Design and Analyse Your Experiment with MINITAB

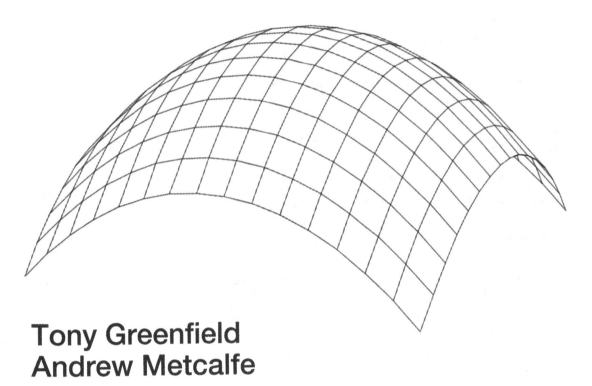

Tony Greenfield
Andrew Metcalfe

T0224267

John Wiley & Sons, Ltd

British Library Cataloguing in Publication Data
A catalogue record for this title is available from the British Library

ISBN: 978-0-470-71114-9

First published 2007
Impression number   10 9 8 7 6 5 4 3 2 1
Year   2012 2011 2010 2009 2008 2007

John Wiley & Sons Ltd, The Atrium, Southern Gate, Chichester, West
Sussex, PO19 8SQ, United Kingdom

For details of our global editorial offices, for customer services and
for information about how to apply for permission to reuse the copyright
material in this book please see our website at www.wiley.com.

Typeset by Charon Tec Ltd (A Macmillan Company)

Contents

Preface

Curiosity is in the nature of all animals. 'Curiosity is, in great and generous minds, the first passion and the last', said Samuel Johnson. 'It is one of the most permanent and certain characteristics of a vigorous intellect.' Curiosity is driving your research. But your research is more than simple curiosity: it is formalised and structured curiosity. You are testing, observing, measuring, recording, displaying, analysing and interpreting with a purpose.

With this book, we help you to structure your curiosity by teaching you how to design and analyse experiments using a statistical package. And, throughout the book, we have tried to help you to understand the technicalities of design and analysis of experiments by using an uncomplicated style and by describing research studies in many areas of inquiry.

The methods we teach are just as applicable in social studies as in physics and chemistry, in linguistics, history and geography as much as in engineering and marketing. You doubt this claim? Then challenge us. Write to us about your research and we believe, almost surely, that we can show how your problems can be resolved with the help of well-designed experiments.

Industry and business keep moving forward, developing new products, doing new things. Research and development are essential for such progress and for continual improvement of products and processes. Experiments, well designed, analysed and interpreted, will help you to find correct, timely and profitable results. Our case studies will convince you of this, whatever field of business or industry you are in.

Although the methods of experimental design and analysis have existed for a long time, they have become more accessible through the development of computers and, especially, the development of statistical analysis packages. MINITAB is one of many suites of programs that are available in almost all, perhaps all, universities across the world, in most research and development laboratories and in many industrial and business companies. With MINITAB you can analyse almost any set of data you are likely to meet. MINITAB will also help you to design your experiments.

MINITAB is compatible with Microsoft Excel so that Excel spreadsheets (as **.xls** files) can be imported to MINITAB worksheets. In places we use functions available in Excel, such as Solver.

So, in this book, we teach you how to design and analyse your experiments with the help of MINITAB. You might like to skim-read it to get a flavour of it, but we urge you to study it thoroughly, working through every case, learning the MINITAB commands, understanding the methods and the interpretations. Retrieve each data set from the website and paste it into a MINITAB worksheet. Look also at the data sets and the examples provided by MINITAB.

We thank MINITAB for the support of the Author Assistance Program, for providing us with copies of the software and for personal support. We also thank Stephen Halder, our editor at Arnold, for his patience and encouragement while we took longer than we had intended to write this book: a book that needed to be constructed with great care for detail because we intend that readers will learn by assiduously following every step. We are also very grateful to Katharina Hyland for dealing so efficiently with the production of the book, and to Richard Leigh for his thorough final editing. Richard checked every procedure from the viewpoint of the reader and suggested several improvements. Nevertheless, we are responsible for any lurking errors and will post a note on the website if any appear.

Complete the course and you will have power in your research and the evidence to prove to others that you can deliver valuable results.

Tony Greenfield
Andrew Metcalfe

Notation

We want to make MINITAB instructions stand out clearly, so that you can follow them easily. So, although we include them in the body of the text, we change the type font to **sans serif** like this: **Stat > Basic Statistics > Display Descriptive Statistics**.

When we introduce new technical terms with which you may not be familiar, we show them in *italics*.

Website for data sets

The full data sets for most of the case studies are available on www.greenfieldresearch.co.uk/doe/data.htm The data set for each case study is in an Excel spreadsheet. Copy and paste the data into a MINITAB worksheet.

Sources of data

Most cases are based on our own consulting experience. Others are based on excerpts of data from published papers and examples from textbooks. We thank people for allowing us to use their data and we list attributions, to the best of our knowledge, below. Regardless of the data source, we are responsible for any errors of analysis or interpretation in the cases presented in this book.

Case 2.1   Department for Education and Skills (UK)

Case 2.5   Department for International Development (UK)

Case 4.4   Gorn (1982)

Case 5.1   Martin Bland's website (originally Close *et al.* 1986)

Case 5.5   From DASL website (originally Frisby & Clatworthy 1975)

Case 5.6   East Malling Research, New Road, East Malling, Kent, England.

Case 6.2   East Malling Research, New Road, East Malling, Kent, England.

Case 6.3   Sadras *et al.* (2004)

Case 7.1   Singh *et al.* (1994)

Case 7.2, 8.2, 9.4, 10.2, 11.2   Greenfield Research (2003)

Case 9.2   Greenfield Research WinDEX

Case 9.3   Wu & Hamada (2000) and Morris *et al.* (1997)

Case 9.5   Norman and Naveed (1990)

Case 12.1   Hanna *et al.* (2005)

Case 12.3   Graham and Martin (1946)

Case 13.1   Miller and Freund (1977)

Case 14.2   Montgomery (2004)

Case 14.3   Clarke and Kempson (1997)

Case 14.4   Anderson & McClean (1974)

Case 15.2   Montgomery (2004) and Kowalski *et al.* (2002)

1

Guide

1.1 Introduction

Experimental design and analysis is an essential part of understanding and improving processes. Every experiment should be well-designed, planned and managed to ensure that the results can be analysed, interpreted and presented. If you do not do this, you will not understand properly what you are doing and you will face the hazard of failing to reach your research goals, and of wasting great effort, time, money and other resources in fruitless pursuits.

We shall follow, throughout this book, a trail of experiments from the simplest to the complicated. The definition of an experimental design is the same at every level. It is *the specification of the conditions at which experimental data will be observed.*

Experimental design is a major part of applied statistics and it has applications in every branch of industry, business and government as well as in the research and development provided by universities in support of industry, business and government. We shall illustrate the different types of experimental design with cases from three organisations to represent the range of applications. One of these is a university: the University of Erewhon or UoE. The other two are manufacturing companies: AgroPharm, which is active in the agricultural and biological sciences, and SeaDragon, which is active in the engineering and physical sciences. The motives for research and development by both these companies are the same: to improve their products and production processes so that they can sustain and increase their profits in an ethical manner. We shall describe these three organisations in more detail later in this chapter.

The purpose of any experiment, whether well designed or not, is to yield data that can be analysed so as to improve understanding of the characteristics of a product or a process (Figure 1.1). Improved understanding usually comes from a comparison of those characteristics, either against standard or nominal values, or against the characteristics of a similar product or process. We usually express the understanding of characteristics algebraically, as a mathematical model, so that we can predict

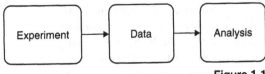

Figure 1.1

the values of characteristics resulting from changes to features of the product or process. Our aim is often to discover the values of those features that can provide the best possible values of the characteristics (Figure 1.2). Those features that we can change are called *control variables*. Those characteristics that respond to changes in the control variables are called *response variables*. We shall say more about these later.

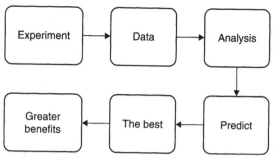

Figure 1.2

Analysis of data used to be daunting, especially when there were more than just a few variables.

Now we are fortunate: we have computers and computer programs to make the analysis easy, provided we know how to use them.

MINITAB is one of many suites of programs that are available in almost all, perhaps all, universities across the world, in most research and development laboratories and in many industrial and business companies. With MINITAB, you can analyse almost any set of data you are likely to meet. MINITAB will also help you to design your experiments. As we move through this book, following a trail of experiments by our three enterprises, we shall show you those functions that we think will be most useful to you. You can access all of these by clicking on buttons in special toolbars, so we shall introduce those toolbars in the relevant chapters. We shall also show you some features of MINITAB that will help you to use it easily and effectively, to switch rapidly between your worksheets of data, your charts, your session window, your history folder where most of your recent commands are stored, your ReportPad and other documents. We assume that you have the latest version of MINITAB which, as we write this book, is version 14, although most of our demonstrations can be applied in earlier versions.

1.2 Starting MINITAB

MINITAB is now fully implemented for Windows with graphic user interfaces. When you first open MINITAB you will see three windows assembled together as in Figure 1.3. The top window is similar to the main window of any other Windows program, with menu for choosing commands and a standard toolbar with buttons for commonly used functions. The session window contains all text output, such as tables of statistics, according to commands that you make during a session and of graphs that you create. You can also use the session window to enter commands instead of using menus and buttons. The data window contains columns and rows

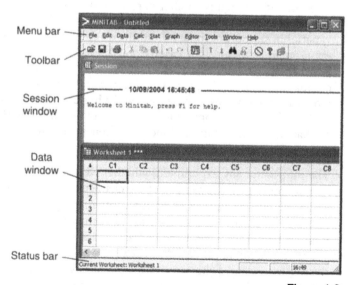

Figure 1.3

of cells in which you enter, edit, and view the data for each worksheet. Since you can have several worksheets open at a time, you can have several data windows open at a time.

We have introduced the word 'worksheet'. What is it? All the data associated with a particular data set are contained in a worksheet. A project can have many worksheets; the number is limited only by your computer's memory. Worksheets are not visible, but you can view your data in the data and session windows, and in the worksheet folder in the project manager which we shall describe later. Each worksheet will have its own data window. There is only one session window, but it can display information on any open worksheet. Any command you use works on the current worksheet. The current worksheet is the worksheet associated with the active data window. You make a window active by clicking on it, choosing it from the window menu, or by right clicking on the worksheet folder and bringing it to the front. If no data window is active, the command acts on the data window that was most recently active. In MINITAB 14, a worksheet can contain up to 4000 columns, 1000 constants, 100 matrices, and 10,000 cells.

Other windows that some users of MINITAB keep visible are the graph windows and the project manager. We prefer to avoid clutter and to hide these windows until we need to see them. Indeed, you can reduce the clutter even further by hiding the session window. To do this, just click on the maximise button at the top right of the data window (Figure 1.4).

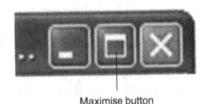

Maximise button

Figure 1.4

So now that we have hidden several windows, how do we see them easily and quickly? We use the project manager toolbar.

The menu at the top of the data window includes a command **Tools**. A click on this will reveal a drop-down menu pointing to **Toolbars**, another drop-down menu in which **Standard** is already ticked (Figure 1.5). Click on **Project Manager** and a new toolbar will appear in the data window. Use the mouse to drag this toolbar to dock with the standard toolbar (Figure 1.6).

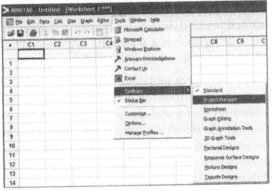

Figure 1.5

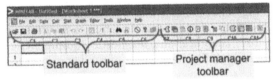

Standard toolbar   Project manager toolbar

Figure 1.6

Although other toolbars are available, including four for different types of experimental designs, your work will be easy if you start with just these two. MINITAB also allows you to design your own toolbars.

We describe the buttons of the project manager toolbar in Figure 1.7. Use this toolbar to access project manager folders and information. Click a project manager toolbar button to open the corresponding folder and any related windows. To return to your previous window set-up, click the appropriate project manager button again.

1.3 Three enterprises

Many people are surprised and even sceptical when we claim that statistical methods, especially the design and analysis of experiments, can be applied across the full range of research and development. Show an example from, say, a continuous chemical process to a machine tool manufacturer and hear this response: 'That's excellent in those circumstances but of course we are different. The method would be of no use to us.' Examples are needed to prove otherwise. That is why we decided to illustrate the different types of experimental design with cases from three very different organisations to represent the range of applications.

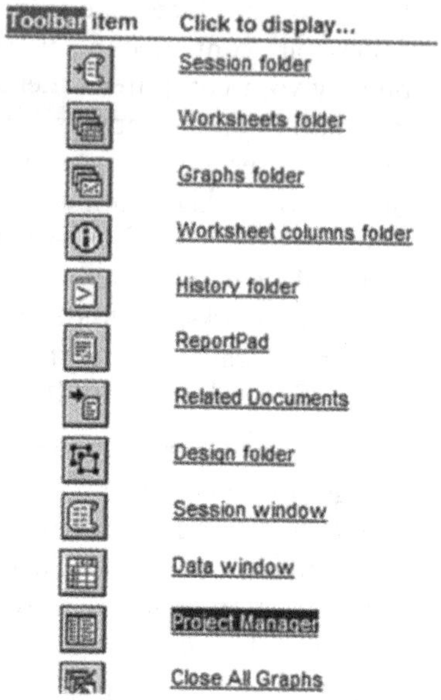

Every university has many departments and a huge range of research and development projects: far too wide to describe in this book. The *University of Erewhon* is typical so we decided to restrict our attentions to one department: the Department of Social Studies. This department has a long tradition of research on development issues such as poverty, oppression, sustainability, conflict between state and civil society, governance and ethnic and religious tensions. Researchers focus on the economics of sustainable development at company, regional, national and global levels, on human resources and population studies, on rural development and the environment.

Toolbar item	Click to display...
	Session folder
	Worksheets folder
	Graphs folder
	Worksheet columns folder
	History folder
	ReportPad
	Related Documents
	Design folder
	Session window
	Data window
	Project Manager
	Close All Graphs

Figure 1.7

We chose two global manufacturing companies, partly because their operations may relate to UoE projects and partly because they offer such a wealth and diversity of examples. Remember, though, that companies with just a few employees can profitably apply the topics of this book and, in our experience, have done.

AgroPharm is active in the agricultural and biological sciences. Its products include seed and plant varieties, fertilisers and pesticides, pharmaceuticals and medical devices for human and veterinary use, analytical and diagnostic equipment, garden and horticultural wares. Its activities include research for the support of rural communities and health services across the world.

SeaDragon has more than 3000 mining, manufacturing, research and development, and sales and distribution premises round the world to support its activities in the engineering and physical sciences. Its products include motor vehicle parts, fire prevention and control equipment, plastics and adhesives as well as raw materials such as cut and polished stone, cement and lime, steel and titanium. The company is now implementing Six Sigma and Lean Manufacturing in all divisions with a view to examining and permanently improving processes to eradicate waste, defects and variability. An important tool of Six Sigma is the design and analysis of experiments.

1.4 More about design and analysis of experiments

The understanding of products and processes demands the analysis of data. This applies whether or not the data came from a designed experiment. Indeed, many data sets have been generated without designed experiments – sometimes for good reasons, but often because those in charge were unaware of their missed opportunities. We shall take you through the analysis of some such data sets. MINITAB provides all the tools you will need for analysis. You do not need to understand all the mathematical theory that supports the techniques of analysis, but you do need to appreciate a little about the techniques because they provide the justification for the methods of experimental design. Figure 1.8 illustrates this.

Many people have written about the scientific method. Its relationship with statistical analysis still raises controversy. We give a brief overview in the context of our cases.

Scientific method is nicely summarised by Wudka (1998) as:

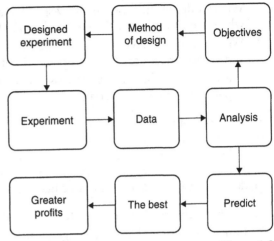

Figure 1.8

1. Observe some aspect of the universe.
2. Invent a tentative description, called a *hypothesis*, that is consistent with what you have observed. [A hypothesis is a working assumption.]
3. Use the hypothesis to make predictions.
4. Test those predictions by experiments or further observations and modify the hypothesis in the light of your results.
5. Repeat steps 3 and 4 until there are no discrepancies between theory and experiment and/or observations.

He adds:

> When consistency is obtained the hypothesis becomes a *theory* and provides a coherent set of propositions that explain a class of phenomena. A theory is then a framework within which observations are explained and predictions are made.

Notice that there is no suggestion that a theory is true in any absolute sense. Newton's theory of gravitation that he proposed in 1666 is an outstanding example of the success of this approach. It remains the basis for mathematical models that are used to describe and predict most physical phenomena. In 1905 Einstein proposed his theory of relativity in which he argued that Newton's laws of motion were invalid at speeds approaching that of light. Einstein's theory of relativity was shown to give excellent agreement with observation during the solar eclipse of 1909. But the fact that mathematical models cannot describe precisely an industrial process for printing patterns on sheets of metal is not due to any relativistic effects. It is due to differences between idealised mathematical models and physical machinery and inevitable random variation. Statistical analysis aims to describe both these aspects. The underlying models for statistical analysis are generally quite simple and fitting them conforms closely to the scientific method. An example is the etching of tin sheet.

1. We have observed that acid etches metal.
2. Our hypothesis is that the depth of etch increases in proportion to the concentration and temperature of the acid within feasible ranges of concentration and temperature. Specifically:

$$\text{depth} = a + b \times \text{concentration} + c \times \text{temperature}$$

where a, b and c are parameters of the model. Suppose we start with plausible guesses for a, b and c.

3. We now design an experiment by specifying values of concentration and temperature at which we will measure the depth of etch. We use the hypothesis to make predictions.

4. We compare our predictions with the results of the experiment and modify the values of a, b and c, and if necessary the algebraic form of the model in the light of our results.

5. We decide whether or not it is reasonable to attribute discrepancies between hypothesis and experiment to random variation.

So the fitting of statistical models conforms with scientific method. There is an alternative approach to statistical analysis, known as *hypothesis testing*, which can be a useful aid to decision making. We consider two examples and discuss how well they conform with scientific method. The first is whether the average tensile strength of a particular alloy of steel, used in the manufacture of turbine blades, is at least equal to some specified value.

1. We have observed our process for making the alloy turbine blades.

2. Our hypothesis is that the average tensile strength of all the blades that would be produced by this process, if it continues on its present settings, is greater than or equal to the specified average value of 620 MPa.

3. We predict that the average value of tensile strength for a random sample of 20 blades will be greater than 610 MPa.

4. We take a sample of blades, measure the tensile strengths, and calculate the sample mean. If the sample mean exceeds 610 we retain the hypothesis. If the sample mean is less than 610 we will reject the hypothesis in favour of an alternative hypothesis that the mean of all the blades that would be produced is less than 620.

5. If we reject the hypothesis in step 2 we will adjust our process and repeat step 4 until we are able to retain the hypothesis.

This statistical hypothesis test is compatible with the scientific method. An important detail, that we discuss later, is that the prediction in step 3 has a probability associated with it.

Our second example is a comparison of two drugs for reducing systolic blood pressure of hypertensive patients. One hundred patients will be randomly assigned to either drug A or drug B, subject to a restriction that exactly half are given drug A. A strategy of asking all the patients to try both drugs had been considered but was ruled out because of the length of time needed to monitor the effect of each drug and the possibility of carry-over effects.

1. We have observed the ability of drugs to reduce blood pressure.

2. Our hypothesis is that drug A and drug B deliver identical average reductions for the population of all possible patients. We call this the null hypothesis because our hypothesis is that there is no difference.

3. We predict that the difference in average reductions in systolic blood pressure between the 50 patients on drug A and the 50 on drug B is less than 5 mmHg.

4. We do the experiment. If the difference in sample means is less than 5 mmHg we retain the null hypothesis. If the difference exceeds 5 mmHg we reject the hypothesis in favour of an alternative hypothesis that the average reduction for drug A is greater or less, according to the sign of the difference, than that for drug B.

Again, the prediction in step 3 has a probability associated with it. The controversy is over the status of the hypothesis in step 2. In the scientific method the hypothesis under test usually has some chance of being accepted as a theory. In the last example, the hypothesis of exact equality is quite unrealistic. Nevertheless, the underlying principle of testing ideas against experimental observations is certainly common to both examples. In statistical hypothesis testing the hypothesis under test, specified in step 2, is known as the *null hypothesis*.

When we have a hypothesis and we know how we intend to analyse the data, we can choose a method to design the experiment. In general the procedure is as follows. State a *hypothesis* about the behaviour of the system of interest. Formulate the hypothesis as a *model* that can be represented in terms of *measurable variables*:

- *control* variables (also called *independent* or *predictor* variables);
- *response* variables (also called *dependent* variables);
- *concomitant* variables (also called *covariates*), which may influence the response variables but which we cannot control, even though we can identify and measure them.

The relationship between a control variable and the response is known as an *effect*. There are also variables that cannot be measured:

- *time-dependent errors* which we cannot identify but which show their presence by trends in the observed values of response variables;
- *random errors* which represent other unidentified variables which taken together show no pattern or trend.

We characterise the errors by a probability distribution. The *statistical objectives* of experimental design are to specify:

1. the number of observations;
2. the values of the control variables at every observation;
3. the order of the observations;

with a view to:

1. ensuring that all effects in the model can be estimated from the observed data;
2. testing the reality of those effects by comparison with random variation;
3. ensuring that all effects can be estimated with greatest possible precision (reducing the influence of random variation);
4. ensuring that all effects can be estimated with the least possible bias, or greatest accuracy (reducing the effects of time-dependent errors);
5. suggesting improvements to the model;
6. keeping within a budget of effort and cost.

1.5 Research never ends

The experience of all researchers is that ideas keep flowing: whenever one experiment ends, the current hypothesis changes and a new cycle of research begins. We can see the complete cycle in practical terms in Figure 1.9. We describe it as a research spiral to emphasise that we do progress as we keep going round the cycle.

We shall return to parts of this cycle in later chapters as we present methods and cases from our three enterprises.

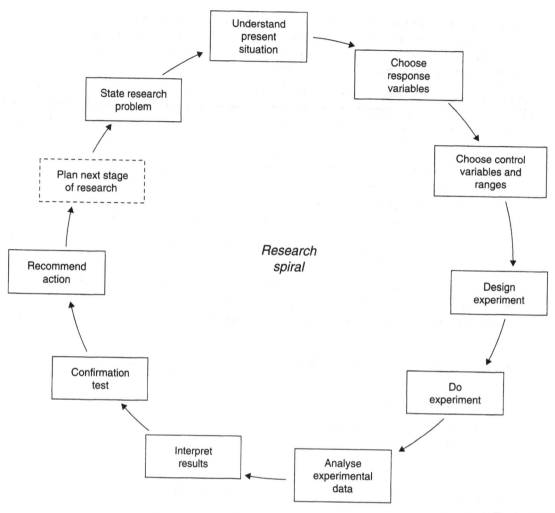

Research spiral

Figure 1.9

1.6 Randomisation

All processes include some random variation, and replicates will not be identical. The precise volume of olive oil in one-litre cans will vary slightly. The porosity of aluminium wheels made by the same process will vary. The yields of tomato plants of the same variety from the same nursery will differ. Participants in a programme that aims to help people to control problem gambling will react in different ways. Patients respond in different ways to the same treatment.

Random variation may be due to inherent variation in the population, variation in operations, measurement error, or some combination of all of these. We aim to reduce it as much as possible, but we cannot eliminate it.

If we are to improve processes we must distinguish the effects of our intervention from random variation. Experimental design and analysis is the means for doing this. It is an essential part of understanding and improving processes.

We will continue with the example of the tensile strength of turbine blades. We can measure the tensile strengths on no more than a sample of blades for at least two very good reasons. The first is that the population of all blades that would be made, if the process continued on its present settings indefinitely, is an imaginary infinite population. The second is that each test of tensile strength destroys the blade. We would like our sample to represent the population. We can never guarantee this but, if we know nothing about the population, the best we can do is to take a random sample.

The assumption of a random sample enables us to make statements about our likely accuracy and precision. But how should we take a random sample from an imaginary infinite population? We have to compromise. A practical proposition might be to take a random sample from one day's production. We can do this by allocating a number to each blade, or at least establishing some rule that enables us to identify a blade corresponding to a particular number. We can make the draw with MINITAB.

Suppose there are 300 blades and we want a random sample of 20. Select **Calc** in the MINITAB menu. The drop-down menu shows **Random Data** from which you can select **Uniform** (Figure 1.10).

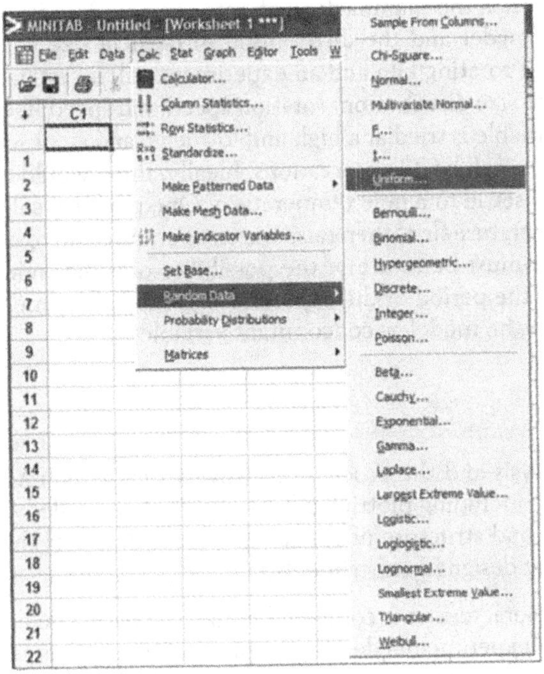

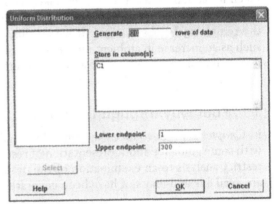

Figure 1.10

Figure 1.11

A window appears in which you can ask for 20 rows of data to be generated in the range 0.5 to 300.5 and to be stored in column C1 (Figure 1.11).

The generated numbers have three decimal places so you must change them to integer values by rounding. Select **Calc** in the MINITAB menu. The drop-down menu shows **Calculator**, which opens the window shown in Figure 1.12. Enter **ROUND**, or choose it from the list of **Functions**, with the parameters **C1, 0** indicating that the numbers you want to be rounded are in column **C1** and that you want them rounded with no decimal places.

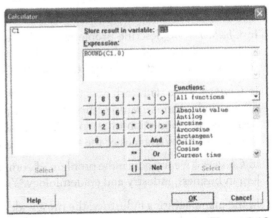

Figure 1.12

If we know something about our population, such that turbine blades are made on two sites but with a single process, we can divide our population into sub-populations and take random samples from each.

Now think about the drugs for reducing blood pressure. Ideally, we would take a random sample of all possible patients. In practice we will have people who are patients at the time of the study who are willing to enter a clinical trial. We hope to remove any systematic bias by the random allocation of patients to the two drugs. When we interpret the results, we assume that the difference between the two drugs with these patients will be similar to the difference between drugs for all other patients, even if the patients are rather different. Doctors with specialist knowledge may be able to comment on the medical grounds for this assumption. Multi-site trials extend the validity of studies more objectively.

A third example that raises issues of randomisation is the manufacture of antimony oxide, a paint whitener and flame retardant. This is a white powder and the distribution of particle size is an important characteristic. The powder is made in a rotating kiln and an experiment will investigate the effects of five control variables on the powder size distribution: rotation speed; fan speed; feed rate; flu venting; and kiln temperature. If each variable is tried at a high and low level and an observation is taken at every possible combination, there will be 32 observations. Ideally, these would be taken in random order. But the kiln takes days to settle to a new temperature! One practical solution is to take the 16 observations at the low temperature first, then raise the temperature and make the remaining 16 observations. If we do this, we must be aware of the possibility of some trend, such as an increase in ambient temperature, over the period of the experiment. Provided we monitor such changes, we can at least include them in the model as concomitant variables.

1.7 Your way through the book

In *Chapter 2* we introduce the principles of analysis and the procedures provided by MINITAB, with some guidance about presentation of results, including plotting data. In the first examples we restrict analysis to an examination of the nature and structure of the data. We use data sets that are available and may not have been generated by designed experiments.

In *Chapter 3* we discuss experiments in which a characteristic is compared with a standard value. It is here that we introduce two important aspects of experimental design: sample size calculations and randomisation. We also introduce and discuss tests of statistical significance and apply them to the examples. We shall show how MINITAB can help you to improve the presentation of your results.

In *Chapter 4* we describe problems in wine tasting, compulsive gambling, smashing of TV tubes and musical tastes of students to show how to compare treatments when the response is simply that one item is better or worse than another, or that an item is satisfactory or not.

In *Chapter 5* you will learn how to compare the means of two populations from which samples are drawn: independent samples first and then paired samples. Again, an important feature of the design is the calculation of sample size. In the independent samples case we shall show you how to compare the variances of the two populations from which the samples were drawn.

In *Chapter 6* we introduce the problem of comparing more than two means. This is a common problem in business, industry and epidemiology. You may find this hard study but it is worth the effort.

In the manufacturing industries, the most widely used class of design is the factorial experiment. This is used when there are several control variables, also called factors. Two-level factorial experiments, in which each factor is set at two levels, high and low, are widely used during development studies. We introduce this class of design in *Chapter 7*, where we also give examples of regression analysis, the most widely used method of model fitting analysis.

A difficulty that occurs with factorial experiments is that the more factors there are, the larger the experiment. This can be a financial and an operational embarrassment, so in *Chapter 8* we show you a special technique for designing fractional factorial experiments.

In the final stage of a development study, when you are seeking the conditions (such as the values of composition and process variables) that will yield the best value of a material property (such as the highest value of tensile strength), additional points must be added to factorial experiments so that the curvature of the response can be estimated. These designs are known as augmented or composite designs, and we describe them in *Chapter 9*.

If you think of a response surface as a mountain, perhaps with many more than three spatial dimensions, you may want to find the top of the mountain. In *Chapter 10*, on hill-climbing, we illustrate ways to do this.

In the experiments mentioned so far our aim may have been to discover the settings of the control variables that will give the best value of the response variable, or even the best values of several response variables. A difficulty with this is that, when the product is subsequently in use, it may not be possible to stick exactly to the optimal control settings. Also the responses may be influenced by ambient conditions, such as temperature, humidity, inattention of the user or ageing of the product. We need to guard against these situations, so robust designs have been developed. They are extensions of factorial and composite designs. We introduce these and the corresponding analysis with industrial examples in *Chapter 11*.

Variables in some industrial processes are nested. Their settings may differ between production batches, for example. We explain these in *Chapter 12*.

We return to factorial experiments in *Chapter 13*, where we discuss factors at more than two levels. Then, in *Chapter 14*, we add further complications with crossed as well as nested factors and with split-plot experiments.

Many processes in the food and drinks industry, and in the pharmaceutical and chemical industry, involve blending ingredients. The proportions of ingredients in the mixture are crucial, and blenders rely on experiments to obtain the best results. Mixture designs deal with this, and we describe them in *Chapter 15*.

In the final chapter, *Chapter 16*, we discuss the design and analysis of experiments in which the response variables are discrete: they have distinct values, such as whole numbers or categories. In particular, we introduce *logistic regression*, a versatile model for analysing the results of experiments in which the response is categorical. We also propose an informal method for estimating the sample size needed for a satisfactory experiment. This method includes the simulation of an experiment and of the variation that we expect in an experiment. To do this, you will need to write a short program in MINITAB's macro language, and we show you how to do this.

2

Descriptive statistics and plotting

Believe it or not, but this can happen. You are presented with a table of data, from whatever source and not necessarily from a designed experiment, and asked 'What can you make of that?' You might reply 'What do you want me to make of it?' or 'Why have the data been collected and assembled into a table?' A reply that we have had is 'Just look through it and see if you can spot any relationships'. You want to help, so you start by examining the data.

This is what you should do with any data set, even if a clear purpose for its collection and analysis has been specified, even if it has been generated from a well-designed experiment: examine the data.

A temptation offered by access to a computer and a statistical analysis package is to load the data and embark immediately on elaborate analysis. Be warned! Eagerness for results can lead to false and costly mistakes. A simple examination of the data can reveal errors such as missing or incorrect values that would make your analysis wrong. A common error is that missing values are entered as zeros so that automatic analysis gives strange associations or perhaps no associations where some useful relationships really exist. Your preliminary examination will also identify any extreme values. You should check these carefully. If they are valid you must retain them. Perhaps they are of particular interest, but they may influence your choice of analysis.

There are two approaches to the initial simple examination of data. We recommend that you should use both with every data set that you want to analyse before you embark on an elaborate analysis. These are *descriptive statistics* and *plotting*. We shall use both these approaches with some examples and explain the MINITAB procedures.

▽ Case 2.1 (UoE)

Greta Green, a lecturer in Department of Social Studies at UoE, is writing an article on absence from primary schools in England. The Department for Education and Skills provides statistics on absence, as recorded by local education authorities (LEAs) in England and Wales, for six school years up to 2000/2001 on its website www.dfes.gov.uk/performancetables/natabs_01/primary.shtml.

For each LEA, the data provided are:

- number of day pupils of compulsory school age;
- percentage of half days missed due to authorised absence;
- average number of half days missed per authorised absent pupil;
- percentage of half days missed due to unauthorised absence;
- average number of half days missed per unauthorised absent pupil.

Greta has copied the data for 2000/2001 into a MINITAB spreadsheet, and will start her analysis by drawing some graphics and calculating some summary statistics. We will follow her example.

Obtain the Excel spreadsheet from the website (www.green-fieldresearch.co.uk/doe/data.htm), copy and paste it into a MINITAB worksheet and maximise the data window (see Figure 2.1). Note that the project manager toolbar is in position as we advised in Chapter 1 (Figure 1.6). When you paste a spreadsheet into MINITAB with spaces as delimiters, it splits the names and creates extra columns. You can avoid this by replacing, in the spreadsheet, underscores for the spaces in names – for example: **day_pupils** and **Barking_and_Dagenham**.

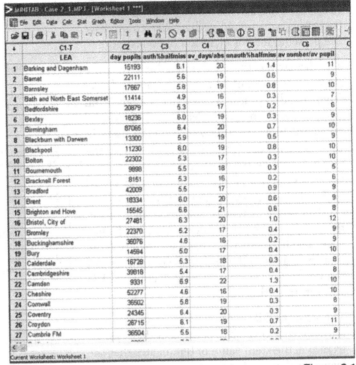

Figure 2.1

The first rows of data are shown in Figure 2.1. When you examine the table you will see that the final row contains the sums of the columns. You should delete this row before doing the calculations and plots. Read through the table, entry by entry, and look for strange values that might be wrong. Just reading the data will give you some idea of the nature of values.

One of the simplest plots available in MINITAB is the stem-and-leaf. Click on **Graph** in the MINITAB window. A drop-down menu of available graphs styles appears (Figure 2.2). Choose **Stem-and-Leaf**. A window appears for you to specify which variables you want to plot (Figure 2.3). We could select as many of the available variables as we like but, for easy explanation, we begin by investigating just one: the percentage of half days missed due to unauthorised absence, which is

Figure 2.2

a measure of truancy. The column that contains these values is headed **unauth% halfmiss**. Highlight this in the list of available variables and click on **Select**, then on **OK**.

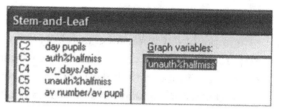

Figure 2.3

The data window is immediately replaced by the session window, which contains the image you see in Figure 2.4. This needs some explanation. Also, you would perhaps like to improve the presentation for inclusion in your report.

The display has three columns: the leaves, the stem and the counts. Each value in the *leaves* (right-hand) column represents a digit from one observation. The 'leaf unit' (declared above the plot) specifies which digit is used. In the example, the leaf unit is 0.01. Thus, the leaf value for an observation of 1.40 is 0 while the leaf value for an observation of 1.45 would be 5 if there were such an observation. There is no data point in which there is a digit, other than zero, in the second decimal place. Hence every leaf in the diagram has a value of 0. Note that the values in the left-hand column are not the counts of leaves in the right column. We will return to this below.

The *stem* (middle column) represents the digit immediately to the left of the leaf digit. In the example, the stem value of 0 indicates that the leaves in that row are from observations with values greater than or equal to zero, but less

```
Stem-and-Leaf Display: unauth%halfmiss

Stem-and-leaf of unauth%halfmiss N = 150
Leaf Unit = 0.010

   1    0  0
   6    1  00000
  28    2  000000000000000000000000
  62    3  00000000000000000000000000000000000
 (20)   4  00000000000000000000
  68    5  00000000000000
  54    6  00000000000000
  40    7  000000
  34    8  00000000000
  23    9  0000000
  16   10  00
  14   11  000
  11   12  00
   9   13  00
   7   14  0
   6   15  0
   5   16  0
   4   17  0000
```

Figure 2.4

than 0.1. The stem value of 1 indicates observations greater than or equal to 0.10, but less than 0.20. A stem value of 14 indicates an observation greater than or equal 1.4 but less than 1.5. So any value in the middle column, when multiplied by 0.1, is the percentage of unauthorised absences.

The *counts* are in the left-hand column. If the *median* value (the middle of the data if they are put into ascending order: half the observations are less than or equal to it) for the sample is included in a row, the count for that row is enclosed in parentheses. The values for rows above and below the median are cumulative. The count for a row above the median represents the total count for that row and the rows above it. The value for a row below the median represents the total count for that row and the rows below it. The median for the sample is 0.4, so the count for the fifth row is enclosed in parentheses (20). We can see that about half the leaves are less than or equal to the median of 0.4. The longest column corresponds to the most common percentage, reported by 34 LEAs, of 0.3. This is the *modal* value. It is another definition of a typical value, but is applicable only if there is a discrete set of numbers that the variable can take.

Not much can be done to improve the presentation of a stem-and-leaf diagram as it appears in the session window. You can change the font. On the main menu go to **Tools > Options**, then double-click on **Session Window** and single-click on **I/O Font**. Few choices are available. The default font is

Courier but you may think the diagram looks better in **Lucida console**. However, you can cut and paste to your Word document for better formatting.

When you come to more advanced graphs, you will find that there is a great deal of choice. Go to **Tools > Options > Graphics** and explore the options. You can change the fonts, style and colours of frames, of lines, of plotting symbols, of fills, of tick marks and of grid lines. We shall show some of the possible variations in following examples.

Now let us return to the current example. In the project manager toolbar at the top of the session window, click on the data sheet button (Figure 2.5). The data sheet window immediately replaces the session window.

Data sheet button

Figure 2.5

Another simple character display of the distribution of the data is the dotplot: **Graph > Dotplots > Simple**. It is similar to the stem-and-leaf plot, but is orientated in the conventional way (Figure 2.6). We have used the **Tools > Options > Graphics** commands to set the style of the plot: font, dot size, fill colour, borders. Note that, unlike Excel , changes to style are not applied to the existing figure and you need to redraw it.

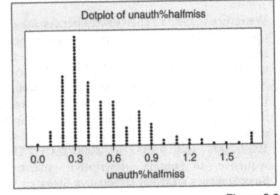

Figure 2.6

When the distribution has a tail of large values, to the right in this orientation, we say it is *positively skewed*. The dotplot clearly shows that there are some extreme values in the right-hand tail of the distribution. You may learn something useful if you can identify those cases (the LEAs) with extreme values. MINITAB has a tool called *brushing* that can do this for you.

Figure 2.7

Make sure that the dotplot is on the screen. You can do this by clicking on the **Show Graphs Folder** button in the project manager toolbar (Figure 2.7). Click on **Editor** in the menu and you will see the drop-down menu of Figure 2.8. You can also obtain this menu by positioning the mouse pointer in the graph area and then right-clicking. From this you can choose the brushing tool. You will perhaps use the brushing tool often so it is worth putting its button in the main toolbar. From the main menu follow the commands **Tools > Customize > Commands > Brushing Editor Menu**. You will see the brush tool button displayed. Move the mouse pointer to it. Click and hold down the left button and drag the button onto to the main toolbar. It will

Figure 2.8

generally be greyed out but will appear in colour when a graph is active (Figure 2.9).

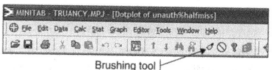

To use the brushing tool, click on it in the drop-down menu or on the toolbar, then point the mouse at the graph and drag out a rectangle over a set of points. These will change colour and a list of row numbers corresponding to those points will appear (Figure 2.10). You may change the colour in **Tools > Options > Graphics > Data View > Symbol**.

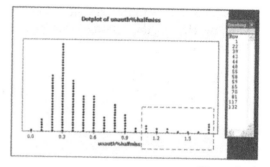

Figure 2.9

As well as producing the list of row numbers, brushing will mark the rows in the current data sheet and will highlight corresponding points in other charts that are open. MINITAB enables you to focus your analysis on the brushed points. With data options (**Graph > Dotplot > Simple > Data Options**) you can restrict further plotting to the brushed points (Figure 2.11).

We return to Greta's analysis and her consideration of unauthorised absence. The ideal percentage, although it is, perhaps, an unrealistic goal, is zero. Nevertheless, one LEA did report 0.0%, to one decimal place. This was the Isles of Scilly, with only 119 primary school pupils. The largest LEA, Kent, with 102,142 primary school pupils, reported 0.4%. A potential basis for comparison is Scotland, where unauthorised absence in primary schools was 0.29% in 1999/2000 and 0.35% in 2000/2001. The MINITAB sequence to obtain descriptive statistics is **Stat > BasicStatistics > Display Descriptive Statistics**.

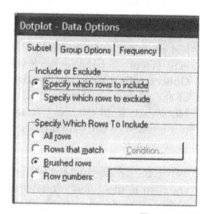

Figure 2.10

Select **unauth%halfmiss** then **Statistics**. You will have a choice of 24 descriptive statistics (Figure 2.12). A number of statistics are selected by default – we will use these, so you need only click **OK**.

Figure 2.11

Note also that there is an option for **Graphs** so that you could use only one sequence of menus to obtain both descriptive statistics and graphs, except that dotplots are not included.

The descriptive statistics appear in the session window, as follows. You can copy this table directly into the **ReportPad** folder: place the cursor anywhere in the text, right click and choose **Append Section to Report**.

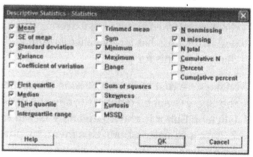

Figure 2.12

```
Descriptive Statistics: unauth%halfmiss
```

Variable	N	N*	Mean	SE Mean	StDev	Minimum	Q1	Median
unauth%halfmiss	150	0	0.5293	0.0296	0.3624	0.000000000	0.3000	0.4000

```
Variable            Q3        Maximum
unauth%halfmiss     0.7000    1.7000
```

By clicking on **ReportPad** (see Figure 1.7), you can fully edit all output in the **ReportPad** folder, and you also can add your own text (such as user comments and notes) and titles to the report.

You can save the contents of the **ReportPad** folder as an RTF (rich text format) file that you can transfer to other applications, or as an HTML file to use as a Web page or in Web-based applications. From the **ReportPad** folder, you can edit and print the reports, or open and edit them in a word-processing program for enhanced formatting options.

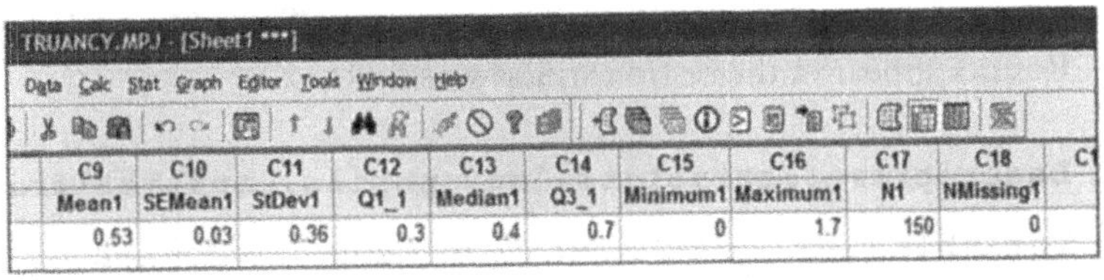

You may want to edit some of the numeric values. For example, you may want to change values of **Minimum**, **Q1**, **Median**, **Q3** and **Maximum** to one decimal place and values of **Mean**, **SE Mean** and **StDev** to two

Figure 2.13

decimal places. An easy way to deal with this in MINITAB 14 is to use **Stat > Basic Statistics > Store Descriptive Statistics**. Note that the default statistics are now just **Mean** and **N nonmissing**, so you select the other statistics you want. This will save the values of the chosen statistics as new variables in the data sheet so that you can define the data format. For each data column in turn, select (highlight) a value in the column, or the complete column or a set of columns, then follow the sequence **Editor > Format Column > Numeric**. If you do not select a value in the column, the word numeric will be greyed out, indicating that it is unusable. In each case, a window will appear where you can specify the number of decimal places for the column's values. The results are shown in Figure 2.13.

From here, you can copy and paste into an Excel worksheet and preserve the tabular structure for further editing into a report. Note that pasting into a Word document will not preserve the tabular structure.

When we study the details of the table, we see that there are 150 LEAs and percentages range from 0 to 1.7. We calculate the mean by adding the percentages for the 150 LEAs, and then dividing this sum by the number of LEAs. The mean is synonymous with 'average', although the latter term is sometimes used rather loosely to include other summary statistics that give some sort of typical value. One of these is the median. You can find it by sorting the percentages into ascending order, and taking the one in the middle. However, with 150 data, and any other even number, no one datum is exactly in the middle so we take the average of the 75th and 76th , in the ascending order, which is 0.4. It follows that half the LEAs have percentages below or equal to 0.4, and half have percentages above or equal to 0.4. Kent is on the median, but below average. This may seem strange, but it arises because the mean (average) is influenced by the few large values, evident in the dotplot of Figure 2.6, where the distribution is positively skewed. The mean of a positively skewed distribution is greater than its median. You can see why if you consider the mean and median of {1, 2, 3, 4, 10} which are 4 and 3, respectively. If you are familiar with the concept of moments, the mean is the balance point of the distribution of points, as shown in Figure 2.6 for example, if they all have equal mass.

We shall explain the other statistics and discuss whether the mean percentage over LEAs is a fair summary for the national percentage later.

If you go back to the descriptive statistics you will notice Q1 and Q3. These are the lower and upper quartiles, respectively. A quarter of the data lie below the lower quartile. It is computed by sorting the data into ascending order and taking the $[(n + 1)/4]$th datum, where n is the number of data. In general, this will be a fraction so we, or MINITAB, interpolate between the two data either side in the ascending order. Similarly, three-quarters of the data lie below, and one-quarter lie above, the upper quartile. It is computed as the $[3(n + 1)/4]$th datum. We shall explain the remaining statistics in the context of two other cases.

We now look at dotplots and descriptive statistics for the other three variables. To bring it all together let us select all variables and present the results in a single table. Following **Stat > Basic Statistics > Display Descriptive Statistics**, select all variables and choose **N nonmissing, Mean, Median,** Results appear in the session window as follows:

```
Descriptive Statistics: day pupils, auth%halfmis, av_days/abs, ...

Variable              N       Mean    Median
day pupils          150      24942     18737
auth%halfmiss       150     5.7127    5.6500
av_days/abs         150     18.507    18.000
unauth%halfmiss     150     0.5293    0.4000
av number/av pup    150      8.927     9.000
```

Most absence is authorised, and the average number of half days lost for unauthorised absentees is about half of that for the authorised absentees. If we are to compare unauthorised absence in England with that in Scotland we should allow for the different sizes of LEAs. For each LEA, we need to multiply the percentage of half days missed by the number of pupils to obtain the total number of half days missed (per half day). To calculate the average per pupil in England, add these totals over LEAs and divide by the total number of pupils in England. In MINITAB: **Calc > Calculator : sum(c5*c2)/sum(c2)** and put the result in **'pupilaverage'** on the data sheet, by typing this name in the **Store Result in Variable** box.

The result, an example of a weighted mean, is 0.4870% and is slightly less than the mean of the 150 LEA percentages (0.5293%). It is, nevertheless, substantially higher than the percentage for Scotland.

If several variables are measured for each member (or *case*) of the sample, we should investigate whether or not there appears to be any association between them. Simple plots of one variable against another are a valuable first step in any analysis, but we should be aware that these plots will not necessarily uncover relationships between three variables, for example.

First we plot the percentage of half days missed due to unauthorised absence against the size of the LEA. Select **Graph > Scatterplot > Simple,** click **OK,** and then put day pupils in the **X variables** box and unauth%halfmiss in the **Y variables** box. The result is shown in Figure 2.14. There does not appear to be any tendency for the percentage to increase or decrease with the

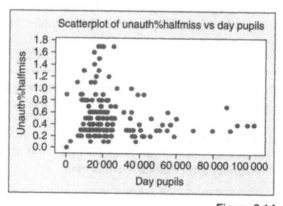

Figure 2.14

size of the LEA, but the percentages seem to vary more for the smaller LEAs.

The next plot (Figure 2.15) is the percentage of half days missed due to unauthorised absence against the percentage of half days missed due to authorised absence. You might think that there is a slight tendency for the percentage of half days missed due to unauthorised absence to increase with the percentage of half days missed due to authorised absence. The lowest right-hand point, which seems somewhat anomalous, is the Isles of Scilly.

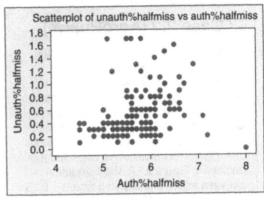

Figure 2.15

Correlation measures how closely points are scattered about a straight line. If the points all lie on a straight line which slopes downwards, the correlation is −1; if they lie on a straight line sloping upwards the correlation is +1. If they are closely scattered about a straight line, the correlation is close to −1 or +1 according to the slope. If there is no hint of a tendency for one variable to increase or decrease with the other, the correlation is close to 0. The correlation for the pairs in Figure 2.15 is obtained in MINITAB from **Stat > Basic Statistics > Correlation**. This appears in the session window as:

```
Pearson correlation of Unauth %halfmiss and Auth%halfmiss = 0.289
P-Value = 0.000
```

The correlation of 0.289 indicates a slight linear association. The P-value is the probability that we obtain by chance a sample correlation as large in absolute value as 0.289, or larger, if the two variables are independent in the corresponding population. A P-value of 0.000 (which doesn't mean 0, but less than 0.0005) tells us that a correlation of 0.289 is most unlikely to have occurred by chance *with a sample of 150 observations* if there is no relationship. Therefore we have strong evidence of a slight tendency for the unauthorised percentage to increase with the authorised percentage, but the effect is too small to be of much practical interest.

The relationship in the next plot (Figure 2.16), of the average number of days missed per unauthorised absent pupil against the percentage of half days missed due to unauthorised absence, is far more substantial. There is a clear tendency for the number of days missed to increase with the unauthorised absence percentage. This suggests that increases in the percentage of half days missed is partly due to the

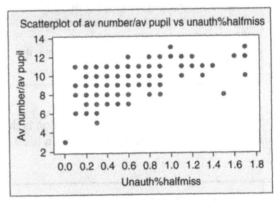

Figure 2.16

unauthorised absences lasting longer. Some LEAs might consider a policy of following up unauthorised absence more quickly.

Correlation measures only linear association and a correlation near zero does not imply that there is no relationship between the variables. Always check by plotting the data.

☐ Case 2.2 (AgroPharm)

Gerard Grey is a research chemist in the Coatings Division. He performed an experiment to investigate the relationship between the amount of additive added to a varnish, used in the electronics industry, and the drying time. The concentration of additive ranged from 0 to 10 mg/g in steps of 1 mg/g. The results of his experiment can be seen in Figure 2.17. The correlation between additive and drying time is −0.25, but this is not at all helpful. His aim is to minimise drying time and it seems that about 5 mg/g additive would be optimal.

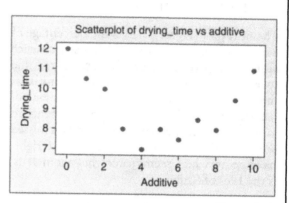

Figure 2.17

☐ Case 2.3 (SeaDragon)

Belinda Bistre is a geologist working in the Prospecting Division. She has recorded the locations of kimberlite pipes, containing diamonds, in part of the Flinders Ranges of South Australia. The locations in metres north (y) and east (x) of an origin are shown in Figure 2.18; the data can be found at www.greenfieldresearch.co.uk/doe/data.htm. The correlation between x and y is −0.08 but it seems that the extent in the north–south direction is greatest towards the middle of the range in the east–west direction. The diamond deposit seems to be in an approximately star-shaped region. Again, the correlation is not a helpful statistic.

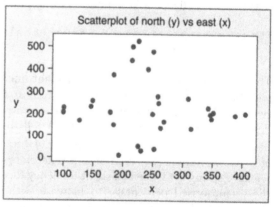

Figure 2.18

Another important feature of correlation is that it is just a measure of association, and an association between two variables does not, in itself, imply that one causes the other.

☐ Case 2.4 (UoE)

Greta Green is giving her class examples of non-causal correlations, sometimes called *spurious*. The data for the first example are minimum ozone layer depth (Dobson units) and tourist spending in Northumbria (NE England) at 1989 prices for the eight years 1989–94, 1996 and 1997, and they are plotted in Figure 2.19; the full data set is available at www.greenfieldresearch.co.uk/doe/data.htm. The correlation is −0.66, but it is hard to imagine any causal link between the two trends.

Although the years are shown in Figure 2.19, a time series plot, obtained from the sequence **Graph > Time Series Plot > Multiple**, shows more clearly that over this period the general trend has been for expenditure to increase and for the minimum ozone depth to decrease.

A somewhat different example of a non-causal correlation is the high positive correlation between the number of firemen and the extent of bush fires. In this case, the severity of the fire accounts for both the extent and number of firemen needed to bring it under control.

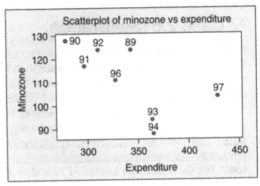

Figure 2.19

□ Case 2.5 (UOE)

Paul Pink is preparing for an economics tutorial class. He has asked his students to prepare for a discussion, centred on a British politician's claim that the per-capita development assistance, from the UK, is above the average for the Development Assistance Committee (DAC) countries. The aid from these 22 countries, in millions of pounds sterling and as a percentage of gross national income (GNI), for the five years from 1998 to 2002 is from data supplied by the UK Department for International Development.

The MINITAB spreadsheet (Figure 2.20) shows the 2002 data, together with estimates of the populations in the 22 countries. The full data set is also available at www.greenfieldresearch.co.uk/doe/data.htm.

The United Nations target is set at 0.7% GNI, and five countries do better than this. However, the politician's claim was in terms of aid per capita. The mean and median of the aid per capita in column 5 are 76.2 and 51, respectively. The mean for the UK is 54, which is less than 76.2 and it may seem that the politician was mistaken. However, if we allow for the different sizes of the DAC countries by calculating a weighted mean given by **sum(c5*c2)/sum(c2)** we obtain 44.0 per capita. This is apt, because the total aid is given by the product of the weighted mean and the total population. The politician's claim is correct in as much as inhabitants of the UK do contribute more than the average of all the inhabitants in DAC countries, although you might think it is not the most appropriate comparison. In general, it is unwise to rely on any single statistic to summarise complex economic data.

MINITAB - DAC.MPJ - [DAC.mtw ***]

File Edit Data Calc Stat Graph Editor Tools Window Help

	C1-T	C2	C3	C4	C5
	country	population	aid	%GNI	aid/capita
1	Australia	19732	641	0.25	32
2	Austria	8188	316	0.23	39
3	Belgium	10289	707	0.42	69
4	Canada	32207	1342	0.28	42
5	Denmark	5384	1087	0.96	202
6	Finland	5191	311	0.35	60
7	France	60181	3454	0.36	57
8	Germany	82398	3571	0.27	43
9	Greece	10666	197	0.22	18
10	Ireland	3924	264	0.41	67
11	Italy	57998	1542	0.20	27
12	Japan	127214	6145	0.23	48
13	Luxembourg	454	95	0.78	209
14	Netherlands	16151	2251	0.82	139
15	NewZealand	3951	83	0.23	21
16	Norway	4546	1164	0.91	256
17	Portugal	10102	188	0.24	19
18	Spain	40217	1072	0.25	27
19	Sweden	8878	1169	0.74	132
20	Switzerland	7319	622	0.32	85
21	UK	60095	3275	0.31	54
22	USA	290343	8598	0.12	30

Figure 2.20

▽ Case 2.6 (SeaDragon)

SeaDragon has been commissioned to produce a report on noise levels near the Tyne and Wear Metro, a light railway, and Brenda Black in the Transport Division is the project leader. Preliminary noise measurements have been made at 46 sites located at a distance of 50 m from the track in residential areas. The data (courtesy of TORG, University of Newcastle upon Tyne) are noise level (decibels) exceeded 10% of exposure time. There is no absolute standard for a maximum level but many guidelines. For example, two useful websites are www.defra.gov.uk/environment/noise/health/page07.htm and www.lhh.org/noise/decibel.htm. The data set is available at www.greenfieldresearch.co.uk/doe/data.htm. A noise level of 70 dB would be typical for a main road. The DOE Advisory Leaflet AL72 states that the noise level outside the nearest occupied room should not exceed 70 dB in suburban areas. A useful comparison is provided by a discussion document for Midland Metro in which it is proposed that noise insulation should be provided if daytime noise levels exceed 68 dB (www.centro.org.uk/metrotwa/WtoB/).

The following descriptive statistics for noise were selected:

N	Mean	Median	TrMean	StDev	SE Mean	Minimum	Maximum	Q1	Q3
46	60.35	59.15	60.01	5.18	0.76	52.3	75.3	57.23	61.08

A histogram is a more refined version of the stem-and-leaf plot and dotplot. You can obtain it at the same time as the descriptive statistics if you use the sequence **Stat > Basic Statistics > Display Descriptive Statistics**; and then choose **Graphs** as well as **Statistics**.

If you choose the **density** option the histogram will be scaled so that the total area under it equals one, and its area between any two noise levels will be the proportion of the data between those two levels. You may choose the density option through the sequence **Tools > Options > Individual Graphs > Histograms**. The density is defined for the grouping intervals, known as bins, as the number of data in the bin divided by both the total number of data and by the width of the bin. We see from the histogram (Figure 2.21) that the distribution of these data is positively skewed, with a few rather high values.

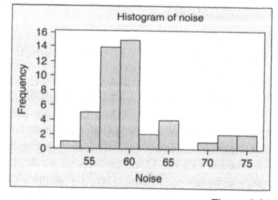

Figure 2.21

We now explain two more of the descriptive statistics. MINITAB calculates a trimmed mean (**TrMean**) by discarding the smallest 5% and largest 5% data and averaging the rest. This is well explained in **Help**, but it is not a commonly used statistic in design of experiments. In contrast, the standard deviation is crucial for both design and analysis of experiments, so it is important that you understand what it is. Here is a procedure to calculate the standard deviation, and its square which is known as the variance, for a set of data.

1. Calculate the mean.
2. Subtract the mean of the data from each datum. The sum of these differences will equal 0 and these modified data are sometimes referred to as 'mean corrected'.

3. Square each of the differences in step 2.
4. Sum the squared differences in step 3.
5. Divide the sum of squared differences in step 4 by one less than the number of data. This quantity is known as the variance.
6. The standard deviation is the square root of the variance.

We will usually rely on MINITAB to do the arithmetic, but we give one brief example. The following five data were randomly selected from the 46 sites and have been rounded to the nearest integer: 74, 61, 55, 66, 64. Following the six steps:

1. The mean is $(74 + 61 + 55 + 66 + 64)/5 = 64$.
2. The mean corrected data are: $10, -3, -9, 2, 0$.
3. The squared differences are: $100, 9, 81, 4, 0$.
4. The sum of squared differences is 194.
5. The variance is $194/4 = 48.5$.
6. The standard deviation is $\sqrt{48.5} = 6.96$.

MINITAB **Help** gives the mathematical formula for calculating the standard deviation. The practical interpretation of the standard deviation is that roughly two-thirds of the data will lie within one standard deviation of the mean. This is a useful approximation for most distributions of data, and if the data have an approximate bell-shaped distribution it is quite accurate. Also, if the data have a bell-shaped distribution, about 95% will lie within two standard deviations of the mean and about 99.7% will lie within three standard deviations of the mean. The standard deviation calculated from all 46 sites is given as 5.184. The mean noise level is 60.354, so the mean minus one standard deviation is 55.17 and the mean plus one standard deviation is 65.54. These numbers are slightly below Q1 and above Q3 respectively, as we might expect given that 50% of the data lie within Q1 and Q3. If you sort the data in MINITAB you will see that three data lie below 55.17 and six lie above 65.54. It follows that 80% (37/46) lie within one standard deviation of the mean for this data set.

You may wonder why we divide by one less than the number of data in step 5. There are several justifications for this, and all suppose that we are calculating the standard deviation from the sample data to estimate the standard deviation in the corresponding population. In this case the corresponding population is, in principle, all sites at 50 m from the track within residential areas. Now imagine taking millions of random samples of 46 sites from this hypothetical population of all possible sites. For each sample of 46 sites you calculate a variance, using the denominator $(46 - 1)$. The average of these variances will equal the population variance and the estimation procedure, which we perform only for our particular sample, is said to be unbiased for the population variance. If we divided by 46, rather than by 45, on average, over imagined repeated samples, we would obtain a value that is slightly too low. This is a common justification for dividing by one less than the sample size, but it is not entirely convincing because it applies to variances rather than standard deviations. A more intuitive explanation follows from considering a sample of size 1. Formally, the mean is the single observation and the deviation from the mean is zero. Our estimate for the population standard deviation is $\sqrt{[0/(1 - 1)]}$, which is an undefined quantity. This is reasonable since a sample of size 1 can tell us nothing about the variability of the corresponding population. If we divide by the sample size the standard deviation would be $\sqrt{[0/1]}$, which is misleading as an estimate of the population standard deviation. In each case, the zero in the numerator arises because we have to estimate the population mean before we can estimate the standard deviation. This is described as the loss of one *degree of freedom*, a term borrowed from mechanics. A more descriptive term for the standard deviation, root mean squared error (RMSE), is often used by engineers, but it is defined with the sample size as the denominator rather than the sample size less 1. In large samples it makes little difference, so you may wonder why we have discussed this point in such detail. The reason is that the concept of degrees of freedom is needed for the analysis of all experiments.

SeaDragon's final report will include contours of noise levels, such as are commonly produced for airports. Nice examples can be found using Google, searching for 'noise contours'.

▽ Case 2.7 (UOE)

A student in Greta's first-year economics class collected data on 32 movies released in 1997–1998. If you are using MINITAB 14 you will find the worksheet (**movies.MTW**) in the folder **studnt14**; locate it and double-click on it to open it. In earlier versions it will be in **studnt12**. If you don't have it, it is also available at www.greenfieldresearch.co.uk/doe/data.htm. Copy it first into a new Excel spreadsheet and from there load it into a MINITAB worksheet.

There are five columns in the original MINITAB file:

1. movie title;
2. gross receipts in the first weekend after release in millions of dollars;
3. total budget for making the movie in millions of dollars;
4. whether or not the movie has a superstar;
5. whether or not the movie was released in the summer.

One of the objectives of the experiment was to investigate the distribution of the proportion of the budget that was earned over the opening weekend. We need to start by calculating the proportion. Follow **Calc > Calculator >** to open the window of Figure 2.22. Enter the expression **'Opening'/'Budget'** by selecting the names of columns to obtain the proportion. Enter **'PropEarn'** as the name of the variable for storing the results. Now calculate descriptive statistics of the proportion earned. MINITAB displays these in the session window as follows:

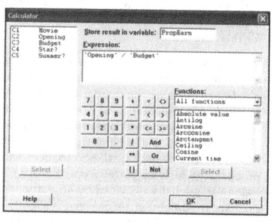

Figure 2.22

Descriptive Statistics: PropEarn

Variable	N	Mean	SE Mean	StDev	Minimum	Q1	Median	Q3	Maximum
PropEarn	32	0.619	0.201	1.139	0.0974	0.223	0.310	0.520	6.568

The median proportion earned is 0.31 while the mean is 0.619, and the standard deviation is 1.139.

The fact that the mean is considerable larger than the median, about a third of a standard deviation, tells us that the distribution is strongly right skewed. This means that there is a long tail to the right;

the distribution is not symmetric and is certainly not near normal. This feature of the distribution is displayed well on the boxplot in Figure 2.23, obtained by following **Graph > Boxplots > Simple**, selecting **PropEarn**, clicking on **Scale** and checking the **Transpose** box.

It is even more striking on the normal probability plot in Figure 2.24, obtained from **Graph > Probability Plot > Single** and selecting **PropEarn**. The default **Distribution** is normal, so you needn't bother with this button, just click **OK**.

Random samples from normal distributions are scattered about a notional line. You can try this with one or two random samples of size 32 from a normal distribution. If you accept the default you get the standard normal distribution with a mean of zero and a standard deviation of one. Use **Calc > Random Data > Normal**.

For some purposes it is convenient to have a transformation of the data so that they are near normal. Common choices, in order of ability to remove right skewness, are: square root, natural logarithm and reciprocal. In the present case, we found that the log transformation was the best.

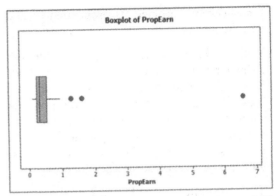

Figure 2.23

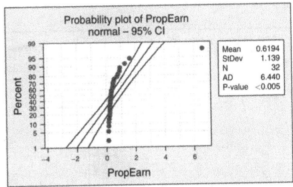

Figure 2.24

Use the **Calculator** window (Figure 2.25) to create a new variable, **LnPE**, as the logarithm of the **PropEarn** values. A boxplot (Figure 2.26) of the transformed data, **LnPE**, is more symmetrically distributed except for one extreme point. This is for the movie *Chasing Amy* which had by far the lowest budget so that, despite having the smallest

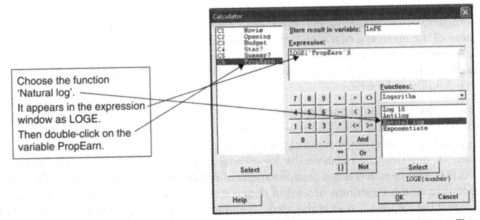

Choose the function 'Natural log'.
It appears in the expression window as LOGE.
Then double-click on the variable PropEarn.

Figure 2.25

first weekend earnings, its proportion earned was the highest.

With brushing (see Figure 2.9) we can identify a single point, or a group of points, on a chart. Select the brush tool from the toolbar and move the pointer over the extreme point. A list will appear showing that the extreme point represents the datum in row7 of the data table, which is for the movie *Chasing Amy*. That row will also be marked in the data window.

The probability plot in Figure 2.27 shows that the transformed data are closer to a normal distribution but the outlying point, representing *Chasing Amy*, is still evident.

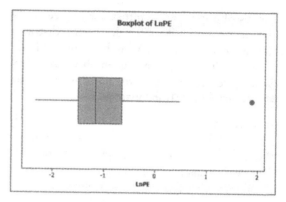

Figure 2.26

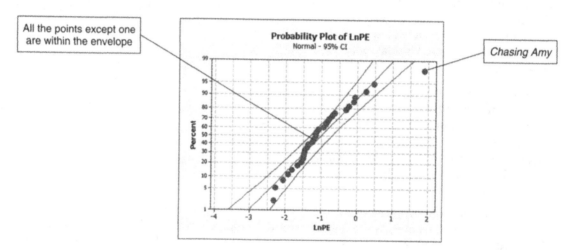

Figure 2.27

Why are we interested in transforming the data to near normality? One reason is that we know a lot about normal distributions. For example, 95% of observations will lie within two standard deviations of the mean and nearly all (99.7%) will lie within about three standard deviations of the mean. In this case, if we assume the normality of the log variable we shall predict nearly all of the log variable values within three standard deviations (sd = 0.866) of the mean (−1.008), that is, between −3.605 and 1.589.

If we back-transform, by taking the exponential, we get a prediction interval for the original variable: between 0.027 and 4.900. This works fine for prediction intervals but not for mean values, because the exponential of the mean of the logarithms of any data is less than the mean of the original data. In this case the mean of the logarithms of **PropEarn** is −1.008, and the exponential of −1.008 is 0.3649 which is less than the mean of **PropEarn** (0.619). In contrast, the exponential of the median of the logarithms of a set of data is equal to the median of the original data. This is because half the data are less than the median, and it follows that logarithms of those data are less than the logarithm of the median and logarithms of data greater than the median are greater than the logarithm of the median. In this case the median of the logarithms of the data is -1.173 and taking the exponential gives 0.3094.

A small numerical example may help make all this clearer. The numbers 1, 10, 100 have median 10 and mean 37. The logarithms, to base 10, of these three numbers are 0, 1, 2 , with median 1.0 and mean 1.0, and 10 raised to the power of 1 is 10.

Given the success of *Chasing Amy*, you may think that the proportion earned decreases with the budget. We leave you to draw scatterplots of **PropEarn** and **LnPE** against the budget, and to calculate correlations. There is some evidence that the proportion earned does decrease with the budget, but the effect is slight and producers shouldn't be too discouraged from big budget films.

3

Single sample experiments

In a single sample experiment, we measure a characteristic of a product or a process several times. The number of times is called the *sample size*. Let's call it N.

We need to decide on a sensible sample size N before doing the experiment. This is the experimental design problem when we wish to use a sample to test whether or not a characteristic of a product or a process is satisfactorily close to a standard specification. The decision relies on some knowledge of the population standard deviation. You may think we are unlikely to have such information when we do not even know the mean. However, nobody does any research in a vacuum of ignorance. We are usually able to use some information from past studies, or from physical argument, to make at least a rough estimate of the population standard deviation. Alternatively, we can take a pilot sample to obtain a first estimate.

We shall present a procedure based on the MINITAB routine **Stat > Power and Sample Size > 1-Sample Z** or **1-Sample t**. We shall explain where they differ in step 5.

1. *State the null hypothesis and the alternative hypothesis.* The hypothesis to be tested is that the mean of the characteristic is precisely equal to a single specified value. This is known as the *null hypothesis*, where the 'null' refers to no difference between the population mean and the specified value. After analysing the data, the conclusion of our experiment will be either that we have insufficient evidence to reject the null hypothesis or that we reject the null hypothesis in favour of the alternative hypothesis. There are three options for the alternative hypothesis and you can choose which one you want by clicking on **Options**.

   * The default is **Not equal**. This is appropriate if the specified value is the ideal target value. The alternative hypothesis is that the population mean is not equal to the specified value. If we reject the null hypothesis, we have evidence that the population mean is too high or too low, according to whether the sample mean is greater than or less than the specified value. An example is a requirement that the average fluoride content of the water supply should be 0.7 parts per million (ppm).
   * If, however, the specified value is a minimum requirement for the population mean, then you should choose **Less than**. Only sample means sufficiently less than the specified mean will be evidence against the null hypothesis. An example is a specified minimum mean

strength for strands that make up wire rope of 700 MPa. The formal null hypothesis is that the mean strength is 700, though in effect we are testing a hypothesis that the population mean is at least 700. The alternative hypothesis is that the mean strength is less than 700.

- Finally, if the specified value is a maximum requirement for the population mean, then **Greater than** is the appropriate choice. Only sample means sufficiently greater than the specified mean will be evidence against the null hypothesis. An example is a requirement that mean polycyclic aromatic hydrocarbon levels in drinking water should be less than 150 ng/l. The null hypothesis is that the mean is 150. The alternative hypothesis is that the mean is greater than 150.

2. *Significance level.* The significance level is a probability value that will be used in the subsequent data analysis to determine whether or not the difference between the sample mean and specified value may be regarded as statistically significant. It is the probability that the difference will be so regarded if in fact there is *no* difference between the population mean and specified value. Thus, it must be a small value and usually it is taken as 0.05. If you use a smaller value, such as 0.01, you may be more confident in reaching the correct decision but you will need more items in your study. With a larger value, such as 0.10, you would be less confident in your decision but you would need fewer items.

3. *Technically significant difference.* The technically significant difference (TSD) is the minimum difference between the population mean and the specified value that the investigator regards as being of practical importance. If the difference is greater than this, then it is technically important and we would hope to reject the null hypothesis and claim evidence of a difference. The TSD must be specified before the study starts.

4. *Power.* The power is the probability of rejecting the null hypothesis if the difference between the population mean and the specified value equals the TSD. We naturally want the power of any experiment to be as high as possible. Indeed, if the power is low, there is little chance of detecting an important difference. It would be a waste of money and time.

5. *Standard deviation.* The standard deviation of the distribution of the characteristic in sample units is needed for the calculation of sample size. We may be willing to assume some specific value for this standard deviation, based on experience of the process. In such cases, we use **1-Sample Z** to determine the sample size and for the analysis. More commonly, we will not know the standard deviation. Then we use **1-Sample t** for the determination of the sample size and for the analysis. But, for the determination of the sample size, we still have to provide a value for the standard deviation. This will have to be a best guess based on experience, similar studies, or a pilot study.

Once we have gone through steps 1–5, we can enter the numbers into the MINITAB routine and obtain a sample size.

So far, our discussion has been confined to a specification for a mean value. A specification should also set limits on variability. We will end our first case study by explaining how to test a hypothesis about a standard deviation.

▽ Case 3.1 (AgroPharm)

Some water companies add fluoride to the water, as a public health measure to reduce the incidence of tooth decay. According to the British Dental Health Foundation, about 10% of the supply in the UK has fluoride added – see www.dentalhealth.org.uk/faqs/leafletdetail.php. The scientific consensus is that fluoridation has been successful in reducing tooth decay, and that it does not have harmful side-effects at the concentration used. Nevertheless, there is still controversy about the addition of fluoride to the public supply. For example, on 28 April 2004, *The Sentinel* (Stoke-on-Trent) reported the case of a man objecting to fluoridation under the headline 'Water rebel willing to face court'. Nevertheless, the medical consensus is that it is beneficial to add fluoride to bring the concentration up to 1 part per million (ppm). This is well below some naturally occurring concentrations. However, some variation about a target value is inevitable. It is important that this variation is kept as low as possible, and people responsible for water supplies are also wary about exceeding 1 ppm. So, a typical current recommendation is a target value of 0.7 ppm (equivalent to 0.7 mg/l) and an explicit requirement for dosing precision is that 'Instantaneous measured fluoride concentration must be within 0.2 mg/l of the target' – see, for example, www.wrc.org.za.

AgroPharm supplies fluoride, and the dosing equipment, to many water companies in the USA, Australia, South Africa and Europe. The company has a commercial interest in working with water companies to demonstrate that the system of dosing is on target and within the allowed range. Bill Brown has been given the job, and his first collaboration is with a water company called AquaBleu which supplies water to an area in Brittany. Routine daily monitoring of the fluoride concentration, and other indicators of water quality, is managed at the treatment works. But, as well as this, Bill intends to monitor the water supply at people's houses about twice a year. He will ask AquaBleu to draw a random set of houses from the billing register, fill a litre jar at the kitchen tap in each house, and measure the fluoride concentrations. He supposes that the hypothetical distribution of fluoride concentrations in all jars that could be filled has a mean μ and a standard deviation σ. The target value for the mean is 0.7 but the standard deviation is not specified explicitly.

He decides to make a further assumption that fluoride concentrations have a normal distribution. The normal distribution has a distinctive bell shape and is appropriate if we imagine that deviations from the target, known as *errors*, are the sum of a large number of mini-errors. In this case the mini-errors might be: slight variation in the concentration of the hydrofluorosilicic acid (H_2SiF_6) used as the source of fluorine; slight variation of water flow through the treatment works; temperature variation; measurement error when determining the natural level of fluorine in the water; slight error in the amount of acid dispensed; measurement errors when determining the concentration of fluorine in the jars; and so on.

The dosing precision requirement is that measurements must be within 0.2 of target. For a normal distribution, the probability of being more than three standard deviations from the mean is between 0.2% and 0.3%. MINITAB can provide a more precise value. Follow **Calc > Probability Distributions > Normal**, click on **Cumulative Probability** and enter **3** in the **Input Constant** box. The default mean of 0 and standard deviation of 1 correspond to a standard normal random variable, generally referred to as Z. MINITAB returns the probability that a standard normal variable is less than 3.0, which is 0.9987. This scales for any other normal distribution, and the probability that a normal variable is less than three standard deviations above the mean ($\mu + 3\sigma$) also equals 0.9987. Hence, the probability that a normal variable is more than three standard deviations above the mean is $1.0 - 0.9987 = 0.0013$. It follows from the symmetry of the normal distribution that the probability of being more than three standard deviations below the mean is also

0.0013. So, the probability of being more than three standard deviations from the mean is 0.0026. A reasonable, but not stringent, interpretation of the dosing precision requirement is to take the, hoped for, maximum discrepancy of 0.2 as 3σ, corresponding to $\sigma = 0.067$. Bill thinks a population standard deviation of 0.06 is realistic for the AgroPharm system and that this will suffice to meet the precision requirement. He now needs to decide how many jars to fill. This is the crucial first step in the design of the experiment. We follow the five steps:

1. *State the null hypothesis and the alternative hypothesis.* The hypothesis to be tested is that the mean for the imaginary population of all possible jars is 0.7. The alternative hypothesis is the MINITAB default: **Not equal**.
2. *Significance level.* Bill makes the common choice of 0.05, which is the MINITAB default.
3. *Technically significant difference.* Bill would like to know if the mean is more than 0.035 from target of 0.700. He specifies the TSD as 0.035.
4. *Power.* He decides to specify this as 0.90.
5. *Standard deviation.* He assumes a value of 0.06. Therefore, he uses **1-Sample Z**.

Bill uses the MINITAB procedure **Stat > Power and Sample Size > 1-Sample Z** and enters **Differences: 0.035, Power values: 0.90, Standard deviation: 0.06**. A sample size of 31 is returned, with a power of 0.9011. He had originally budgeted for a sample of 30 jars, and he finds this corresponds to a power of 0.891. Bill settles on a sample size of 30. The result is:

Power and Sample Size

1-Sample Z Test

```
Testing mean = null (versus not = null)
Calculating power for mean = null + difference
Alpha = 0.05 Assumed standard deviation = 0.06
```

	Sample	
Difference	Size	Power
0.035	30	0.891601

A few months later he has the results from his experiment; these can be found at www.greenfieldresearch.co.uk/doe/data.htm. He enters the measurements in a single column (**C1**) and makes the stem-and-leaf plot shown in Figure 3.1. He obtains the following descriptive statistics:

N	Mean	SE Mean	TrMean	StDev
30	0.732	0.02	0.73	0.09
Minimum	Q1	Median	Q3	Maximum
0.59	0.65	0.7055	0.8058	1.017

Stem-and-leaf of fluoride		
N = 30		
Leaf unit = 0.010		
1	5	9
8	6	2233444
12	6	6689
(4)	7	0000
14	7	555667
8	8	012334
2	8	5
1	9	
1	9	
1	10	1

Figure 3.1

You should now be familiar with all except the SE Mean. Imagine our experiment being repeated millions of times. Each time it is repeated we can calculate a sample mean, and these means vary from experiment to experiment. But, they don't vary as much as individual values. If the experiments are independent inasmuch as the data in any one experiment are not influenced by the data that arose in any of the others, then the standard deviation of the means is the standard deviation of the individual data values divided by the square root of the number of data in the experiment:

$$\text{standard deviation of the means} = \frac{\text{standard deviation of the individual data values}}{\text{square root of the number of data in the experiment}}$$

The standard deviation of the means is commonly called the *standard error of the mean*, denoted by SE Mean in the MINITAB summary. You can check that $0.0941/\sqrt{30} = 0.0172$.

Bill has an impression from the stem-and-leaf plot that the data are not from a normal distribution. The mean of 0.732 doesn't give any cause for alarm, but the standard deviation of 0.0941 is rather high. He proceeds, using **Stat > Basic Statistics > 1-Sample Z**, to test the null hypothesis that the mean is equal to 0.7 against an alternative that it is not equal to 0.7 at the 5% level on the basis that he knows the population standard deviation is 0.06. This procedure does not require an entry of the significance level (5%) because it calculates the P-value and leaves the judgement to Bill.

However, he clicks on **Options** where he is asked for **Confidence level**. The default for this is 95% and it is needed to estimate a corresponding confidence interval. **Options** also asks for the alternative hypothesis, and Bill chooses **Not equal**. The result is:

```
One-Sample Z: fluoride

Test of mu = 0.7 vs not = 0.7
The assumed standard deviation = 0.06


Variable   N      Mean     StDev  SE Mean        95% CI          Z      P
fluoride  30  0.732000  0.094097 0.010954 (0.710530,  0.753470)  2.92  0.003
```

The test is referred to as a Z-test because if the null hypothesis is true, the ratio of the difference between the hypothesised mean and sample mean to the assumed known standard error of the mean is a standard normal variable. In this case SE Mean is $0.06/\sqrt{30}$ which equals 0.011, and the ratio is $(0.732 - 0.700)/0.011$ which equals 2.92. The P-value is the probability that a standard normal variable exceeds 2.92 or is less than -2.92, and is 0.003. The P-value of 0.003 is less than 0.05, so on the basis of this test he could claim evidence that the mean is greater than 0.7. But, Bill is wary about assuming that the population standard deviation is 0.06 when the sample estimate is 0.0941, and proceeds as follows.

In many applications we do not wish to assume a value for the population standard deviation, and it seems natural to replace it by the sample estimate. If the null hypothesis is true, the ratio of the difference between the hypothesised mean and sample mean to the sample estimate of SE Mean has a Student t-distribution. In our case the estimate of the standard error of the mean is $0.0941/\sqrt{30}$ which equals 0.0172.

Student's t-distribution was originally derived by W. S. Gossett, who worked for the Guinness brewery in Dublin and published his work under the pseudonym 'Student' in *Biometrika* in 1908. The distribution looks similar to the standard normal distribution but the peak is lower and the tails are rather higher. The exact shape depends on the degrees of freedom available for estimating the standard deviation. The t-distribution is close to the standard normal distribution if the degrees of freedom exceed about 30. The MINITAB syntax is **Stat > Basic Statistics > 1-Sample t**, using the same options as for the Z-calculation. The result is:

```
One-Sample T: fluoride

Test of mu = 0.7 vs not = 0.7

Variable    N    Mean   StDev   SE Mean      95% CI          T      P
fluoride   30   0.732   0.094   0.017   (0.697, 0.767)   1.86   0.073
```

The P-value is now 0.073 and we no longer claim evidence against the hypothesis that the population mean is 0.7 ppm at the 5% level. The difference between 0.700 and 0.732 can reasonably be attributed to chance.

The MINITAB tests include 95% confidence intervals (CI) for the population mean. If we assume the population standard deviation is 0.06 we are 95% confident that the population mean is between 0.7105 and 0.7535. A loose practical interpretation is to interpret this as a 95% chance that the mean is between the two values. An explanation for the construction of the confidence interval relies on the notion of millions of imagined repeated samples which lead to a distribution of sample means. Then:

1. The sample mean has an approximate normal distribution even if the individual values do not. The approximation improves as the sample size increases.
2. The distribution of the sample mean has a mean equal to the population mean μ and a standard deviation, **SE Mean**, equal to $\sigma/\sqrt{N}$ where σ is the standard deviation of individual measurements and N is the sample size.
3. 95% of a standard normal distribution lies between -1.96 and $+1.96$.
4. Therefore there is a 95% chance that the sample mean lies within $1.196\sigma/\sqrt{N}$ of μ.
5. There is a 95% chance that μ is within $1.96\sigma/\sqrt{N}$ of the sample mean.
6. If we write $\bar{x}$ for the sample mean, a 95% confidence interval for the population mean μ is from $\bar{x} - 1.96\sigma/\sqrt{N}$ up to $\bar{x} + 1.96\sigma/\sqrt{N}$.

Notice that the confidence interval becomes narrower as the sample size N increases, but the reduction is only proportional to its square root. If, for example, we wish to reduce the width of a confidence interval by a factor of one-half we need to increase the sample size by a factor of 4.

If we do not know the population standard deviation, σ, we replace it by our sample estimate s which is made with $N - 1$ degrees of freedom. We also need to modify our '95% of a standard normal distribution' (3 above) to 95% of a t-distribution with $N - 1$ degrees of freedom lies between $-L$ and L, where L is the number such that 2.5% of the t-distribution lies above it. You can find this number from MINITAB, but you will not usually need to. For our case, N equals 30 and the degrees of freedom are 29. Follow **Calc > Probability Distributions > t**, select **Inverse cumulative probability** and enter **Degrees of freedom 29** and **Input constant 0.975**. The result is as follows:

```
Inverse Cumulative Distribution Function

Student's t distribution with 29 DF

P(X <= x)        x
   0.975     2.045
```

Notice that the value is close to 2.0. The ensuing 95% CI is part of the **One-Sample T** output.

There is a duality between the hypothesis test and the confidence interval. We reject the null hypothesis that $\mu = 0.7$ at the 5% level if the 95% CI excludes 0.7. If the 95% CI includes 0.7 we have no evidence to reject the null hypothesis at the 5% level. We recommend the CI approach because you can see at a glance whether the lack of evidence is attributable to a small sample or is because the population mean is close to the specified mean.

Bill thought that the stem-and-leaf plot did not look like a random sample from a normal distribution. A normal probability plot allows us to investigate this more thoroughly. The sequence to follow is **Graph > Probability Plot > Normal; Single** will be highlighted, so click **OK.** Select **fluoride** as **Graph variable**, click on **Distribution** and choose **Normal;** leave historical parameters blank. The graph is in Figure 3.2.

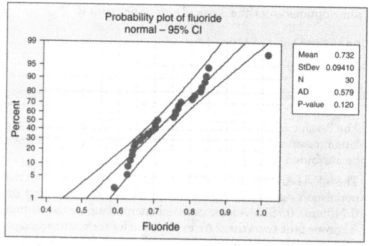

Figure 3.2

The points are plotted by sorting the data into ascending order and then plotting the ith point corresponding to the ith datum on the horizontal axis and a probability of $i/(N + 1) \times 100\%$ on the vertical Percent axis. For example, the 4th datum in ascending order for our case is 0.633, and the corresponding percent is $4/31 \times 100 = 12.9\%$.

If the data are sorted into ascending order they are known as *order statistics*.

The curves either side of the fitted line provide a 95% CI for the line corresponding to a population assumed to be normally distributed. Individual points lying beyond these lines are not necessarily strong evidence against the null hypothesis being from a normal distribution. A formal test is given by **Stat > Basic Statistics > Normality Test > Anderson-Darling**. However, the legend with the graph of Figure 3.2 already includes the results of this test: AD = 0.579, with an associated P-value of 0.12. Despite appearances, we have only weak evidence against the null hypothesis using this test. The Ryan–Joiner test statistic of 0.961 has a P-value of 0.04 (**Stat > Basic Statistics > Normality Test > Ryan-Joiner**) and provides more substantial evidence that the data are not from a normal distribution.

Now that we have some evidence against an assumption that the data are from a normal distribution, we should consider the consequences. It has little effect on our hypothesis tests about the mean or our confidence intervals. This is because they rely on the sample mean having a normal distribution rather than individual values necessarily having a normal distribution. It does, however, have an effect on our claim that only 2.6 in a thousand individual fluoride measurements will be more than three standard deviations from the mean. If a distribution is positively skewed we are likely to get more points beyond three standard deviations above the mean. This is undesirable when water companies are keen to demonstrate that they maintain the fluoride in the water supply below 1.0 ppm.

The sample standard deviation (s), which equals 0.094, is considerably higher than the value of 0.06 which Bill had assumed. He is concerned about this difference and wonders whether it can reasonably be attributed to chance or whether it provides evidence that the AgroPharm system is less precise than he thought. Before we follow this up, we review the concept of confidence intervals.

A point estimate of a population parameter, such as the mean, is a single best estimate and is usually the value of the corresponding sample statistic. So the point estimate of the population mean (μ) is the sample mean ($\bar{x}$). Similarly, the point estimate of the population standard deviation (σ)

is the sample standard deviation (s) which is calculated with the denominator ($N - 1$). An interval estimate of a population parameter is known as a *confidence interval* (CI) and provides an estimate of the precision with which the population parameter has been estimated. For example, when making decisions, we can act as if there is a 95% chance that the 95% confidence interval for the mean does include the population mean. Less formally, we can be fairly confident that the 95% confidence interval includes the corresponding population parameter.

In this chapter, the experiment is a comparison with a standard or specified value. If our 95% confidence interval excludes the specified value, we have evidence, at the 5% level, that the population mean differs from the specified value. If our 95% confidence interval includes the specified value, we have insufficient evidence, at the 5% level of significance, that the population mean differs from the specified value. However, the width of the confidence interval, and hence our conclusions, depend on the sample size. If we have a very large sample size, the confidence interval will be narrow and we may find that a small difference between the sample mean and the specified value is sufficient evidence to reject the hypothesis that the population mean equals the specified value. But we must consider whether a small difference is of practical concern. At the other extreme, if we have a very small sample the confidence interval will be wide and even a large difference between the sample mean and the specified value will not provide any convincing evidence that the population mean has moved away from the specified value. Specifying the technically significant difference is crucial for determining a suitable sample size.

MINITAB provides confidence intervals for standard deviations as part of its **2-variances** procedure. However, the procedure requires two columns of data so first copy the data from **c1** into **c2**. Use **Stat > Basic Statistics > 2 Variances > Storage** and tick both **Upper confidence limit for sigmas** and **Lower confidence limits for sigmas**.

It's a Bouferroni procedure, so alpha is halved for an individual sigma. Using MINITAB, a 97.5% confidence interval for the standard deviation of the process is [0.073, 0.132]. The lower bound of the interval (0.073) is greater than the assumed value of 0.06 so there is evidence that the process is too variable to meet the requirement for dosing precision.

△ Review of Case 3.1

To summarise the situation so far, Bill Brown filled litre jars of water at a random sample of 30 kitchen taps, to check compliance with a specified mean of 0.70 ppm and standard deviation of 0.06 ppm. The sample mean was 0.73 ppm and this is compatible with an assumed population mean of 0.70. However, the sample estimate of the population standard deviation is 0.094 and this does suggest that the process is too variable. He is also concerned that the largest value in the sample exceeded 1.0 ppm. His next move is to check the procedures at the treatment works.

In some experiments the response is restricted to one of two possible values that are conveniently referred to as 'success' and 'failure'. The object of the experiment will be either to estimate the proportion of successes in the corresponding population or to test whether the population proportion differs from a specified value.

▽ Case 3.2 (SeaDragon)

The Electronics Division of SeaDragon manufactures scientific calculators. The designer, Sam Silver, has modified the casing and wishes to investigate the durability of his new design.

Sam decides to test the design by dropping calculators, with the case open, from a height of 1 metre onto a concrete floor. Sam hopes that the probability of the calculator being damaged will be less than 0.1. If the probability is as high as 0.2 he would reconsider the design. To decide on a sample size he uses MINITAB, with a choice of 0.8 for the power: **Stat > Power and Sample Size > 1 Proportion**, entering **Hypothesized p: 0.1, Alternative value of p: 0.2** and **Power value: 0.80** MINITAB returns a sample size of 86:

Power and Sample Size

```
Test for One Proportion

Testing proportion = 0.1 (versus not = 0.1)
Alpha = 0.05

Alternative      Sample      Target      Actual
 Proportion       Size       Power        Power
  0.200000         86        0.8000       0.8020
```

Sam performs the experiment and finds that 17 of the calculators are damaged by the fall. He uses MINITAB to test the hypothesis that the proportion in the corresponding population is 0.1 and to construct a 95% confidence interval for this proportion. The procedure is **Stat > Basic Statistics > Proportion**, click on **Summarized data** and enter **Number of trials: 86, Number of events: 17**, then select **Options** and enter **Test Proportion: 0.1**. The result is:

Test and CI for One Proportion

```
Test of p = 0.1 vs p not = 0.1

                                                              Exact
Sample     X    N    Sample p          95.0% CI             P-Value
1          17   86   0.197674   (0.119578, 0.297521)         0.011
```

Sam does have evidence, at the 5% level, that the proportion differs from 0.1 since the P-value of 0.011 is less than 0.05. The 95% confidence interval [0.12, 0.30] is more useful. It not only provides Sam with evidence that the population proportion exceeds 0.1, but also warns him that it might be as high as 0.3. In most cases the only damage was that hinges on the front cover broke, so he decided to review this aspect of the design. Sam might have claimed that a one-sided alternative, that the sample proportion exceeds 0.1, is more appropriate because only then would he review the design. In **Options** change **not equal** to **greater than** to obtain the following:

Test and CI for One Proportion

```
Test of p = 0.1 vs p > 0.1

                                                      Exact
Sample     X    N    Sample p   95.0% Lower Bound    P-Value
1          17   86   0.197674        0.130046         0.005
```

The P-value is halved, and we are given a one-sided confidence interval. Sam can be 95% confident that the proportion exceeds 0.13. The lower limit of a one-sided 95% CI is also the lower limit of a (two-sided) 90% confidence interval, [0.13, 0.28].

In Case 3.2, the confidence interval is the best summary of Silver's experiment, but the notional hypothesis test was useful for determining a sample size. In the next case the hypothesis test corresponds to US Environmental Protection Agency (EPA) standards. The EPA has established an enforceable lead concentration action level for public water supplies. A public water system exceeding 15 µg/l in more than 10% of sampled homes is required to take action to reduce lead levels. Typical sampling by, for example, the City of New York Department of Environmental Protection is of 1300 water samples per month.

▽ **Case 3.3 (AgroPharm)**

AquaBleu have recently won a contract to operate the water supply system in a US city and surrounding suburbs. AquaBleu must therefore demonstrate compliance with EPA standards. The EPA will monitor, and occasionally run checks on, Aqua Bleu's sampling strategy. Bill Brown has been retained to advise. The manager responsible for the contract thinks that the company should aim for less than 4% of homes exceeding 15 µg/l. The manager will report the results of tests each month, and he asks Bill what the power of a test that the proportion is 0.04, at the 0.05 level with a two-sided alternative, based on 200 water samples per month would be if the actual proportion is 0.1.

```
Power and Sample Size

Test for One Proportion

Testing proportion = 0.04 (versus not = 0.04)
Alpha = 0.05

Alternative  Sample
 Proportion   Size    Power
   1.00E-01    200   0.9392
```

The manager wonders what would be the effect of increasing the number of samples each month to 250:

```
Power and Sample Size

Test for One Proportion

Testing proportion = 0.04 (versus not = 0.04)
Alpha = 0.05

Alternative  Sample
 Proportion   Size    Power
   1.00E-01    250   0.9701
```

The probability of concluding that the proportion exceeds 0.04, if the population proportion is as high as 0.10, has increased from 0.94 to 0.97 with the increase in the sample size from 200 to 250. The manager must decide if this is worth the extra cost. Before doing so, he asks about using a one-sided alternative with the sample size of 200:

```
Power and Sample Size

Test for One Proportion

Testing proportion = 0.04 (versus > 0.04)
Alpha = 0.05

Alternative   Sample
  Proportion    Size      Power
        0.1      200   0.960286
```

△ **Review of Case 3.3**

Using MINITAB, Bill follows **Stat > Power and Sample Size > 1 Proportion**. He enters **200** in the **Sample sizes** box, **0.1** in the **Alternative values of p** box, and **0.04** in the **Hypothesized p** box. Clicking on **Options** he confirms that the **Alternative Hypothesis** is **Not equal** and that the **Significance level** is **0.05**. He clicks **OK, OK**: The manager decides to keep to a sample size of 200 and consider the one-sided alternative hypothesis, because AquaBleu will take action only if the proportion appears too high.

4

Comparing proportions

In this chapter, we compare treatments when the response is simply that one item is better or worse than another, or that an item is satisfactory or not. If items corresponding to the two treatments can be paired, this usually leads to a more precise comparison. However, this is not always practical. In our first case study, visitors to a vineyard are asked to express a preference between an Australian sparkling white wine and a popular French champagne. The two glasses of wine presented to each taster are a pair, and the aim of the pairing is to remove variability among visitors in their assessment of champagne-style wines. The second case concerns the effect of a talk, in which the gambler's fallacy is explained, on a facet of compulsive gambling behaviour known as chasing losses. It is also a paired experiment, in which the pairing is of behaviour before and behaviour after the talk for each participant. The intention is that the pairing will remove the variation in gambling behaviour between participants. In the third case study, we compare resistance to impact of samples of TV screens from two suppliers. Here the samples are assumed to be independent because there is no advantage to be gained from pairing. The analysis of the fourth case study also assumes independent samples.

▽ Case 4.1 (AgroPharm)

Dionysius Darkred is the winemaker at Wombat Wines, a vineyard owned by AgroPharm. Wombat make a sparkling Chardonnay using the remouage method. The bottles are turned so that the sediment settles in the neck of the bottle. This increased production by ten times over the traditional method and it brought the price down to where the average person can enjoy champagne. The wine sells well in Australia and the company is keen to start exporting it to the UK. However, Dionysius thinks they should first compare it with a popular non-vintage French champagne, which has the largest share of the English market and which would remain the main competitor. He intends to do this by running an experiment in which he asks visitors, who are interested in taking part, to taste both wines and give their preference. He will present the wines in a random order, subject to each being presented first on an equal number of occasions.

Dionysius next needs to decide how many visitors to ask to be tasters. He thinks he can find six visitors a day, over a 12-day period, who would be willing to compare the sparkling Chardonnay with the champagne. Each day, he will aim to present the Chardonnay and champagne first on the same number of occasions. To make a specific sample size calculation, he supposes that 50 of the 72 visitors, might express a preference. If there is no difference in the quality of the wine, he would expect a proportion of 0.5, of those who could tell a difference, to prefer the champagne. If the proportion preferring the champagne was as low as 0.3 he would like to know about it. He can use MINITAB to evaluate the probability of detecting this. He follows: **Stat > Power and Sample Size > 1 Proportion** and enters **Sample size: 50; Alternative value of p: 0.3; Hypothesised p: 0.5.** The result is:

```
Power and Sample Size

Test for One Proportion

Testing proportion = 0.5 (versus not = 0.5)
Alpha = 0.05

Alternative   Sample
Proportion     Size      Power
      0.3        50    0.828326
```

The power is 0.83.

The result of the experiment was that 53, of the 72 visitors, detected a difference, with just 17 of these preferring the champagne. Dionysius used **Stat > Basic Statistics > 1 Proportion** with the **Summarized data** option. The number of trials is 53 and the number of events is 17:

```
Test and CI for One Proportion

Test of p = 0.5 vs p not = 0.5

                                                      Exact
Sample     X     N  Sample p         95.0% CI        P-Value
1         17    53  0.320755  (0.199187, 0.463161)    0.013
```

The sample proportion preferring the champagne is 0.32 and a 95% confidence interval for the proportion of the population of all visitors who can tell a difference preferring the champagne is [0.20, 0.45]. Since the entire interval is below 0.5, there is clear evidence that a higher proportion of the population of visitors expressing a preference prefer the Australian sparkling Chardonnay.

△ Review of Case 4.1

Dion asked 72 visitors to taste both the sparkling Chardonnay and the champagne. Half were randomly selected to taste the Champagne first, and the other half tasted the sparkling Chardonnay first. About three quarters (74%) stated a preference and 36 out of these 53 (68%) preferred the sparkling Chardonnay. Dion is encouraged by this result, although he realises that he can expect only half of his potential customers to prefer the sparkling Chardonnay and that this estimate is far from precise. It is also possible that Australian and British tastes may differ, but he thinks this is unlikely. He decides to go ahead with the plan to export the sparkling Chardonnay, provided he can keep its retail price below that of the champagne.

▽ Case 4.2 (UoE)

Chasing losses (or *chasing* for short) is one of ten criteria listed by the American Psychological Association for identifying compulsive gamblers. The diagnosis is made if someone exhibits five or more of these criteria. Ariella Auburn, in the Department of Social Studies,

		After presentation	
		Not chase	Chase
Before presentation	Not chase	118	33
	Chase	57	28

Table 4.1

has decided to run an experiment to investigate the effect of a half-hour talk on the gambler's fallacy – the mistaken notion that the odds for something with a fixed probability increase or decrease depending upon recent occurrences – on students' gambling behaviour. She asked for student volunteers from those students who gamble at least once a month, and had 236 responses. The students were asked whether or not they had chased losses during the past month and then asked to attend the talk. One month later they were again asked whether or not they had chased losses during the past month. The results are summarised in Table 4.1.

The analysis concentrates on those students who have changed their behaviour after hearing the talk. If the talk has no effect we would expect half of these to change from not chasing losses to chasing losses. In this case 33 out of 90 did so. A 95% confidence interval for the proportion in the corresponding population can easily be found from MINITAB:

Test and CI for One Proportion

Test of p = 0.5 vs p not = 0.5

Sample	X	N	Sample p	95.0% CI	Exact P-Value
1	33	90	0.366667	(0.267522, 0.474851)	0.015

There is strong evidence that the talk has discouraged students at UoE from chasing losses.

This an example of an experiment that is analysed by McNemar's test. It is similar to Case 4.1 except that there is now a before and after aspect to the experiment. There will be differences in students' interpretation of chasing losses, but this is not crucial because students report only the change in their own gambling behaviour. It is important to realise that the results of Ariella's experiment are formally restricted to an imaginary population of all possible UoE students. It may be reasonable to extend this to all students, but we have no evidence it applies to the population as a whole.

△ Review of Case 4.2

Ariella recruited 236 volunteers from the students at UoE who gamble at least once a month. Fifty percent denied chasing losses, and 12% had chased losses and intended to continue doing so. This left 90 students whose behaviour changed. Of these 63% did not chase losses in the month after the talk. There is evidence that the talk has had some effect, but Ariella had hoped that the results would be more dramatic. She decides to discuss the content of the talk with a colleague in the Statistics Department, with the aim of making it more compelling.

▽ Case 4.3 (SeaDragon)

A subsidiary company of SeaDragon manufactures television sets. TV tubes contain a vacuum and implode if broken, so an important safety feature of a TV set is its resistance to impact if, for example, a child falls against it. Valerie Violet is the production engineer and she wishes to compare the impact resistance of the glass bowls, which are the basis of the tube, from two suppliers, A and B.

The tests are done in a sealed room. The glass bowl is held in a jig and an iron ball is swung against it under standard conditions. The response is whether or not the bowl breaks. Valerie wishes to determine an appropriate sample size to take from each supplier. She expects, from experience, about 40% of tubes from A to break. If the percentage from B that break is only 30% she would like a probability of 0.5 of detecting the improvement, testing at the 5% level with a two-sided alternative. Using MINITAB, she finds that the sample size needs to be 175:

Power and Sample Size

Test for Two Proportions

Testing proportion 1 = proportion 2 (versus not =)
Calculating power for proportion 2 = 0.3
Alpha = 0.05

Proportion 1	Sample Size	Target Power	Actual Power
0.400000	175	0.5000	0.5005

This is unreasonably large. If she expects the difference in proportions to be greater and is prepared to take this as the TSD, then she could test a smaller sample with a higher power. She can do this by taking samples of 100 and the percentage of breakages from B as low as 25%. The power is 0.62, and after discussion with members of her team decides to go ahead with the experiment. The results were that 35 out of 100 bowls from A broke and 22 out of the 100 from B broke. She used MINITAB to analyse the results:

Test and CI for Two Proportions

Sample	X	N	Sample p
1	35	100	0.350000
2	22	100	0.220000

Estimate for p(1) − p(2): 0.13
95% CI for p(1) − p(2): (0.00618053, 0.253819)
Test for p(1) − p(2) = 0 (vs not = 0): Z = 2.06 P-Value = 0.040

The 95% confidence interval for the difference in proportions is [0.01, 0.25] which excludes zero, so there is some evidence that the proportion of bulbs from manufacturer B that fail is less than the proportion of those from manufacturer A. Since the confidence interval excludes zero, we know the P-value will be less than 0.05, and it is given as 0.040.

▽ Case 4.4 (UoE)

Crispin Chartreuse is in the Management School at UoE. As part of a study of the effects of music in advertising, Crispin divided 244 undergraduates into four groups. Group 1 listened to music they liked during which a light blue pen was advertised, and Group 2 listened to music they disliked during which the light blue pen was advertised. Groups 3 and 4 were treated similarly except that a beige pen was advertised. A pilot study had indicated that students were indifferent about the pen colour, and that most liked the music from *Grease* and disliked Indian music. At the end of the advertisement students were asked to choose a free pen from either a box of blue pens or a box of beige pens, and to express their reaction to the music on a scale from 1 to 5. In the pilot study the average score for the music from *Grease* and the Indian music had been 4.3 and 1.5, respectively. Students were dropped from the analysis if they did not respond 1 or 2 to the Indian music and 4 or 5 to the music from *Grease*. You may be surprised to hear that this left 195 students. The results are summarised in Table 4.2.

	Chose advertised pen	Chose non-advertised pen	Totals
Liked music	74	20	94
Disliked music	30	71	101
Totals	104	91	195

Table 4.2

The analysis is easy using **Stat > Basic Statistics > 2 Proportions** with the **Summarized data** option:

```
Test and CI for Two Proportions

Sample      X      N   Sample p
1          74     94   0.787234
2          30    101   0.297030

Estimate for p(1) − p(2): 0.490204
95% CI for p(1) − p(2): (0.368604, 0.611805)
Test for p(1) − p(2) = 0 (vs not = 0): Z = 7.90 P-Value = 0.000
```

There is strong evidence that the music affected the consequences of the advertising. The corresponding population in this case is rather restrictive, but it does indicate that music can have a subtle effect on people's choices. For more on this, see Gorn (1982), on which paper this case is based.

△ Review of Case 4.4

Ninety four students listened to music from *Grease*, during which a pen was advertised, and responded that they liked the music. Of these, 77 (79%) chose a pen of the advertised colour. One hundred and one students listened to Indian music, during which a pen was advertised, and responded that they did not like the music. Of these only 30 (30%) chose a pen of the advertised colour. The music has a strong effect on the students' choices when they were initially indifferent about the colour. We might surmise that the background music is a crucial part of an advertisement.

5

Comparing two treatments

5.1 Paired experiments

In Chapter 3 we studied an experiment in which the mean of a single population was compared against a standard or target value, using the mean of a sample. In Case 3.1 the target value for the fluoride mean was 0.7 ppm, and the null hypothesis was that the mean equalled this target. There is a special case of such an experiment: when the null hypothesis is that the mean equals zero. This happens when two populations are being compared but we are able to pair every observation in a sample from one population with a corresponding observation from the other population. The null hypothesis is that the means of the two populations are equal. These experiments are called *paired experiments*. They offer valuable economies in research provided certain criteria are met. Here are some examples:

1. *Wear of bearings made from two metals.* Two bearings (A and B), one made from white metal and the other made from an alternative soft alloy metal, are fitted to a shaft. In normal use the bearing and shaft are separated by a pressurised oil film and the wear should be negligible over a few days. In the experiment, no lubrication was used to induce rapid wear. The shaft is rotated, under a fixed load, for a constant number of turns. Despite our best efforts the fixed load may vary somewhat if we repeat this trial. However, this need not affect our comparison because we can analyse the differences in wear of the two bearings from each trial. For each trial the wear of each bearing is measured as a weight loss (W_A = wear of A, W_B = wear of B). The difference between the two bearings is computed ($W_A - W_B$). We repeat this with, say, 29 more pairs, giving a sample size of 30, so we have 30 values of a variable that we can call *difference*. We now compute the mean of difference and compare it with a value of zero, using the procedure introduced in Chapter 3. If the mean is significantly different from zero we conclude that the two types of bearing do not wear the same amount under the same conditions. There may also be a biasing effect: one end of the shaft may experience slightly more load than the other end. We overcome this by randomly allocating the bearings of each pair to the shaft ends, subject to each metal being used an equal number of times at each end of the shaft.

2. *Effect of humidity on adhesion of sprayed paint.* Suppose some metal tiles are each sprayed with paint. Half of them are subjected to 90% humidity and half to 60% humidity and the surface

conditions are measured. We are interested in the effect of humidity on the surface condition. But we have a source of variation that we have not controlled: the amount and quality of sprayed paint will have varied between the metal tiles, no matter how hard we tried to keep them constant. A paired experiment enables us to control the effect of this variation. After all the tiles have been sprayed, cut each one into two and label the halves A and B. Subject those labelled A to 90% humidity and those labelled B to 60% humidity. There may also be a biasing effect: the left half of each tile may in every case receive slightly more or less paint than the right half. We overcome this by randomly labelling the two halves of each tile, subject to the right and left halves receiving the same numbers of A and B labels.

3. *Effect of storage on brightness of a pigment.* A woven material may be treated with a standard pigment and we should like to know if the pigment will fade more over a year in a store in Carlisle than a similar store in Cardiff. The procedure is similar to that in the paint adhesion study.

4. *A cross-over clinical trial to compare the effects of two treatments.* Each patient receives both treatments, one after the other, and the difference between the responses is tested against the null hypothesis that the difference in a large population would be zero. This has the great advantage that far fewer patients are needed in the trial than in a randomised parallel treatments clinical trial, in which patients are assigned at random to just one of the two treatments. The advantage arises because the variation between patients is eliminated. The order in which each patient tries the two treatments should be randomised, subject to each treatment being used first the same number of times. There are several difficulties with cross-over trials such as carry-over effects, which may perhaps be explored in a more sophisticated analysis, and they cannot always be used.

The outstanding question for all of these paired experiments is how big the sample size should be. This question is answered by following the procedure described in Chapter 3 for comparison of a single mean against a target value. But in these cases we compare the *mean of differences* against a hypothesised value of zero.

▽ Case 5.1 (AgroPharm)

AgroPharm is developing a new drug, Lowprex, for the management of hypertension. It will market it as an occasional treatment for people with mild or moderate hypertension who are willing to monitor their own blood pressure. The marketing manager, Omar Ochre, is keen to demonstrate that home monitoring of blood pressure can be simple and reliable. He will assess the performance of several commercially available instruments, including a new oscillometric device for measuring blood pressure at the wrist.

To start with, Omar surveys the literature on devices for measuring blood pressure. In an article in the *British Medical Journal*, Close *et al.* (1986) compared blood pressure measurements made with an inflatable cuff around the upper arm, the standard method, and a finger monitor (Minimonitor, Tripod Industries) on 881 patients. Omar analyses data from a random subsample of 200 patients. The data are available on the internet and are reproduced at www.greenfieldresearch.co.uk/doe/data.htm.

There are three variables, all measured in millimetres of mercury (mmHg): arm systolic pressure (**armsys**), finger systolic pressure (**fingsys**); and the difference (**diff**), obtained by subtracting **armsys** from **fingsys**.

Apart from assessing whether there is evidence of a difference in the means for the two methods, we can investigate whether there is a tendency for the differences to depend on the underlying blood pressure over a wide range of pressures. For example, the finger method might overestimate relatively low blood pressures and underestimate relatively high blood pressures. Then the differences would tend to be positive for low blood pressures and negative for high blood pressures. We assume that the standard arm cuff method is considerably more precise than the finger method and plot the difference against the arm (Figure 5.1). There is no clear tendency for the difference to depend on the arm blood pressure reading. There is considerable variability in the differences, which range from around −50 to a little above +50. We might guess that the sample mean difference is positive, but this seems less of an issue than the background variability.

To investigate whether there is a systematic difference between the arm and finger measurements we use the 200 differences to construct a 95% confidence interval for the average difference in the corresponding population. Follow the MINITAB procedure **Stat > Basic Statistics > Paired t** (naming the first sample as **fingsys**) as in Figure 5.2. The results are as follows:

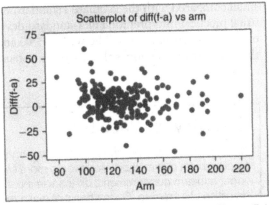

Figure 5.1

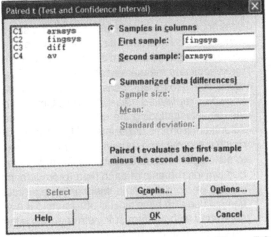

Figure 5.2

Paired T-Test and CI: fingsys, armsys

```
Paired T for fingsys - armsys

              N      Mean     StDev   SE Mean
fingsys     200   132.815    25.648     1.814
armsys      200   128.520    23.288     1.647
Difference  200   4.29500  14.58636   1.03141

95% CI for mean difference: (2.26110, 6.32890)
T-Test of mean difference = 0 (vs not = 0): T-Value = 4.16 P-Value =
0.000
```

Since the 95% confidence interval for the population mean difference is all positive, and hence the P-value is less than 0.05, there is evidence that the finger measurement is systematically higher, on average, than the arm measurement. Our best estimate of this mean difference is 4.3, but this is small compared with the estimated standard deviation of the differences, which is 14.6. Using our usual practical interpretation of a standard deviation, we can predict that approximately two-thirds of finger measurements will lie within -10 and 19 mmHg of the arm measurement. Omar thinks this is too wide and hopes that modern instruments will give closer agreement.

△ Review of Case 5.1

Omar estimates that the finger monitor will, on average, give readings 4 mmHg higher than the standard arm cuff method. However, a more serious limitation is that the differences between finger and arm measurements on the same patient on the same day have an estimated standard deviation of 14 mmHg. On the basis of these results he could not have recommended the finger monitor for monitoring blood pressure. But his experiment will be to assess more recently developed instruments. Nevertheless, he has some outstanding questions, whatever instruments he is comparing. The first is how much measurements, made with the same instrument on the same person, vary if they are made on the same day and on different days. The second is whether the difference in measurements made on the same patient with different instruments is relatively constant for that patient. If so, alternative blood pressure measuring instruments should be calibrated against the arm cuff method for individual patients. To answer his questions, Omar must make replicate measurements on patients. The best way to determine the precision of a measuring technique is to repeat measurements on the same patient within a short period of time. These are known as *replicated measurements*. Omar decides that he will compare the precisions of instruments, including the arm cuff type, by making replicate measurements on himself and a few volunteers from his department. He will ask volunteers not only to make replicate measurements at much the same time, but also to repeat this exercise on different days. From these results he hopes to distinguish measurement error from day-to-day variation in human blood pressures. Also, he will buy two instruments of each type to provide some check of an assumption that instruments of the same type from the same supplier have practically equal precision. Any instrument types that are easy to use and have similar precision to the arm cuff, which is not well suited to home use, will be compared with the arm cuff instrument on a larger sample to investigate any systematic difference. He will ask participants to make comparisons on several occasions to indicate whether or not any difference is approximately constant for them.

As an aside, we mention that if another instrument has a similar precision to the arm cuff instrument it would be better to plot the difference in measurements against their average than against the standard arm cuff measurement. The reason is that the difference, alternative minus standard, will be positively associated with the alternative and negatively associated with the standard.

Omar proceeds to find three types of instrument that appear suitable. He decides to use the two instruments of each of these types and the two arm cuff instruments to measure blood pressure on volunteers from the AgroPharm site. He will ask each volunteer to replicate the eight measurements twice on four days. Each set of eight measurements will be made in a random order. This generalisation of the paired comparison procedure is an example of a *randomised block design*, which we cover in more detail in the next chapter. Ideally the subjects in the experiment would be a random sample from the population of all people with mild or moderate hypertension. It is not practical for Omar to obtain such a sample, so he will have to assume that any systematic difference between instruments is independent of the population on which they are used.

▽ Case 5.2 (SeaDragon)

SeaDragon has been using oxy-propane gas welding torches for cutting plain carbon steel plates, but now wishes to compare this fuel with oxy-natural gas. A slight hardening at the cut surface of the plates is unavoidable, with any gas cutting method, and material has to be removed by a non-thermal process such as grinding. Oxy-natural gas has advantages for some types of work, provided the hardening problem is not made worse. The workshop manager, Reginald Rose, wishes to compare cuts made with the two gas types.

Reginald has found eight off-cut plates, of slightly different thicknesses which may have been from made from different carbon steels, for a pilot experiment. Given the small number of plates, and the variation in the plates' characteristics, the only feasible design is to cut each plate with both gas types. Although the cuts would be well separated, Reg thought that the order of cutting might make a difference to the hardness measurements. He therefore randomised the order of oxy-

Plate number	Oxy-propane	Oxy-natural gas	Difference
1	370	333	37
2	333	336	−3
3	330	299	31
4	306	294	12
5	314	297	17
6	322	373	−51
7	290	304	−14
8	303	299	4

Table 5.1

natural gas and oxy-propane cutting over the eight plates, subject to a restriction that each gas combination would be used to make the first cut on four plates. He measured the hardness of the cut edge by grinding away 1 mm of the heat-affected zone and then making 36 Vickers hardness measurements along the edge. The maximum hardness along the edge is the critical characteristic when welding plates together, so Reg decided to use the maximum of the 36 hardness measurements along an edge in the analysis. The results of the experiment are in Table 5.1 The data set is also available at www.greenfieldresearch.co.uk/doe/data.htm.

Using the same MINITAB procedures as in Figure 5.2, Reg's analysis produced the following results:

Paired T-Test and CI: propane, natural_gas

Paired T for propane - natural_gas

```
              N    Mean      StDev     SE Mean
propane       8  321.000    24.454     8.646
natural_gas   8  316.875    28.028     9.909
Difference    8  4.12500   27.90001   9.86414
```

95% CI for mean difference: (−19.19999, 27.44999)
T-Test of mean difference = 0 (vs not = 0): T-Value = 0.42 P-Value = 0.688

Since the 95% confidence interval for the mean of the population of such differences includes zero, he has no evidence of a systematic difference in the hardness of the cut edges for the two gas types. But the sample size is small and the confidence interval for this mean difference has a width of 47 VH10 units. Reg would like a confidence interval with a width of about 20 VH10. If the standard deviation is known, the width of a confidence interval decreases in proportion to the square root of the increase in the sample size:

$$\frac{\text{width new interval}}{\text{width old interval}} = \sqrt{\frac{\text{old sample size}}{\text{new sample size}}}$$

This is equivalent to

$$\text{new sample size} = \left(\frac{\text{width old interval}}{\text{width new interval}}\right)^2 \times \text{old sample size}$$

This is a useful general principle for calculating the sample size from a pilot study, but it can only give approximate results because the value of the sample standard deviation will be somewhat different in the follow-up experiment. We cannot tell in advance whether this will lead to a wider or narrower interval than that planned, but we will have more degrees of freedom for the t-distribution which will tend to reduce, slightly, the width of the confidence interval predicted by our approximate formula. Reg calculated a sample of size for the follow-up experiment of:

$$\left(\frac{47}{20}\right)^2 \times 8 = 44$$

It is useful to know how to calculate a sample size from the ratio of the width of a confidence interval in a pilot study to the desired width, because this is a general principle. For this experimental design we could also use **Stat > Power and Sample Size > 1-Sample t** in MINITAB. An equivalent requirement to the width of the 95% confidence interval being 20 is a power of 0.5 of detecting a difference of absolute magnitude equal to half the width of the interval.

Power and Sample Size

1-Sample t Test

Testing mean = null (versus not = null)
Calculating power for mean = null + difference
Alpha = 0.05 Assumed standard deviation = 27.9

Difference	Sample Size	Target Power	Actual Power
10	32	0.5	0.501881

The MINITAB sample size of 32 is considerably less than 44 and you may think there is some mistake. The difference is that MINITAB has used a t-value with 31 degrees of freedom, 2.04, whereas the ratio of confidence intervals was based on a t-value with only seven degrees of freedom, which at 2.36 is substantially greater than two. Both calculations rely on the unrealistic assumption that the estimate of the population standard deviation made from the new sample will be the same as that in the pilot study. Sample size calculations provide an indication rather than an exact value.

There is no suggestion that the hardening is worse with oxy-natural gas cutting, but the pilot experiment is too small to be confident about this. The hardening is a vital issue because it can lead to cracks developing in welds on structures fabricated from the plates. Typical structures are ships, offshore platforms, and pressure vessels. The cost of each run in the experiment, cutting a single off-cut plate with both fuel combinations and grinding the cut edges, is low compared with the consequences of adopting a cutting technique that leads to increased hardening. Reg decides to base his follow-up experiment on 50 test plates. To represent SeaDragon's range of work, the plates will be of five different thicknesses, and from two suppliers. Reg will treat thickness and supplier as concomitant variables and these can be included in the analysis using regression.

▽ **Case 5.3 (UoE)**

The sociology department at UoE has been running a programme to help people who admit they have a problem with gambling. Participants are invited to social meetings in the department every two weeks and to have consultations with psychologists who specialise in problems related to addictive behaviour, once a month. Qualitative research indicates that the programme is successful, and the programme is attracting increasing numbers of participants. Ariella's boss, Valerie Violet, wants to augment the programme with a 24-hour telephone help-line and intends to apply for research funds to do this. She thinks that a quantitative assessment of the benefits of a telephone help-line will add to her research proposal.

Valerie's proposal needs to be based on precise definitions. The United States National Council on Problem Gambling gives this definition: 'Problem gambling is gambling behavior which causes disruptions in any major area of life: psychological, physical, social or vocational. The term "Problem Gambling" includes, but is not limited to, the condition known as "Pathological", or "Compulsive" Gambling, a progressive addiction characterized by increasing preoccupation with gambling, a need to bet more money more frequently, restlessness or irritability when attempting to stop, "chasing" losses, and loss of control manifested by continuation of the gambling behavior in spite of mounting, serious, negative consequences.' These definitions are still not easy to turn into measurable quantitative variables.

Valerie outlines her design for the experiment in the proposal. She will start by asking programme participants if they are willing to take part in a trial of the help-line. Those who are willing will be matched as pairs as closely as is possible according to current weekly stake, sex, age group, and whether they live in urban or rural areas. All participants will be told that their objective is to reduce their current weekly stake. For each matched pair, one will be chosen at random to have access to the help-line. Both will continue with the programme of meetings and consultations. Over a six-month period all participants will record their weekly stakes. At the end of this period, for each participant, the ratio of the average weekly stake during the trial period to the average stake before the trial began will be calculated. The ratios will be analysed with the MINITAB **paired t** procedure. Valerie has no experience that she can quote to justify a particular sample size, but she knows it is important to demonstrate that the experiment will have some discriminatory power. She presents the following argument. There are now 48 people in the programme and she thinks they will all be willing to participate. This will give 24 pairs and, as she expects recruitment in the first three months to exceed the dropout rate, she is fairly confident of a sample size of at least 24 over a nine-month period.

Given that the existing programme is successful, the population of all possible ratios, with or without the help-line, should have means less than 1.0, and can vary from zero upwards. It is uncommon for the standard deviation of a non-negative variable to exceed its mean, so a fairly pessimistic assumption is that ratios have a standard deviation as high as 1.0. The difference between two independent ratios would have a standard deviation of me $\sqrt{(1^2 + 1^2)} = 1.414$.

But the purpose of the pairing is to reduce this. She first supposes the standard deviation of the paired differences would be reduced to 1.0. She considers that a difference in ratios of 0.4 would undoubtedly justify the phone-line. The MINITAB power calculation for testing at the 10% level (alpha = 0.10) with an assumed standard deviation (sigma) of 1.0 and with a two-sided alternative gives:

Power and Sample Size

1-Sample t Test

Testing mean = null (versus not = null)
Calculating power for mean = null + difference
Alpha = 0.1 Sigma = 1

	Sample	
Difference	Size	Power
0.4	24	0.6015

The power of the proposed study is only 60%. This means that, if there is an important difference of 0.4 between ratios of those who use the help-line and those who do not, the probability of producing evidence of a reduction is only 0.6. The power depends crucially on the unknown standard deviation, so if she were to make a highly optimistic supposition that the standard deviation of differences is only 0.5 then MINITAB gives a power of more than 98%:

Power and Sample Size

1-Sample t Test

Testing mean = null (versus not = null)
Calculating power for mean = null + difference
Alpha = 0.1 Sigma = 0.5

	Sample	
Difference	Size	Power
0.4	24	0.9844

Unfortunately, she does not know and certainly cannot control the standard deviation. She could, however, increase the power of the study by letting it run beyond nine months, and she discusses this possibility in her proposal.

Valerie was awarded a grant, despite one referee suggesting that it might not amount to more than a pilot study, and ran the experiment. The data are in Table 5.2 and at www.greenfieldresearch.co.uk/doe/data.htm. The first two columns are the proportions, and we start by drawing a boxplot of their differences (Figure 5.3).

help-line	no help-line	ln(h-l)	ln(no h-l)	help-line	no help-line	ln(h-l)	ln(no h-l)
1.82	1.07	0.599	0.068	0.02	0.25	−3.912	−1.386
0.01	0.47	−4.605	−0.755	0.02	0.13	−3.912	−2.04
0.57	1.65	−0.562	0.501	0.46	0.84	−0.777	−0.174
0.01	0.19	−4.605	−1.661	0.99	0.87	−0.01	−0.139
0.4	0.65	−0.916	−0.431	0.08	0.5	−2.526	−0.693
0.79	0.83	−0.236	−0.186	0.23	0.85	−1.47	−0.163
0.4	0.48	−0.916	−0.734	0.1	0.14	−2.303	−1.966
0.06	0.83	−2.813	−0.186	0.05	0.37	−2.996	−0.994
2.09	1.46	0.737	0.378	0.06	0.8	−2.813	−0.223
1.84	1.33	0.61	0.285	1.66	1.07	0.507	0.068
0.12	0.1	−2.12	−2.303	0.22	1.14	−1.514	0.131
0.61	0.48	−0.494	−0.734	0.31	0.29	−1.171	−1.238

Table 5.2

To draw this boxplot, create a new variable, **difference = help-line – no help-line**. Follow **Graph > Boxplot**. Choose **One Y** and **Simple**. Select **difference** as the graph variable. Click on **Scale** and tick **Transpose value and category scales**. The median is clearly below zero, but there is considerable variation in the differences.

More formal analysis of the differences of the first two columns, following the procedure in Figure 5.2, together with **Options > Confidence level 90**, gives the following results:

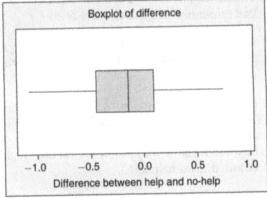

Figure 5.3

Paired T-Test and CI: help-line, no help-line

Paired T for help-line - no help-line

	N	Mean	StDev	SE Mean
help-line	24	0.538	0.657	0.134
no help-line	24	0.700	0.437	0.089
Difference	24	−0.1612	0.481	0.098

90% CI for mean difference: (−0.3295, 0.0070)

```
T-Test of mean difference = 0 (vs not = 0):
T-Value = -1.64 P-Value = 0.114
```

The results are promising, but far from convincing: the P-value is 0.114 and it follows that even a 90% confidence interval straddles zero. There is insufficient evidence to reject the null hypothesis at the 10% level.

A colleague suggests analysing the logarithms of the ratios. So Valerie computed the logarithms of the first two columns following the MINITAB procedure shown in Figure 2.25. These values are shown in Table 5.2. She then calculated the differences of the logarithms of the ratios as **dif-logs**. A boxplot of these differences is quite striking (Figure 5.4).

She then repeated the analysis to give the following result:

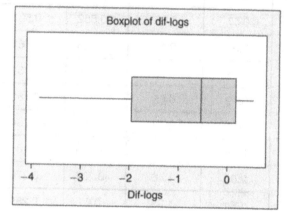

Figure 5.4

```
Paired  T-Test  and  CI:  ln(h-1),
ln(no h-1)

Paired T for ln(h-1) - ln(no h-1)

                  N      Mean     StDev    SE Mean
ln(h-1)          24    -1.592     1.660     0.339
ln(no h-1)       24    -0.607     0.799     0.163
Difference       24    -0.985     1.288     0.263

90% CI for mean difference: (-1.436, -0.534)
T-Test of mean difference = 0 (vs not = 0): T-Value = -3.75 P-Value =
0.001
```

She has substantial evidence that the help-line has reduced the mean of the logarithm of the ratios. The reason for this difference in the strength of evidence against the null hypothesis is that the logarithms give relatively greater weight to the dramatic reductions achieved with the help-line (see boxplot in Figure 5.4). There are no such high reductions amongst the participants who were randomised to no-help line.

△ Review of Case 5.3

There is substantial evidence that the help-line has had some beneficial effect for people on the programme. The estimated reduction in the mean is 0.14. Valerie expects that these results will at least justify the continuation of the trial. However, given the evidence that the help-line has some effect, she will suggest offering the help-line to all participants and monitoring its success by frequency of use and comments from people who use it.

5.2 Independent samples

It is not always possible to use a paired comparison design. An example (see Case 5.4 below) is the manufacture of high-voltage switches for use in the electricity supply industry. One of the important characteristics of these switches is the open and close times. A modified design of switch, with fewer moving parts, is being compared with the standard design, and the engineer responsible wishes to ensure that there is no substantial increase in the open and close times. In this case, times for the new switches will have to be compared with times for the old ones. We have no choice but to base the comparison on independent samples from the two populations.

In other cases the feasibility and advantages of using a paired comparison design may be less clear. But it is generally better to use a paired comparison for comparing means, if it is possible to do so. The slight drawback is a reduction in degrees of freedom for the t-distribution but, despite this, the increase in precision, for the same amount of experimental effort, can be considerable.

▽ Case 5.4 (SeaDragon)

Harriet Henna is responsible for the design of high-voltage switches in SeaDragon's heavy electrical machinery division. A modification to the design leads to a switch with fewer mechanical parts than the base model, which should be even more reliable in service. However, it is important to investigate whether the performance is compromised.

The open and close times of the switch are the most important characteristics, and their lower and upper specification limits (LSL and USL) are:

	LSL	USL
Open time (ms)	18	26
Close time (ms)	90	105

Switches are hand-built and all are tested before being shipped to customers, many of whom are overseas. From past records Harriet knows that the means of open and close times for the base model are close to 23 ms and 100 ms respectively, and that their standard deviations are about 0.9 ms and 1.8 ms respectively. Harriet is prepared to market the modified switch only if she can be reasonably confident that the mean open and close times have not increased. Small decreases in either of these mean times would be a benefit. There have been no changes in production methods over the past two months, during which time 38 switches have been made. She has decided to manufacture ten switches with the modification to compare with the 38 made to from the past two months. She will then review the modified design, and if necessary increase the sample size.

The results of her experiment are available at www.greenfieldresearch.co.uk/doe/data.htm. Harriet begins by analysing the open times (variables **BasOpen** and **ModOpen**). Although Harriet's experiment has been described in terms of means, she would not wish to go ahead with the modified design if the open and close times have become more variable. She starts her analysis with a MINITAB routine that compares the sample variances and provides boxplots in a single figure. The procedure **Stat > BasicStatistics > 2 Variances > Samples in different columns** gives the results in Figure 5.5.

The upper half of the figure gives 95% Bonferroni confidence intervals for the two population standard deviations. We can be at least 95% confident that the two Bonferroni confidence intervals

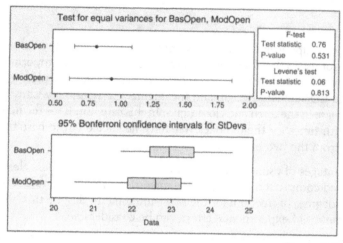

Figure 5.5

value of 0.531 tells her that there is insufficient evidence to reject the hypothesis that the population variances are equal. Levene's test leads to the same conclusion. Levene's test generalises for comparing variances from more than two populations. The null hypothesis that population variances are equal is identical to a null hypothesis that the standard deviations are equal. The former phrase is more commonly adopted because the test is explicitly based on the ratio of sample variances.

The analysis for means is given by selecting **Stat > BasicStatistics > 2-Sample t**:

Two-Sample T-Test and CI: BasOpen, ModOpen

Two-sample T for BasOpen vs ModOpen

	N	Mean	StDev	SE Mean
BasOpen	38	23.021	0.807	0.13
ModOpen	10	22.462	0.923	0.29

Difference = mu BasOpen − mu ModOpen
Estimate for difference: 0.559
95% CI for difference: (−0.138, 1.256)
T-Test of difference = 0 (vs not =):
T-Value = 1.75 P-Value = 0.106 DF = 12

Harriet can be 95% confident that the difference in population means is between an increase of 0.14 and a decrease of 1.26. She is quite happy with this range. The sample standard deviation is slightly greater for the modified design, and Harriet will monitor the variability during production. Her point estimates of the mean and standard deviation of open times for the modified design are 22.46 and 0.923 respectively, and suggest that the specification of 18 to 26 should be met if the distribution of open times is approximately normal. The usual requirement for meeting a specification is that the process mean should be at least three standard deviations from the specification limits – the rationale being that if the characteristic has a normal distribution this corresponds to less than three in 1000 items being outside the specification.

She now analyses the close times (variables **BasClose** and **ModClose**) with the following results and the plots of Figure 5.6.

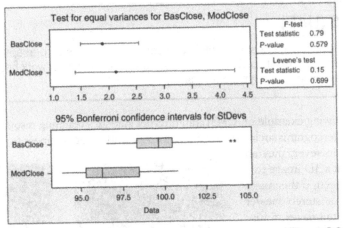

Figure 5.6

```
Two-Sample T-Test and CI: BasClose, ModClose

Two-sample T for BasClose vs ModClose

              N      Mean     StDev    SE Mean
BasClose     38     99.64     1.89      0.31
ModClose     10     96.85     2.12      0.67

Difference = mu BasClose — mu ModClose
Estimate for difference: 2.791
95% CI for difference: (1.198, 4.384)
T-Test of difference = 0 (vs not =): T-Value = 3.79 P-Value = 0.002
DF = 13
```

She can be confident that the mean of the close times is lower for the modified design. Her point estimates of 96.85 and 2.12 for the mean and standard deviation of close times again suggest the specification between 90 and 105 can be met. However, Harriett is aware that the sample is small, and also that processes can change, so she will need to keep the test results from all the switches that are shipped and to monitor progress.

When you use **2-Sample t** you should notice an **Assume equal variances** option. If you check this, you get a slight variant on the analysis, based on an assumption that the population variances are equal which usually leads to slightly narrower confidence intervals.

△ Review of Case 5.4

There is no evidence that the design modification has had any effect on the open times, and it appears to have led to a small decrease in the mean close time. This is itself a slight advantage, and Harriet is keen to proceed with the modified design. The best estimates of the standard deviations of open and close times are slightly higher for the modified design than for the basic design, but the increase can reasonably be attributed to chance in the sampling. Since all switches are tested before shipping, the performance of the modified design can be closely monitored.

▽ Case 5.5 (UoE)

Ingrid Indigo is beginning her research into the effects of visual cues on perception. Before designing her own experiment she surveys the literature to find out what work has already been done and to obtain an idea about typical variability.

She found the following example on DASL (an internet statistics teaching resource). The experiment used random dot stereograms such as those in Figure 5.7. Both images appear to be composed entirely of random dots. However, they are constructed so that a 3D image (of a diamond) will be seen, if the images are viewed with a stereo viewer, causing the separate images to fuse. Another way to fuse the images is to fix your gaze on a point between them and defocus your eyes. This technique takes some effort and practice. The aim of the experiment was to determine whether knowledge of the form of the embedded image affected the time required for subjects to fuse the images. One group of subjects (group NV) received no information about the shape of the embedded object. A second group (group VV) were shown a drawing of the object.

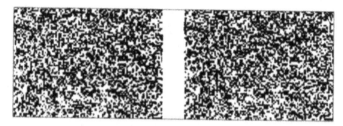

Figure 5.7

The data (time in seconds) are at www.greenfieldresearch.co.uk/doe/data.htm. There are three columns of data for the variables **time, cue** and **ln(time)**. **Cue** is a string variable: it is represented by alphabetic characters instead of by numbers. The values of **cue** are **NV** and **VV**. These are known as the subscripts of the sample variable to be analysed. Since the data are available, Ingrid decides to check the analysis with MINITAB. She starts with a comparison of variances of **time** for the two samples. Since the samples are all in one column, the MINITAB procedure is not the same as in Case 5.4 (see Figure 5.8 and the result in Figure 5.9).

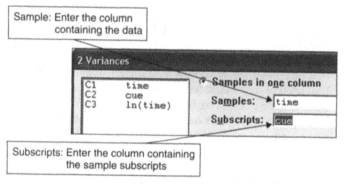

Figure 5.8

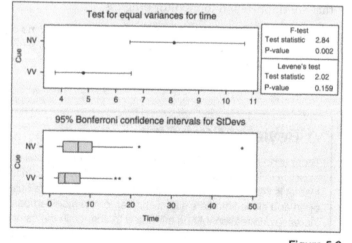

Figure 5.9

There is clear evidence that the **NV** times are more variable. The boxplots indicate that the times for both groups have highly skewed distributions. The **2-Sample t** procedure gives:

Two-Sample T-Test and CI: time, cue

```
Two-sample T for time

cue        N      Mean    StDev   SE Mean
NV        43      8.56     8.09     1.2
VV        35      5.55     4.80     0.81

Difference = mu (NV) - mu (VV)
Estimate for difference: 3.01
95% CI for difference: (0.06, 5.95)
T-Test of difference = 0 (vs not =): T-Value = 2.04 P-Value = 0.045
DF = 70
```

There is evidence that people who are given a visual cue are quicker at fusing the image. With such large samples the confidence interval and hypothesis test are not sensitive to the assumption that the data are from normal distributions. Nevertheless, it is a good idea to investigate what happens

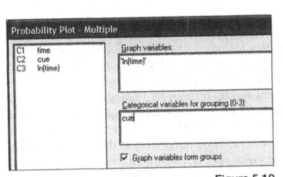

Figure 5.10

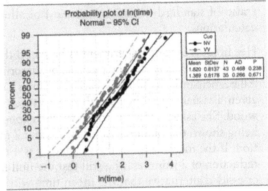

Figure 5.11

if the data are transformed. In this case the logarithms of reaction time can reasonably be assumed to come from a normal distribution. The procedure **Graph > Probability Plot > Multiple** gives the window shown in Figure 5.10. Choose **ln(time)** as the variable to plot grouped by the categorical variable **cue**. This produces the probability plot of Figure 5.11.

She repeats the comparison of variance for the two groups, using the transformed variable **ln(time)**. The result is in Figure 5.12.

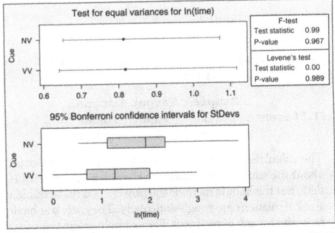

Figure 5.12

The original standard deviations are approximately equal to the means so the logarithms of the variables have variances that are almost equal. The **2-Sample t** procedure provides evidence of a difference in means for the distributions of logarithms of times. We have checked **Assume equal variances** although the confidence interval is unchanged if you do not.

```
Two-Sample T-Test and CI: ln(time), cue

Two-sample T for ln(time)

cue          N      Mean     StDev    SE Mean
NV          43     1.820     0.814     0.12
VV          35     1.389     0.818     0.14

Difference = mu (NV) - mu (VV)
Estimate for difference: 0.431
95% CI for difference: (0.060, 0.801)
T-Test of difference = 0 (vs not =): T-Value = 2.32 P-Value = 0.023
DF = 76
Both use Pooled StDev = 0.8156
```

Ingrid might surmise that variables such as time to recognition have coefficients of variation (ratio of standard deviation to mean) of about one, and use this assumption in her sample size calculations.

Her first perception experiment is based on the children's game of finding hidden objects in a picture. The picture is of an oil painting of Victorian toys, including six rather well-hidden wooden dice. All participants are told that each die shows a different number, but half are randomly selected and given a visual clue. They are shown pictures of the dice, which are made from different types of wood. She asked a few friends to try the experiment and their times for finding all six dice, without being shown the pictures, were 2, 3, 4, and 7 minutes. She makes the following sample size calculation. If the mean without the visual clue is 4 minutes, she would like a power of 0.8 of detecting a reduction of 2 minutes. She will test the null hypothesis, of no difference, at the 5% level against a one-sided alternative that the mean time with the visual clue is less. She assumes a standard deviation of four minutes. In MINITAB, **Stat > Power and Sample size > 2-Sample t** gives:

Power and Sample Size

```
2-Sample t Test

Testing mean 1 = mean 2 (versus >)
Calculating power for mean 1 = mean 2 + difference
Alpha = 0.05 Sigma = 4

              Sample  Target   Actual
Difference     Size   Power    Power
        2        51   0.8000   0.8059
```

The calculation leads to sample sizes of 51, equivalent to 102 participants. Given all the vagaries about the sample size calculation she will aim to recruit 100. She may also analyse logarithms of the data, but this should increase the power. You make think, with some justification, that typical sample size calculations are rough-and-ready. They are, but having some indication of sample size is a prerequisite for any experiment. For example, using this case, just hoping that samples of size 20 will do is likely to be a waste of resources.

□ Case 5.6 (AgroPharm)

Ursula Umbre is responsible for designing horticultural experiments. Fifteen years ago the company had planted a stand of 16 apple trees. Eight of these trees are of root-stock I and the other eight are of root-stock II. The locations of seedlings to plot within the stand had been randomised. An inconvenient feature of experiments with trees is the length of time it takes until the results are available.

gh4yl	gh4yll	extg4yl	extg4yll	gh15yl	gh15yll	wt15yl	wt15yll
111	105	2569	2074	358	409	760	1036
119	117	2928	2885	375	406	821	1094
109	111	2865	3378	393	487	928	1635
125	125	3844	3906	394	498	1009	1517
111	117	3027	2782	360	438	766	1197
108	115	2336	3018	351	465	726	1244
111	117	3211	3383	398	469	1209	1495
116	119	3037	3447	362	440	750	1026

Table 5.3

The data in Table 5.3 (courtesy of East Malling Research) are available at www.greenfieldresearch. co.uk/doe/data.htm. Variables are: trunk girth (mm) above the graft union at four years, **gh4yl** and **gh4yll**; extension growth (cm) at four years, **extg4yl** and **extg4yll**; trunk girth above the ground at 15 years, **gh15yl** and **gh15yll**; and estimated weight of the tree above ground (lbs) at 15 years, **wt15yl** and **wt15yll**.

Ursula used the MINITAB procedure to compare the two root-stocks for each variable:

Two-Sample T-Test and CI: gh4yI, gi4yII

Two-sample T for gh4yI vs gh4yII

```
              N      Mean    StDev   SE Mean
girth4yI   8     113.75    5.82      2.1
girth4yI   8     115.75    5.85      2.1
```

Difference = mu girth4yI − mu girth4yII
Estimate for difference: −2.00
95% CI for difference: (−8.31, 4.31)
T-Test of difference = 0 (vs not =): T-Value = −0.69 P-Value = 0.505
DF = 13

Two-Sample T-Test and CI: extg4yI, extg4yII

Two-sample T for extg4yI vs extg4yII

```
              N     Mean    StDev   SE Mean
extg4yI       8     2977     448      158
extg4yII      8     3109     552      195
```

```
Difference = mu extg4yI - mu extg4yII
Estimate for difference: -132
95% CI for difference: (-675, 411)
T-Test of difference = 0 (vs not =): T-Value = -0.53 P-Value = 0.608
DF = 13
```

Two-Sample T-Test and CI: gh15yI, gh15yII

Two-sample T for girth15yI vs girth15yII

```
              N     Mean    StDev   SE Mean
girth15y      8    373.9     18.8      6.6
girth15y      8    451.5     34.0       12
```

```
Difference = mu girth15yI - mu girth15yII
Estimate for difference: -77.6
95% CI for difference: (-108.2, -47.0)
T-Test of difference = 0 (vs not =): T-Value = -5.65 P-Value = 0.000
DF = 10
```

Two-Sample T-Test and CI: wt15yI, wt15yII

Two-sample T for wt15yI vs wt15yII

```
              N     Mean    StDev   SE Mean
wt15yI        8      871     168       59
wt15yII       8     1281     238       84
```

```
Difference = mu wt15yI - mu wt15yII
Estimate for difference: -409
95% CI for difference: (-634, -185)
T-Test of difference = 0 (vs not =): T-Value = -3.98 P-Value = 0.002
DF = 12
```

There is no significant difference between the two groups for either trunk girth or extension growth after four years, although the means are slightly higher for root-stock II. In contrast, measurements after 15 years of both trunk girth and weight provide evidence that root-stock II is the more vigorous.

The data also provide valuable information about standard deviations. Suppose that Ursula wants to design an experiment that would detect a difference of 3 cm at four years with power 0.9, when testing the null hypothesis at the 5% level with a two-sided alternative. If she assumes a standard deviation of 5.8, on the basis of the *t*-test between **gh4yI** and **gh4yII**, MINITAB gives:

Power and Sample Size

2-Sample t Test

Testing mean 1 = mean 2 (versus not =)

Calculating power for mean 1 = mean 2 + difference
Alpha = 0.05 Sigma = 5.8

```
              Sample   Target   Actual
Difference     Size    Power    Power
         3       80    0.9000   0.9017
```

She would need to plant 80 trees of each root-stock.

△ **Review of Case 5.6**

There is strong evidence that root-stock II is more vigorous than root-stock I , when planted under the growing conditions of the AgroPharm stand. After 15 years, root-stock II is estimated to have about 20% more girth and 50% more weight. However, after just four years there was little difference between the root-stocks. Ursula considers this stand to be typical of English orchard conditions, but she will provide a description of the soil type, with details of how the saplings were tended, with her recommendations to apple growers.

6

Comparison of several means

In Chapter 5 we compared the means of two populations using either a paired experiment or independent samples. In this chapter, we compare more than two populations. We begin, in Section 6.1, with comparisons using independent samples. We continue in Section 6.2 with the randomised block design, which is a generalisation of the paired experiment.

6.1 Comparison based on independent samples

We shall introduce the general principles in the context of comparing the burst strength of four types of filter membrane, which we shall refer to as A, B, C, and D (see Case 6.1). Suppose we have a random sample of N burst strengths for each type. Why should we not use our routine for comparing two means from independent samples, and compare A with B, A with C, ..., C with D? A drawback is that there are six possible comparisons of two types chosen from four types. Suppose an overall null hypothesis, that all four population means are equal, is true and that we test the six individual null hypotheses (mean A equals mean B, ..., mean C equals mean D) at the 5% level. The probability that we wrongly reject at least one of the individual null hypotheses, when the overall null hypothesis is true, is complicated because the comparisons are not independent. For example, a comparison of B with D is not independent of a comparison of C with D. All we can say easily is that the probability is less than the product of the number of comparisons and 0.05, which equals 0.30. This upper limit for the probability is a consequence of the fact that the probability of one or both of two events occurring is less than, or equal to, the sum of the individual probabilities. Only if the two events cannot occur together do we get equality. It follows that the upper estimate of 0.30 is unrealistically high. Even so, it is sometimes used and referred to as the Bonferroni inequality.

If we assume the burst strengths of all four types of membrane are normally distributed with the same standard deviation, there is an exact result for testing the overall null hypothesis. Somewhat surprisingly, this is based on a comparison of two variances calculated from the data. The first of these is an estimate of the variance of burst strengths of the same type of membrane (σ^2) known as the *within-sample variance*. We estimate the within-sample variance by estimating the variance of each type of membrane from each of the four samples, and averaging the results.

As an aside, you should realise that we could add any constant number to all of the data in any of the samples without altering the within-sample variance.

The second variance in the comparison is referred to as the *estimated between-samples variance*. We obtain it by estimating the variance of the four sample means. The sample means have a variance equal to the variance of individual test results divided by N, that is, σ^2/N. If the overall null hypothesis is true, we can estimate the variance of the four sample means from the sample. But if the overall null hypothesis is not true, and there are differences in the population means, this is likely to be an overestimate. The between-samples estimate of variance is the product of this estimate of means and their sample size N. If the overall null hypothesis is true it is an unbiased estimator of σ^2. But, again, if the overall null hypothesis is false, and the population means differ, the expected value of the between-samples estimate of variance is greater than σ^2. A test of the overall null hypothesis is based on the ratio:

$$\frac{\text{between-samples estimate of variance}}{\text{within-samples estimate of variance}}$$

If the null hypothesis is true, we expect this ratio to be approximately one. Values considerably greater than one are evidence against the null hypothesis. Values less than one do not provide any evidence against the null hypothesis and are attributed to sampling variability. If the null hypothesis is true, the ratio has a probability distribution known as the *F-distribution*, after Sir Ronald Fisher who pioneered experimental design at the English agricultural research station at Rothamsted. This enables us to quantify what we mean by 'considerably greater than one', with P-values for the test of the overall null hypothesis. If we can reject it at some reasonable level, 5% or perhaps even 10%, we need to use follow-up procedures to try to identify where the differences lie. The calculations are set out as an analysis of variance (ANOVA). We illustrate these ideas with the following case.

▽ Case 6.1 (SeaDragon)

Christine Crimson is the production manager in a division that manufactures filters for liquids in the pharmaceuticals and food industries. She wished to compare the burst strengths of four types of filter membrane. The first (A) was the company's own standard membrane material and the second (B) was a new material it had developed. The company bought other types (C and D) from external suppliers of filter membrane. She had records from routine measurements of the standard material. The usual procedure was to calculate a burst strength for a batch of filter membrane as an average of fail pressure readings for a random sample of five filter cartridges from that batch. The standard deviation of these burst strengths from different batches of the same type was about 4.3 kPa. There was usually little variation in the fail pressures of the five cartridges that were averaged to obtain the recorded burst strength measurement.

6.1.1 Choosing a sample size

Chris used the MINITAB procedure **Stat > Power and Sample Size > One-way ANOVA** to determine the size of samples to compare the burst strengths of the four membrane types. She thought that a difference of 8 kPa would be important, and requested a probability of about 0.9 of detecting such a difference. She assumed a standard deviation of 4.3 kPa.

```
Power and Sample Size

One-way ANOVA

Alpha = 0.05    Assumed standard deviation = 4.3    Number of Levels = 4

    SS  Sample  Target                        Maximum
  Means   Size   Power   Actual Power        Difference
    32     10     0.9      0.926090               8

The sample size is for each level.
```

This sample size calculation assumes we have the same number of burst strength measurements for each type of filter membrane. This is known as a *balanced design*. The sample size would increase if she decided to compare more types. Chris recommended a sample size of 10 for each type of filter membrane. She realised that this meant ten batches. An increase in the number of filters sampled within a batch would not be a substitute because she believes that the main source of variability in burst strengths is between batches. As production manager, Chris could ensure that ten different batches of material of membrane types A and B were used in the experiment. She would try to obtain material from different batches of types C and D by placing separate orders with some time between them. So, it was several weeks after the experiment had been proposed before

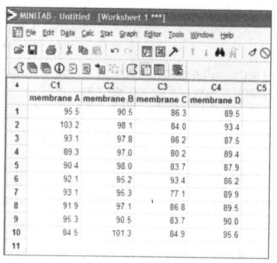

Figure 6.1

all the data were gathered. Chris enters the trial results into a MINITAB worksheet as in Figure 6.1. There is one column for the burst strengths of each membrane type labelled **MembraneA, ..., MembraneD**. The full data set is available at www.greenfieldresearch.co.uk/doe/data.htm.

6.1.2 Looking for evidence of any difference in means

MINITAB offers several procedures for the analysis.

Chris first obtains boxplots for the data (Figure 6.2) by selecting **Graph > Boxplot > Multiple Y's > Simple**. She clicks on **Data View** and chooses **Interquartile range box** and **Outlier symbols**. From the boxplots, it seems as though membrane C may have a lower mean strength.

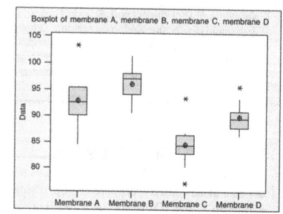

Figure 6.2

Chris feels she could have a better appreciation of the data if she saw a graph with every data point plotted. This is possible with **Graph > Scatterplot** but she can get a better representation with **Graph > Individual Value Plot**. This requires the data to be stacked. She selects **Data > Stack > Columns** to open the window on the left in Figure 6.3. Note that she asks MINITAB to

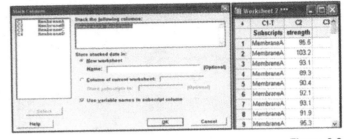

Figure 6.3

store the stacked data in a new worksheet and to use the variable names in the subscript column. The stacked data appear as in Figure 6.3, where she needs to enter a name, **strength**, for the data column.

She selects **Graph > Individual Value Plot**, chooses **With Groups** under **One Y**, specifies **strength** as the **Graph variable** and **Subscripts** as the **Categorical variable**, clicks on **Data View** and selects **Individual symbols** as in Figure 6.4. This produces the graph of Figure 6.5. This graph has several deficiencies. One is that the heading needs to be rewritten. Another is that the columns of dots are not in straight lines. This is because MINITAB has, by default, added jitter to the display.

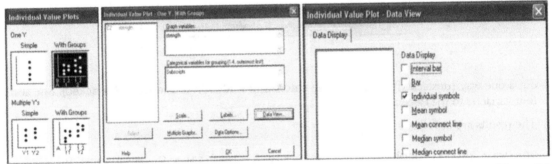

Figure 6.4

Chris was able to remove the jitter. She clicks on a dot in the chart (any will do) and then right-clicks to get a drop-down menu. She chooses **Edit Individual Symbols** and, as in Figure 6.6, she removes the check from **Add jitter**. She clicks on the heading and changes the text to **strength vs membrane**. Then she deletes the **Subscripts** label at the bottom of the graph. Finally, the graph she wants appears as in Figure 6.7.

Chris now proceeds to a more formal assessment of the data with analysis of variance. She returns to the worksheet with the unstacked data to carry

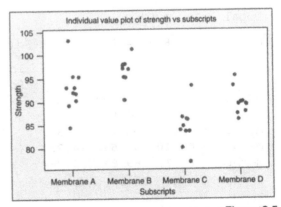

Figure 6.5

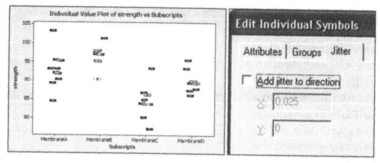

Figure 6.6

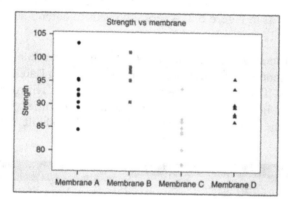

Figure 6.7

out a one-way (unstacked) ANOVA. She enters **Stat > ANOVA > One-way (Unstacked)**. She add all four variables to the **Responses** box.

The results are as follows:

```
One-way ANOVA: membrane A, membrane B, membrane C, membrane D

Source   DF      SS      MS      F       P
Factor    3    709.2   236.4   15.54   0.000
Error    36    547.8    15.2
Total    39   1257.0

S = 3.901    R-Sq = 56.42%    R-Sq(adj) = 52.79%

                                   Individual 95% CIs For Mean Based on
                                   Pooled StDev
Level        N    Mean   StDev   ------+---------+---------+---------+--
membrane A  10  92.840   4.831                       (----*----)
membrane B  10  96.080   3.391                           (----*----)
membrane C  10  84.630   4.287   (----*----)
membrane D  10  89.890   2.764             (----*----)
                                 ------+---------+---------+---------+---
                                   85.0      90.0      95.0     100.0
Pooled StDev = 3.901
```

In the ANOVA table, the first part of the output, **Factor** is a general term for the variable that is being investigated. Here it is the type of membrane material, which has four different levels. The entry in the **Error** row and **MS** (which stands for *mean square*) column is the within-samples estimate of the variance of measurements of burst strength made on individual batches. You can check this from the standard deviations for each of the four samples, which follow the ANOVA table:

$$\frac{4.831^2 + 3.391^2 + 4.287^2 + 2.764^2}{4} = 15.214$$

This within-sample estimate of variance, which is more succinctly called *error mean square*, has 36 degrees of freedom because it is an average of the four variances, each of which has $10 - 1 = 9$ degrees of freedom.

The *pooled standard deviation*, given as $S = 3.901$ at the end of the MINITAB output, is the square root of the within-samples variance.

Returning to the ANOVA table, the entry in the **Factor** row and **MS** column is called either the *factor mean square* or the *between-samples estimate of variance*. It is 10 times the variance of the four sample means:

$$\frac{(92.84 - 90.86)^2 + (96.08 - 90.86)^2 + (84.63 - 90.86)^2 + (89.89 - 90.86)^2}{4 - 1} \times 10 = 236.4$$

This has three degrees of freedom because it is calculated from four means. *Factor mean square* is a better description than *between-samples estimate of variance* because it is only a satisfactory estimator of the variance of burst strengths, within a membrane type, if the overall null hypothesis is true. The ratio of the factor mean square to error mean square is the value of the *F*-statistic. In this case the *F*-ratio is 15.54, and it is much greater than one. The probability of obtaining a value of an *F*-statistic as high as or higher than this if the overall null hypothesis is true is given under the **P** column. It is 0.000 rounded to three decimal places. As this probability is so low, we have overwhelming evidence against the overall null hypothesis. There is evidence that the material types have different mean strengths.

Now that Chris has evidence of differences between population means, she must decide where the differences lie. There is certainly evidence of a difference between the populations with the largest and smallest mean values, B and C. The individual 95% confidence intervals are of limited use. If confidence intervals for different means do not overlap we can be confident there is a difference between those means, but the converse is not necessarily true. This is because the standard deviation of a difference in means is the square root of the sum of the variances of the two means, which is less than the sum of their standard deviations. So, in this case the individual confidence intervals do provide evidence of differences between C and A, B, and D, and between B and D.

Finally, the **SS** column in the ANOVA table is the product of the **DF** and **MS**. We return to this later.

6.1.3 Multiple comparisons

We can obtain confidence intervals for the differences in population means by using a different MINITAB procedure. Before we proceed, recall the relationship between confidence intervals and hypothesis testing. The difference between two sample means is statistically significant at the 5% level – that is, we have evidence to reject the null hypothesis that population means are equal at the 5% level – if and only if the 95% confidence interval for the difference in population means does not include zero. In general, we have evidence to reject the null hypothesis at the $P \times 100\%$ level if and only if the $(1 - P) \times 100\%$ confidence interval does not include zero.

Stack the data for the four different membranes into a single column. Follow **Data > Stack > Columns**; select all four variables to be stacked and store the result in a new worksheet. Head the column **Strength** and put numbers **1** up to **4** to represent membrane materials **A** up to **D** respectively in a second column, **membrane**, using **Calc > Make Patterned Data** (see Figure 6.8 and www.greenfieldresearch. co.uk/doe/data.htm). You can now use **Stat > ANOVA > One-Way** and put **strength** in the **Response** box and **membrane** in the **Factor** box. Click on **Comparisons** and you will see the window shown in Figure 6.9. Here you have a choice of four different procedures for providing confidence intervals for the differences between means, using four different methods: Tukey's, Fisher's, Dunnett's, and Hsu's MCB.

Which method to use depends on the aim of the experiment, and we explain the principles assuming the default of a 5% error rate. If the aim is to compare membrane materials, on an equal basis, then use both the Tukey and Fisher procedures. The Fisher procedure provides individual 95% confidence intervals for all possible differences. Minimum evidence of a difference in population means is that the individual confidence interval for that difference excludes zero. In contrast, Tukey's procedure takes a conservative approach and produces a set of intervals within which we can be 95% confident that all population differences lie. The Tukey 95% intervals are always wider than the Fisher 95% intervals and, with six possible comparisons, are close to Fisher 99% intervals (individual error rate of 1%). If the Tukey interval for a difference excludes zero we shall be more confident about declaring a difference in population means.

If our original intention was to compare other membranes with the company standard, then Dunnett's procedure with the company standard declared to be the control group level is appropriate. It is inefficient to use the Tukey all-pairwise approach when Dunnett's is suitable, because the Tukey confidence intervals will be wider and the hypothesis tests less powerful for a given family error rate. For the same reasons, Hsu's MCB is superior to Dunnett if you want to eliminate levels that are not the best and to identify those that are best or close to the best. We want a confidence interval for the difference between each treatment mean and a control mean, so we choose Dunnett's, and specify the control as the company's standard membrane

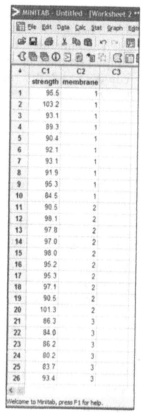

Figure 6.8

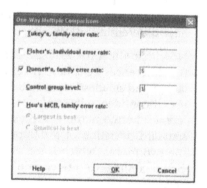

Figure 6.9

material which we coded as 1. Note that the **5** in the text box tells MINITAB to use a 5% test. The results are as follows:

```
Dunnett's comparisons with a control

Family error rate = 0.0500
Individual error rate = 0.0192
Critical value = 2.45
Control = level (1) of membrane
Intervals for treatment mean minus control mean
Level      Lower     Center     Upper   --+--------+---------+--------+----
2         -1.038     3.240      7.518                          (----*----)
3        -12.488    -8.210     -3.932   (----*----)
4         -7.228    -2.950      1.328             (----*----)
                                        --+--------+---------+--------+----
                                        -12.0     -6.0       0.0      6.0
```

There are three comparisons. The family error rate is 0.05. That is, if the overall null hypothesis, that the four population means are identical, is true the probability that at least one interval does not include zero is 0.05. Thus there is a probability of 0.05 that we erroneously claim a difference, or differences, because one or more of the confidence intervals excludes zero. A simpler, and we think more useful, statement is that we can be 95% confident that the three confidence intervals simultaneously include the differences in population means. In our case we can be confident that type C has a lower mean burst strength than type A because the corresponding confidence interval is all negative.

Now suppose the original aim of the experiment was to identify the strongest membrane material. With this scenario, Hsu's procedure is the most appropriate to use:

```
Hsu's MCB (Multiple Comparisons with the Best)

Family error rate = 0.0500

Critical value = 2.13
Intervals for level mean minus largest of other level means
Level      Lower     Center     Upper   ------+--------+-------+--------+-
1         -6.960    -3.240     0.480           (------*------)
2         -0.480     3.240     6.960                      (----*----)
3        -15.170   -11.450     0.000   (-----*--------------)
4         -9.910    -6.190     0.000          (------*-------)
                                        ------+--------+-------+--------+-
                                        -12.0     -6.0     0.0     6.0
```

The interpretation is similar to that for Dunnett's procedure except that comparisons are against the largest of the sample means of the other three types. We can be confident that types C and D have lower mean burst strengths than type B because the corresponding confidence intervals are all negative or zero, and do not extend into the positive numbers.

Finally, let us suppose that the aim of the experiment was to compare the materials, regardless of their origin. We start with:

```
Tukey 95% Simultaneous Confidence Intervals

All Pairwise Comparisons among Levels of membrane
Individual confidence level = 98.93%

membrane = 1 subtracted from:

membrane    Lower    Center    Upper   -----+--------+--------+--------+----
2          -1.460    3.240    7.940                    (----*----)
3         -12.910   -8.210   -3.510     (----*----)
4          -7.650   -2.950    1.750          (----*----)
                                        -----+--------+--------+--------+----
                                           -10        0       10       20

membrane = 2 subtracted from:

membrane    Lower    Center    Upper   -----+--------+--------+--------+----
3         -16.150  -11.450   -6.750   (----*----)
4         -10.890   -6.190   -1.490        (----*----)
                                        -----+--------+--------+--------+----
                                           -10        0       10       20

membrane = 3 subtracted from:

membrane    Lower    Center    Upper   -----+--------+--------+--------+----
4           0.560    5.260    9.960              (----*----)
                                        -----+--------+--------+--------+----
                                           -10        0       10       20
```

The interpretation is similar to that for Dunnett's procedure except that we now make all possible comparisons so the confidence intervals are wider if we maintain the family error rate at 0.05. Using this procedure, we can be confident that: type C has a lower mean burst strengths than the other three types; and type D has a lower mean burst strength than type B.

```
Fisher 95% Individual Confidence Intervals

All Pairwise Comparisons among Levels of membrane

Simultaneous confidence level = 80.32%

membrane = 1 subtracted from:

membrane    Lower    Center    Upper   -----+--------+--------+--------+----
2          -0.298    3.240    6.778                  (----*----)
3         -11.748   -8.210   -4.672     (----*----)
4          -6.488   -2.950    0.588           (----*----)
                                        -----+--------+--------+--------+----
                                          -8.0       0.0      8.0      16.0

membrane = 2 subtracted from:
```

```
membrane     Lower   Center   Upper    ----+--------+--------+-------+----
3          -14.988  -11.450  -7.912    (----*----)
4           -9.728   -6.190  -2.652      (----*----)
                                       ----+--------+--------+-------+----
                                        -8.0      0.0      8.0     16.0

membrane = 3 subtracted from:

membrane     Lower   Center   Upper    ----+--------+--------+-------+----
4            1.722    5.260    8.798               (----*----)
                                       ----+--------+--------+-------+----
                                        -8.0      0.0      8.0     16.0
```

These are individual 95% confidence intervals for the six differences in population means. On the basis of these individual confidence intervals we can be 95% confident that A, B, and D are better than C, and that B is better than D. These claims are based on the corresponding confidence intervals excluding zero. Initially we counselled against individual tests, but we now have evidence against the overall null hypothesis that was the basis for our Bonferroni argument. In this case, the conclusions drawn from the individual 95% confidence intervals given by the Fisher procedure are the same as those from Tukey's procedure with 95% confidence for the family of intervals, but we would usually claim more differences using the former. Chris is particularly interested in the difference between A and B, the company's standard and new material. The confidence interval for the difference, mean A minus mean B, is $[-6.8, 0.2]$. Although this just includes zero, there is a suggestion that B may have a higher mean.

Whether you choose to be guided by the individual or overall error rate depends on the context. In micro-array comparisons of DNA samples, the expression levels of thousands of genes will be compared. This technique is used for medical screening when we hope the overall null hypothesis is true. In this application it is essential to adopt an overall error rate. In contrast, if we are comparing different types of filter membrane we expect differences, and may think individual confidence intervals are more appropriate.

△ Review of Case 6.1

Christine is confident that the company's standard material A, with an estimated mean strength of 92.8 kPa, is stronger than C, and there is no suggestion that the material from D, with an estimated mean of 89.9, is stronger than A. She thinks the new material B shows promise, although she would not recommend a changeover from the standard material to the new material without further testing. The estimated mean strength for B was 96.1, which is 3.5% higher than the estimated mean for A. She recommends a follow-up experiment to compare materials A and B only.

6.1.4 Mathematical model

We can continue the analysis by fitting a mathematical model to the data. Although this isn't necessary for a comparison of several means based on independent samples, it provides an introduction to the models that are a valuable aid to understanding the more complicated designs that we discuss later.

We shall refer to our data by a pair of numbers. The first corresponds to the type of membrane material and is one of the numbers: 1, 2, 3 or 4. The second runs from 1 to 10 to represent the test results for the ten batches of each membrane type. So, for example, strength(3,4) represents the test result for the fourth batch from membrane type C, which turned out to be 80.2 kPa. Our model is:

$$\text{strength}(i, j) = \text{OM} + \text{membrane}(i) + \text{error}$$

where i runs from 1 to 4 to represent the four types of membrane material, A to D respectively, and j runs from 1 to 10 to represent the test results for the ten batches of each membrane type. OM, the overall mean, is the mean of the means of the four populations corresponding to the four membrane types. A corresponding population is the hypothetical population of all batches of membrane that would be produced if the process producing membrane of the given type continued on its present settings indefinitely.

The term membrane(i) is the difference between the mean for type i and OM. It is called the *effect* of type i.

Hence the four population means are OM + membrane(1), OM + membrane(2), OM + membrane(3) and OM + membrane(4) for membrane types A, B, C and D respectively. Since OM is the mean of the four population means we have a constraint that the deviations from OM, the membrane effects, add to give zero. That is:

$$\text{membrane}(1) + \ldots + \text{membrane}(4) = 0$$

The errors allow for deviations from the deterministic part of the model. They allow for variations in the strength of different batches of membrane material of the same type, and error in measuring the mean strength of a batch of material. The errors are assumed to be random and independent with a mean of zero and a standard deviation σ.

The constants in the model (OM, membrane(i), σ^2) are known as the *parameters*. Their values are not known, and one objective of our analysis is to estimate the values of these parameters from our 40 strength(i, j) observations.

As part of the analysis we will also estimate the values taken by the errors, known as the *residuals*, to estimate σ^2 and to check some, but not all, of the assumptions we make about the distribution of the errors. The estimates of the errors are calculated as the differences between observations and their predicted values using the model. These predictions are calculated from the estimated values of the parameters with the error term set at its expected value of zero and are known as the *fitted values*. We summarise this explanation in terms of the mathematical model, which can be rewritten with error on the left-hand side as

$$\text{error} = \text{strength}(i, j) - (\text{OM} + \text{membrane}(i))$$

We replace the unknown (OM + membrane(i)) by our best estimate which is the mean of the ten tests for membrane(i) and define

$$\text{residual}(i, j) = \text{strength}(i, j) - \text{mean of tests}(i)$$

We estimate the variance of the errors by squaring the residuals, adding the squares, and dividing by their number less the number of parameters estimated from the data and used to calculate the residuals. In this case four parameters have been estimated from the data: the OM and three of the four effects. The fourth effect follows from the constraint that the four effects add to zero and is not considered as estimated from the data. This is equivalent to estimating the four population means.

The number of data used to estimate a variance less the number of parameters estimated from the data and used in the calculation is known as the *degrees of freedom*. We met this concept when estimating the variance of a single population in Case 2.6, and return to it below.

We can now provide a more detailed explanation of the ANOVA table by referring to the model. The total sum of squares (SS), 1257.0, in the ANOVA table is calculated by subtracting the mean of all 40 tests, 90.860 (not shown in the output), from the individual test results, squaring these differences and then adding them up. The error SS is the sum of the squared residuals, which are our best estimates of the errors. The factor SS is the sum of squared differences between the mean of tests on membrane(i) and the mean of all tests multiplied by the number of tests for each membrane type (J in general, here 10). The factor SS and error SS add up to the total SS, and the ANOVA table is a breakdown of this Total SS into a proportion that is attributable to the structure of the model, the factor SS, and the unexplained part, the error SS. The better the model, the smaller the error SS.

The degrees of freedom associated with each of the SS is the number of terms in the sum less the number of model parameters that are estimated to calculate that sum. So, for the total SS it is 40–1, since we need to estimate OM by the mean of all 40 tests. For the error SS the degrees of freedom is 40–4, since we need estimates of the four population means. For the factor SS it is 4–1, since this is calculated from just the four membrane means. Notice that the degrees of freedom for the treatment mean square and error mean square add to give the total degrees of freedom, matching the breakdown of the total sum of squares into its components. The mean square is defined as the sum of squares divided by its degrees of freedom.

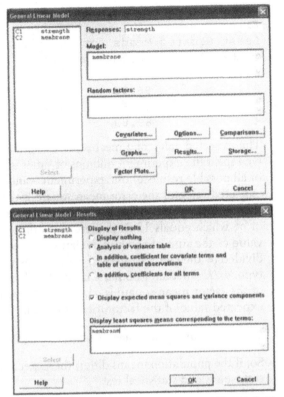

Figure 6.10

Here, we point out that the ANOVA can also be performed with the **Balanced ANOVA** and **General Linear Model** procedures. Their use for a one-way ANOVA is similar, and as the **General Linear Model** procedure is the more versatile we describe its use. The useful additional features it offers are the expected values of the mean squares and residual plots.

To obtain a general linear model, select **ANOVA > General Linear Model**. Put **strength** in the **Responses** box and **membrane** in the **Random factors** box (Figure 6.10). Click **Results** and check **Display expected mean squares and variance components**. In **Display least squares means corresponding to the terms** enter **membrane**. The results are as follows:

```
Source            Expected Mean Square for Each Term
  1 membrane      (2) + Q[1]
  2 Error         (2)
Error Terms for Tests, using Adjusted SS
Source            Error DF    Error MS    Synthesis of Error MS
  1 membrane      36.00       15.22       (2)
Variance Components, using Adjusted SS
```

```
Source            Estimated Value
Error                    15.22
Least Squares Means for strength
membrane          Mean     SE Mean
1                92.84      1.234
2                96.08      1.234
3                84.63      1.234
4                89.89      1.234
```

The expected value of a mean square is the average value of the mean square if we imagine that we repeat the experiment millions of times – that is, an average over this hypothetical population of all possible results of our experiment – and is expressed in terms of the model parameters. The expected value of the error mean square is the variance of the errors, σ^2. The numerical value of the error mean square that we calculate from the results of our one experiment is our estimate, s^2, of σ^2 which equals 15.2. If there is no difference between the population means the expected value of the sum of squared deviations between the mean of tests (i) and the mean of all tests, divided by the factor degrees of freedom, estimates the variance of the means for each membrane type, σ^2/J. The factor SS has a multiplier J so, given equal population means, the expected value of the factor mean square is σ^2. In general, whether or not the population means are equal, the expected value of the factor mean square is:

$$\sigma^2 + \left[[\text{membrane(1)}]^2 + [\text{membrane(2)}]^2 + [\text{membrane(3)}]^2 + [\text{membrane(4)}]^2 \right] / (4 - 1)$$

So, if the population means differ, the expected value of the factor mean square increases and this is the basis of a statistical test.

MINITAB does not provide such detail about the expected value of the mean squares. It provides a useful summary, which is sufficient for our purposes throughout this book. Referring back to the MINITAB output, **Error** is the second listed source of error and **(2)** represents the expected value of the error mean square σ^2. **Q[1]** represents the additional term in the **Factor**, here membrane, mean square. It is zero if the overall null hypothesis is true, and positive if there is a difference in any of the population means. The **F-ratio** is the numerical value of the membrane mean square divided by the numerical value of the error mean square. The **P-value** is the probability of such a large or larger F-ratio if the null hypothesis of no difference between means, equivalent to **Q[1]** equalling zero, is true.

It is worthwhile displaying the expected values of the mean squares because this is the rationale for the F-ratios and P-values calculated by MINITAB.

6.1.5 Investigating the residuals

The residuals can be used to check that some of the assumptions made about the errors are plausible. But we should remember that there are two crucial assumptions, namely that the errors have a mean of zero and are uncorrelated with the predictor factors, that cannot be checked from the residuals. The first of these cannot be checked because if the errors have a mean different from zero, this bias cannot be distinguished from the overall mean. It is an inevitable consequence of the analysis, for any set of data, that the sum of the residuals will equal zero. Similarly, any correlation between errors and the membrane types will be indistinguished from the means for each type. For example, if the errors tended to be positive for material A and negative for material C we will obtain a biased estimate of the difference. Again, it is an inevitable consequence of the analysis that the residuals are uncorrelated with the predictor factors, for any set of data. The practical imperative is that measuring equipment and techniques must be calibrated carefully. Manufacturing companies ensure that micrometers, for instance, are regularly checked and that they can be related back to international standard lengths. We can use the residuals to check that the assumption that errors have a constant standard deviation, although in simple cases we can often judge this from the original plot of the data. We can use a probability plot to check that the residuals appear to come from an approximate normal distribution. The assumption that errors have a normal distribution is more important for constructing prediction intervals than it is for estimation of the parameters or for the calculation of the P-values. Estimators remain unbiased if normality does not hold. Perhaps the most useful aspect of the probability plot is that it highlights any particularly unusual observations.

The normal probability plot of residuals is produced from the **ANOVA** window or from the **General Linear Model** window (Figure 6.10). Click on **Graphs** and then **Normal plot of residuals** (Figure 6.11). There are no clear outlying observations in this case. But the general shape with a concentration of points with a steeper slope in the middle suggests that we might have a mixture from normal distributions with different standard deviations. However, there is no strong evidence to support this. If we calculate the ratio of the largest sample variance (23.34 for membrane A) to the smallest (7.64 for membrane D) we have a ratio of 3.05, which does not exceed the upper 5% point of an *F*-distribution with 9 and 9 degrees of freedom. From **Stat > Basic Statistics > 2 Variances** we get a P-value of 0.112.

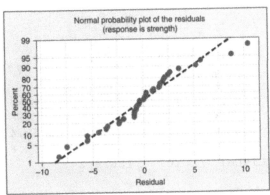

Figure 6.11

The estimates of the population means would remain the same even if the populations did have different standard deviations, and the *F*-tests are not unduly sensitive to the assumption of equal population standard deviations. The practical importance of different standard deviations is that we might prefer a less variable membrane material even if its mean strength were lower. If we want a more precise comparison, we shall need to increase the sample sizes.

We can also check the assumption that errors are uncorrelated with each other if we can put the residuals into time order.

Since we had equal numbers of results for each membrane type, the design is balanced. The theory of the analysis is only slightly more complicated if there are different numbers of results for the different treatments. Although this is no longer a balanced design, all four routines handle such cases. Despite its name, the balanced ANOVA requires balance only if there is more than one factor. The one-way ANOVA is so called because there is a single factor, membrane type in our example, which is applied at different levels, membrane types A, B, C and D in our example. In Section 6.2 we will require a two-way ANOVA.

☐ Case 6.2 (AgroPharm)

Ursula Umbre remembered that there had been many more stands of apple trees before AgroPharm had sold part of the estate to Gigantic Garden Centres about ten years ago. Fortunately, the trunk girth after four years had been recorded before the sale. The data set has been provided by courtesy of East Malling Research and is available at www.greenfieldresearch.co.uk/doe/data.htm.

The trunk girth (in millimetres) was recorded for 13 types of root-stock for apple trees. Ursula draws a boxplot (Figure 6.12) and compares the means in MINITAB using **Stat > ANOVA > One-Way**. There are 78 possible pairwise comparisons, so rather than use the Tukey multiple comparison procedure she uses Hsu's MCB with a family error rate of 0.05:

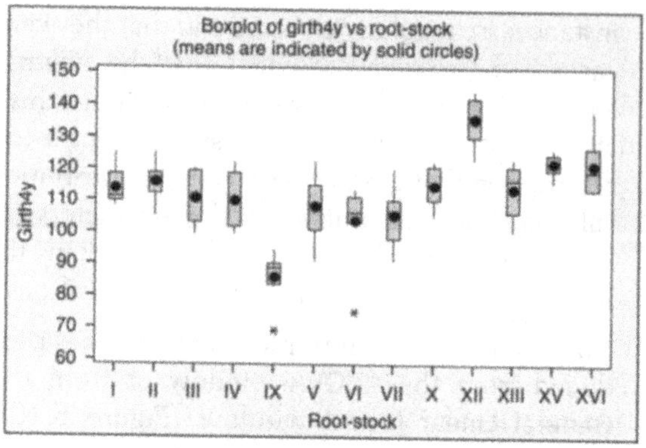

Figure 6.12

One-way ANOVA: girth4y versus root-stock

```
Analysis of Variance for girth4y
Source      DF        SS       MS        F      P
root-sto    12   12748.8   1062.4    16.44  0.000
Error       91    5880.4     64.6
Total      103   18629.2
```

```
                                Individual 95% CIs For Mean
                                Based on Pooled StDev
Level     N      Mean    StDev   ------+---------+---------+-------
I         8    113.75     5.82              (--*--)
II        8    115.75     5.85               (--*--)
III       8    110.75     8.28           (--*--)
IV        8    109.75     8.58           (--*--)
IX        8     85.88     7.70   (--*--)
V         8    108.00     9.96            (--*--)
VI        8    103.63    12.15          (--*--)
VII       8    105.25     9.02          (--*-)
X         8    114.75     5.97              (-*--)
XII       8    135.63     7.31                          (--*--)
XIII      8    113.88     8.29            (--*--)
XV        8    122.00     3.25                 (--*--)
XVI       8    121.38     8.60                (--*--)
                                ------+---------+---------+-------
Pooled StDev = 8.04                   100       120       140
```

```
Hsu's MCB (Multiple Comparisons with the Best)
      Family error rate = 0.0500
Critical value = 2.54
Intervals for level mean minus largest of other level means
Level     Lower    Center    Upper    ------+--------+--------+--------+---
I       -32.098   -21.875    0.000              (---*-------)
II      -30.098   -19.875    0.000               (---*------)
III     -35.098   -24.875    0.000            (---*-------)
IV      -36.098   -25.875    0.000            (---*-------)
IX      -59.973   -49.750    0.000    (---*----------------)
```

```
V      -37.848   -27.625    0.000        (---*--------)
VI     -42.223   -32.000    0.000        (---*---------)
VII    -40.598   -30.375    0.000        (---*--------)
X      -31.098   -20.875    0.000         (---*------)
XII      0.000    13.625   23.848                      (----*----)
XIII   -31.973   -21.750    0.000          (---*--------)
XV     -23.848   -13.625    0.000           (----*----)
XVI    -24.473   -14.250    0.000           (----*----)
                                    ----+--------+--------+-------+----
                                      -50      -25      zero      25
```

Root-stock XII is significantly better than all the others. Root-stock IX is significantly worse than all the others. If she arbitrarily makes comparisons against root-stock I she obtains:

```
Dunnett's comparisons with a control
     Family error rate = 0.0500
Individual error rate = 0.00580

Critical value = 2.83
Control = level (I) of root-sto
Intervals for treatment mean minus control mean
Level    Lower    Center    Upper      ------+---------+--------+--------
II      -9.356    2.000    13.356              (----*----)
III    -14.356   -3.000     8.356            (----*----)
IV     -15.356   -4.000     7.356            (----*----)
IX     -39.231  -27.875   -16.519     (----*----)
V      -17.106   -5.750     5.606           (----*----)
VI     -21.481  -10.125     1.231          (----*----)
VII    -19.856   -8.500     2.856          (----*----)
X      -10.356    1.000    12.356               (----*----)
XII     10.519   21.875    33.231                    (----*----)
XIII   -11.231    0.125    11.481             (----*----)
XV      -3.106    8.250    19.606               (----*----)
XVI     -3.731    7.625    18.981               (----*----)
                                    ------+---------+--------+------
                                        -20       zero      20
```

The only statistically significant differences are between root-stocks I and IX and between I and XII. The conclusions that Ursula can draw from this analysis are the same for both these follow-up procedures.

△ Review of Case 6.2

The mean girth, after four years, of the 104 apple trees was 112.3 mm. Ursula has evidence that root-stock XII is larger, with an estimated mean girth of 135.6 mm, than all the others and that root-stock IX, with an estimated mean girth of 85.9 mm, is smaller. There is no substantial evidence that the other root-stocks differ after four years. These findings relate to the growing conditions on the AgroPharm estate, but horticulturalists consider that they will apply to typical orchards in temperate climates.

6.2 Randomised block design

The principle of a paired comparison of two treatments is that we can remove variation between experimental items by applying both treatments to each item and analysing the differences. The same idea can be used for a comparison of more than two treatments, provided the experimental items are sufficiently large; this is known as a *randomised block design* (RBD). The term *block* for the experimental item comes from the application of RBD to agricultural field trials.

▽ Case 6.3 (AgroPharm)

Chu-Hua Cream is a researcher in the crop science division of AgroPharm who specialises in crop rotation. She wants to investigate a rotation of wheat grown after: faba bean harvested for grain; faba bean incorporated as green manure; ryegrass pasture; or medic pasture. Three farms have volunteered to participate in the experiment and each has allocated a 180 m × 100 m area of a field as a block for the experiment. Each block will be divided into four 90 m × 50 m plots, and one of the four rotations will be assigned at random to each plot.

6.2.1 Choosing a sample size

Chu-Hua has considerable experience of crop yields from her own experiments and from reading reports of others. She thinks the expected wheat yield would be about 3200 kg/ha, with a standard deviation over neighbouring plots of this size of about 160 kg/ha.

Before proceeding with the experiment she needs to determine whether three farms will be sufficient. She can use the MINITAB **One-Way ANOVA** calculator provided she gives a standard deviation corresponding to yields of different plots within a block if they all received the same treatment. She chooses **Stat > Power and sample size > One-Way ANOVA**. She enters four levels, a sample size of 3 (farms), and a power of 0.8 (Figure 6.13). The result is as follows:

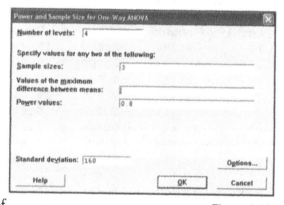

Figure 6.13

Power and Sample Size

```
One-way ANOVA
Sigma = 160  Alpha = 0.05  Number of Levels = 4
Sample                       Maximum
Size    Power     SS Means   Difference
  3     0.8000    1.53E+05    553.5969
```

Chu-Hua has a 0.80 probability of detecting a difference of 554, and considers this adequate as she hopes that the faba bean rotation will increase yields by about 20%. Although the primary objective is to compare yields, she also measures the inorganic soil nitrogen at sowing (kg N/ha) and the grain protein (% by mass) in the wheat crop. Low levels of nitrogen fertiliser were applied to the wheat crop to represent farming practice in the area.

6.2.2 Looking for evidence of any difference in means

We are grateful to Victor Sadras and colleagues at CSIRO for permission to use an excerpt of data from a larger experiment (Sadras *et al.*, 2004). They are shown in Table 6.1 and are available from www.greenfieldresearch.co.uk/doe/data.htm as Case 6.3.

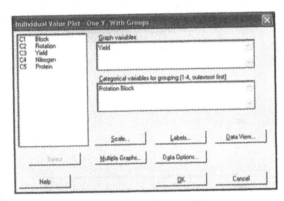

Block	Rotation	Yield	Nitrogen	Protein
1	grass_g	3177	55.915	11
1	faba_gm	4558	111.58	15.3
1	medic_g	3021	76.854	15.9
1	faba_g	4316	103.584	15
2	medic_g	3073	77.685	15.8
2	grass_g	2840	53.165	11.7
2	faba_gm	4437	110.748	15.6
2	faba_g	4247	99.21	14.6
3	faba_g	4541	105.351	14.5
3	faba_gm	4523	113.618	15.7
3	medic_g	3332	81.567	15.3
3	grass_g	3436	62.123	11.3

Table 6.1

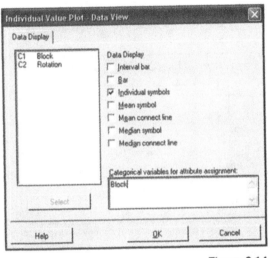

Figure 6.14

As usual, we can learn a lot from the experiment using simple plots. First, Chu-Hua plots yield against the treatment using different plotting symbols to represent the blocks. She uses **Graphs > Individual Value Plot** and selects **With Groups** under **One Y** (Figure 6.14).

She clicks on **Data View**, checks **Individual symbols** and enters **Block** as a **Categorical variable for attribute assignment**. The plot is in Figure 6.15.

The wheat yields after growing faba are substantially higher. Yields on block 3 seem highest and those on block 2 lowest, but this may be accounted for by random variation. She investigates further by plotting **Yield** against **Nitrogen** and **Protein** against **Nitrogen**, using **Graphs > Scatterplot**, enters the variables and selects **Block** as the **Categorical variable for grouping**. The plots are shown in Figure 6.16.

The plot of **Yield** against **Nitrogen** shows the relationship that any farmer or crop researcher would expect. The plot of **Protein** against **Nitrogen** suggests that there may be a nitrogen threshold,

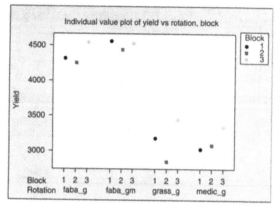

Figure 6.15

around 70 kg N/ha, below which protein content is reduced (the probability that the three lowest yields correspond to the three lowest nitrogen levels, if there is no association, is one in 220).

MINITAB offers several procedures for an analysis using ANOVA. Chu-Hua uses **Stat > ANOVA > Two-Way**, puts **Yield** in the **Response** box, **Rotation** in the **Row factor** box, **Block** in the **Column factor** box, and checks **Display means** for both row and column, which gives the following results:

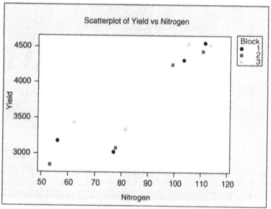

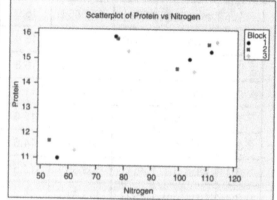

Figure 6.16

Two-way ANOVA: Yield versus Rotation, Block

```
Source      DF        SS        MS         F        P
Rotation     3   5024858   1674953    105.66    0.000
Block        2    194038     97019      6.12    0.036
Error        6     95115     15852
Total       11   5314010

S = 125.9    R-Sq = 98.21%    R-Sq(adj) = 96.72%
```

```
                           Individual 95% CIs For Mean Based on Pooled StDev
Rotation    Mean      ------+---------+---------+---------+----
faba_g      4368                                    (--*---)
faba_gm     4506                                      (--*---)
grass_g     3151      (---*---)
medic_g     3142      (---*--)
                      ------+---------+---------+---------+----
                         3000      3500      4000      4500
```

```
                      Individual 95% CIs For Mean Based on Pooled StDev
Block      Mean     ------+---------+--------+--------+----
1        3768.00            (--------*--------)
2        3649.25    (--------*--------)
3        3958.00                       (--------*--------)
                    --+---------+--------+--------+-------
                    3520      3680     3840     4000
```

The interpretation follows similar lines to that for the one-way ANOVA, except that the total corrected sum of squares is now decomposed into three components: that due to treatments; that due to blocks; and the error. The degrees of freedom for treatments and blocks are one less than their number, for similar reasons to that for treatments in the one-way ANOVA. The degrees of freedom for error, 6, are the number of observations, 12, less one for the OM, less 3 of 4 estimated rotation effects, less 2 of 3 estimated block effects. Note that the degrees of freedom associated with rotation effects are one less than the number of effects because they are constrained to sum to zero; similarly for block effects. The mean square **(MS)** is the sum of squares **(SS)** divided by the degrees of freedom **(DF)**.

The treatment F-ratio (105.66) is the ratio of the treatment MS to the error MS. The null hypothesis for treatments is that there is no difference between the population means for the wheat yields under the four rotation regimes. If this null hypothesis is true Chu-Hua would expect an F-ratio of about one. The P-value is the probability of obtaining a ratio as great as 105.66 if the null hypothesis is true, and it is 0.000 to three decimal places. Therefore she has very strong evidence of a difference in wheat yields under the different rotations. She can tell from the individual confidence intervals that the yield after either faba is much greater than after grass or medic. She will construct confidence intervals for differences later, but, given so much overlap of the grass and medic intervals and of the two faba intervals, she can anticipate that the differences between grass and medic and between the two faba are not statistically significant.

The block F-ratio (6.12) is the ratio of the block MS to the error MS. There are alternative ways to describe the corresponding null hypothesis, depending on whether Chu-Hua considers the blocks to have *fixed effects* or *random effects*. The blocks were on different farms. If she relates the results to these specific farms she is treating blocks as having fixed effects. The null hypothesis for blocks is that there is *no difference between the population means* for the wheat yields on the three farms. As usual, the population is the imaginary infinite population of wheat yields if the experiment was repeated indefinitely. If she considers the three farms to be a random sample of all possible farms then the blocks are random effects. The null hypothesis for blocks is now that *the variance of block effects is zero*.

Whichever form of the null hypothesis she chooses, if it is true we would expect an F-ratio of about one. The P-value is the probability of obtaining a ratio as great as 6.12 if the null hypothesis is true, and it is 0.036. If the blocks are considered as fixed effects we have evidence of a difference in wheat yields from the different farms, and can use follow-up procedures to determine where the differences are likely to be. All we can tell from the individual confidence intervals is that there is evidence the mean yield is higher for block 3 than block 2. This also follows from the significant F-test. If the blocks are considered random effects, we have evidence that the variance between blocks is greater than zero and our follow-up will be to estimate this variance.

6.2.3 Mathematical model

We will refer to our data by a pair of numbers. The first corresponds to the type of crop rotation and is one of the numbers 1, 2, 3 or 4. The second corresponds to the block and is1, 2 or 3. So, for example, yield(1,2) represents the yield after faba bean harvested for grain on block 2, which is 4247 kg/ha. Our model is:

$$\text{yield}(i, j) = \text{OM} + \text{rotation}(i) + \text{block}(j) + \text{error}(i, j)$$

where i runs from 1 to 4 to represent the four rotations, and j runs from 1 to 3 to represent the blocks. The rotation(i) and block(j) effects are measured relative to the overall mean (OM) and are therefore constrained to add to zero. The errors are assumed to be random and independent with a mean of zero and a standard deviation σ. They represent the differences in yields of wheat on the same plot using the same rotation if we imagine the experiment being repeated indefinitely. Since the design is balanced, the OM, rotation(i), and block(j) parameters are straightforward to estimate from the rotation and block means. The estimate of the OM is the mean of all 12 yields, which is also both the mean of the four estimated rotation means and the mean of the three estimated block means, and equals 3791.75. The estimates of the rotation effects are the estimated rotation means less the estimated OM:

$$\text{est-rotation}(1) = 4368 - 3791.75 = 576.25$$

$$\text{est-rotation}(2) = 4506 - 3791.75 = 714.25$$

$$\text{est-rotation}(3) = 3151 - 3791.75 = -640.75$$

$$\text{est-rotation}(4) = 3142 - 3791.75 = -649.75$$

The MINITAB General Linear Model procedure offers a complete analysis including various residual plots (see Figure 6.5). Select **Stat > ANOVA > General Linear Model**. Enter **Yield** in the **Responses** box, **Rotation** and **Block** in the **Model** box, and **Block** again in the **Random factors** box. Now click on **Comparisons**; put **Rotation** in the **Terms** box, and under **Method** make sure that **Turkey** is checked. Click **OK** and then on **Results**. Click on the radio buttonfor **In addition, coefficients for all terms**, check **Display expected value mean squares and variance components**, and in the **Display least squares means corresponding to the terms** box put **Rotation** and **Block**, The results of this analysis are as follows. First, we have a summary of the factors and the ANOVA:

```
General Linear Model: yield versus rotation, block

Factor        Type Levels Values
rotation      fixed      4 faba_g    faba_gm    grass_g    medic_g
block         random     3 1 2 3

Analysis of Variance for yield, using Adjusted SS for Tests

Source      DF     Seq SS     Adj SS     Adj MS       F       P
rotation     3    5024858    5024858    1674953  105.66   0.000
block        2     194037     194037      97019    6.12   0.036
Error        6      95114      95114      15852
Total       11    5314010
S = 125.906   R-Sq = 98.21%   R-Sq(adj) = 96.72%
```

Now we have the estimates of most of the parameters in the model. We can obtain **est-rotation(4)** as the negative of the sum of the first three rotation effects:

$$-(576.25 + 714.25 - 640.75) = -649.75$$

Similarly **est-block(3)** is $+66.25$. The constant is the estimate of the overall mean.

```
Term             Coef  SE Coef        T        P
Constant      3791.75    36.35   104.32    0.000
rotation
faba_g         576.25    62.95     9.15    0.000
faba_gm        714.25    62.95    11.35    0.000
grass_g       -640.75    62.95   -10.18    0.000
block
1              -23.75    51.40    -0.46    0.660
2             -142.50    51.40    -2.77    0.032
```

We specified that blocks were a random effect so in the following statement of the expected values of the mean squares **(2)** represents the variance between blocks. This table is the justification for the F tests.

```
Expected Mean Squares, using Adjusted SS

Source   Expected Mean Square for Each Term
1        rotation  (3) + Q[1]
2        block     (3) + 4.0000(2)
3        Error (3)
```

The next table gives error terms for the F tests which, in this case, are just the error MS in the original ANOVA:

```
Error Terms for Tests, using Adjusted SS

Source       Error DF   Error MS   Synthesis of Error MS
1 rotation      6.00      15852    (3)
2 block         6.00      15852    (3)
```

The next table gives us the estimated variance of the errors, which is the error MS in the original ANOVA, and the estimated variance between blocks 20,292. The estimated standard deviations for errors and between blocks are 126 and 142, respectively.

```
Variance Components, using Adjusted SS

Source    Estimated Value
block           20292
Error           15852
```

Now we have multiple comparisons between the rotation means.

```
Tukey 95.0% Simultaneous Confidence Intervals
Response Variable yield
All Pairwise Comparisons among Levels of rotation

rotation = faba_g subtracted from:

rotation   Lower   Center   Upper    ------+---------+-------+------
faba_gm     -218      138   494.2                      (-----*-----)
grass_g    -1573    -1217  -860.8     (-----*-----)
medic_g    -1582    -1226  -869.8     (-----*-----)
                                     ------+---------+-------+------
                                        -1200      -600       0
```

```
rotation = faba_gm subtracted from:

rotation   Lower    Center    Upper      ------+---------+-------+------
grass_g    -1711    -1355      -999      (-----*-----)
medic_g    -1720    -1364     -1008      (-----*-----)
                                         ------+---------+-------+------
                                           -1200       -600        0


rotation = grass_g subtracted from:
rotation   Lower    Center    Upper      ------+---------+-------+------
medic_g   -365.2    -9.000    347.2                          (-----*-----)
                                         ------+---------+-------+------
                                           -1200       -600        0
```

Tukey Simultaneous Tests
Response Variable yield
All Pairwise Comparisons among Levels of rotation
rotation = faba_g subtracted from:

Level rotation	Difference of Means	SE of Difference	T-Value	Adjusted P-Value
faba_gm	138	102.8	1.34	0.5725
grass_g	-1217	102.8	-11.84	0.0001
medic_g	-1226	102.8	-11.93	0.0001

rotation = faba_gm subtracted from:

Level rotation	Difference of Means	SE of Difference	T-Value	Adjusted P-Value
grass_g	-1355	102.8	-13.18	0.0001
medic_g	-1364	102.8	-13.27	0.0001

rotation = grass_g subtracted from:

Level rotation	Difference of Means	SE of Difference	T-Value	Adjusted P-Value
medic_g	-9.000	102.8	-0.08755	0.9997

The conclusion is unchanged. There is clear evidence that wheat yields are higher after faba has been grown, either for feed or as green manure, than after grass or medic. There is no evidence of a difference between **grass** or **medic**. Neither is there evidence of a difference between faba grown for feed or used as green manure, although the sample mean for the latter is slightly higher. A larger experiment might detect a difference.

The residuals are estimates of the errors. They are calculated by rearranging the model with error(i, j) on the left-hand side, and replacing the parameters by their estimates:

$$\text{residual}(i, j) = \text{yield}(i, j) - (\text{estimate OM} + \text{estimate rotation}(i) + \text{estimate block}(j))$$

For example, the residual corresponding to yield$(1,2)$ is:

$$4247 - (3791.75 + 576.25 + (-142.50)) = 21.5$$

It would be tedious to do this calculation for all 12 yields, but it is easy to obtain the residuals from **MINITAB**. For example, in **General Linear Model** select **Residuals** under **Storage** and **Normal plot of residuals** under **Graphs**.

Since the experiment has not been replicated, the interpretation of residuals as estimates of the errors depends crucially on the additive form of the model being realistic. The model does not allow for the effect of a rotation to depend on the block. Such two-factor effects are known as *two-factor interactions*. A residual that is extraordinarily large in absolute magnitude would be evidence of an interaction effect. We discuss interactions later in the book.

The experiment was performed over three farms. The estimated standard deviation of farm differences, 126 kg/ha, is slightly less than the estimated standard deviation of different replicates, 142 kg/ha. Chu-Hua considers that the findings are applicable to farms in the area. However, extrapolation of the results to farms with different soil types or climatic conditions relies on agricultural knowledge rather than statistical evidence. This is one reason why an understanding of the physical processes involved is so important.

The high yield is largely due to the increased nitrogen in the soil after the faba crop, and Chu-Hua would like to run a follow-up experiment in which she investigates the effects of different levels of nitrogen fertiliser when wheat is grown after grass or medic.

△ Review of Case 6.3

Wheat yields, with low levels of nitrogen fertiliser, are higher after faba has been grown, either for feed or for green manure, than after grass or medic. The estimated mean after faba is 4550 kg/ha, whereas the estimated mean after grass or medic is 3384 kg/ha.

▽ Case 6.4 (UoE)

Twelve junior managers in the head office of SeaDragon asked the personnel director, Karl Khaki, to pay for them to attend a course on leadership skills. This included seminars and outdoor activities over a long weekend in a mountain resort. Karl was sceptical about the value of this expensive course and concerned about setting a precedent of sending people on it. However, he decided to compromise by offering to set up an experiment to evaluate the course. He would compare it with both a one-day course of seminars on leadership skills without any associated outdoor activities and, as a control treatment, no formal training on leadership.

Karl enlisted the support of Ingrid Indigo, a psychologist at UoE, who would assess their leadership skills during the six months following the weekend course. She would ask all the staff who had day-to-day dealings with each manager to complete a questionnaire. She would also interview each manager and award him or her a score, between 0 and 100, based on the interview and questionnaire results. The leadership scores would be held anonymously, by Ingrid, but managers would be given details of their own performances if they wished. Karl would be given only the conclusions drawn from the analysis.

Ingrid had used this method for obtaining leadership scores before and advised that the typical standard deviation for a group of junior managers was 10.0. Karl did not wish to send more than 12 delegates on the weekend course. Before proceeding, he decided to calculate the power of a test to compare the means of three courses with samples of size 12, if the maximum difference in course means was as large as 10, testing at the 5% level with a two-sided alternative:

```
Power and Sample Size

One-way ANOVA
Sigma = 10   Alpha = 0.05   Number of Levels = 3

              Sample                 Maximum
SS Means       Size      Power     Difference
      50         12     0.5430             10
```

The result of the MINITAB calculation is a power of 0.54. Karl hoped he could improve the precision of the experiment by running it as a randomised block design. He persuaded another 24 managers to take part. For each of the 12 original applicants, Karl found two managers from the 24 who were similar in terms of age, education and level of responsibility within the company. This gave 12 groups of three managers. Each person in a group of three similar mangers was assigned at random to one of: no formal training (the control group); the day course; or the weekend course. The randomisation was subject to a restriction that four of the original 12 applicants would be in the control group, four in the day group, and four in the weekend group. This was an important aspect of the experimental design. Karl would have been more popular had he sent all the original applicants on the weekend course, but this would introduce potential bias. It is possible that the enthusiasm of these 12 applicants for the weekend course will be associated with higher leadership scores if they are sent on it. The results of the experiment are given in Table 6.2 and from www.greenfieldresearch.co.uk/doe/data.htm as Case 6.4. Ingrid stacks the data into a MINITAB worksheet as in Figure 6.17.

She starts the analysis with a plot of the data (Figure 6.18). This is a **With Groups** scatterplot with **score** as the **Y variable** and **coursetype** (see Figure 6.17) as the **X variable. group** is specified as a group. Double-click on a data point, and the window to edit symbols appears. Here you

Group_1	control	day	weekend
1	65	63	70
2	60	72	59
3	61	56	52
4	73	59	68
5	58	79	74
6	67	73	82
7	44	58	49
8	68	82	73
9	46	53	57
10	65	72	61
11	59	79	73
12	66	73	64

Table 6.2

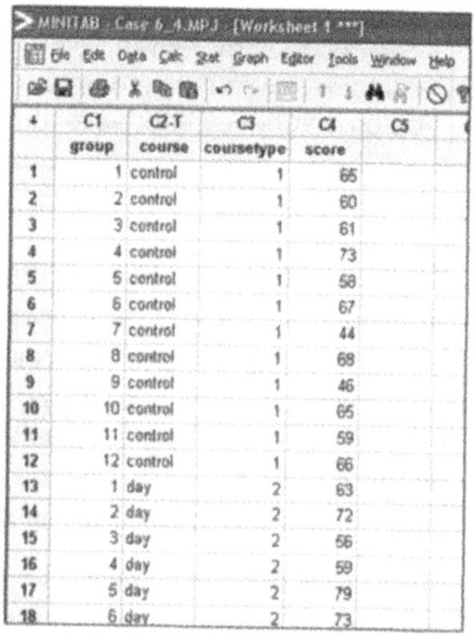

↓	C1 group	C2-T course	C3 coursetype	C4 score	C5
1	1	control	1	65	
2	2	control	1	60	
3	3	control	1	61	
4	4	control	1	73	
5	5	control	1	58	
6	6	control	1	67	
7	7	control	1	44	
8	8	control	1	68	
9	9	control	1	46	
10	10	control	1	65	
11	11	control	1	59	
12	12	control	1	66	
13	1	day	2	63	
14	2	day	2	72	
15	3	day	2	56	
16	4	day	2	59	
17	5	day	2	79	
18	6	day	2	73	

Figure 6.17

can switch off **Jitter** so that all the data points
are stacked in straight columns. The values of
coursetype appear on the X-axis as numbers.
Ingrid edits these labels to identify the names of
the course types. The plot suggests that scores
for the control are generally less than those for
either the day or the week course.

Ingrid uses the MINITAB two-way ANOVA to
produce the following results:

Figure 6.18

Two-way ANOVA: score versus course, group

```
Analysis of Variance for score
Source    DF    SS      MS      F     P
course    2     317.7   158.9   3.93  0.035
group     11    2092.3  190.2   4.70  0.001
Error     22    889.6   40.4
Total     35    3299.6
```

Individual 95% CI

```
course    Mean    ------+---------+---------+---------+--
control   61.0    (--------*--------)
day       68.3                    (--------*--------)
weekend   65.2           (--------*--------)
                  ------+---------+---------+---------+--
                      60.0      64.0      68.0      72.0
```

Dunnett 95.0% Simultaneous Confidence Intervals
Response Variable score
Comparisons with Control Level
course = control subtracted from:

```
course    Lower   Center  Upper    ------+--------+---------+---------+--
day       1.116   7.250   13.38                 (--------*--------)
weekend   -1.967  4.167   10.30    (--------*--------)
                                   ------+--------+---------+---------+--
                                       0.0      5.0      10.0      15.0
```

Dunnett Simultaneous Tests
Response Variable score
Comparisons with Control Level
course = control subtracted from:

```
Level    Difference   SE of               Adjusted
course   of Means     Difference  T-Value  P-Value
day      7.250        2.596       2.793    0.0198
weekend  4.167        2.596       1.605    0.2104
```

This provides evidence that the day course raises leadership scores above what they would be with
no formal training. There is no strong evidence that the day course is better than the weekend

course, or that the weekend course is better than no formal training. There is strong evidence of a difference between mean leadership scores for the groups. This confirms that the strategy of assigning the managers into groups of three with similar attributes has increased the precision of the comparison of courses. If you doubt this, try analysing the data without allowing for groups using a one-way ANOVA. You would infer that there is insufficient evidence of any difference. The better experiment is the one designed to attribute variation to between groups.

Karl had considered making before and after measurements of leadership scores, in which case the responses would have been the differences, but was put off by the additional six months that the experiment would take. Another consideration is that he would then be comparing the courses, and control, in terms of improvement in leadership skills rather than attained leadership skills, and the latter are of more relevance to the company. You may think that the control group, with no formal training on leadership, would not improve, but the anticipated assessment of their leadership skills may have a salutary effect.

△ Review of Case 6.4

Karl has evidence that the day course has improved the leadership skills of his junior managers and is keen to provide this for all such staff. He decides against offering the weekend course.

▽ Case 6.5 (SeaDragon)

Thiobacillus is a bacterium that is both useful and harmful. It is widespread in marine and terrestrial habitats; it oxidises sulphur, producing sulphates useful to plants; in deep ground deposits it generates sulphuric acid, which dissolves metals in mines and corrodes concrete and steel. The dissolving of metal can be turned to advantage.

Ohin Olive runs the extraction process at SeaDragon's copper refinery in Zambia. The extraction process uses bacterial leaching, and he wants to compare the yields from four strains of thiobacillus: A, B, C, and D. The company has three mines and he knows that the copper contents of the ores differ. He decides to take large batches of ore from each mine, to divide each batch into four smaller batches, and to randomly assign these smaller batches to the four strains. The mines are experimental blocks. He knows that the expected value of yields is about 20 kg per tonne, and that the standard deviation of yields from smaller batches at the same mine is about 2 kg/t. He makes a power calculation for detecting a technically significant difference of 5.0.

Power and Sample Size

```
One-way ANOVA
Sigma = 2      Alpha = 0.05    Number of Levels = 4
                Sample          Maximum
SS   Means    Size    Power   Difference
     12.5      3     0.5044        5
```

The power is rather low, but he decides to proceed with the experiment because the only way he can increase the sample size is to run it several times. He will consider this possibility after this first performance. The results of the experiment are in Table 6.3 and as Case 6.5 at www.greenfield-research.co.uk/doe/data.htm.

thiobacillus	mine	yield
A	1	22
A	2	28
A	3	25
B	1	31
B	2	33
B	3	26
C	1	20
C	2	25
C	3	24
D	1	25
D	2	28
D	3	26

Table 6.3

Ohin carries out an analysis using the general linear model:

```
Factor      Type  Levels   Values
mine        random    3    1 2 3
thiobaci    fixed     4    A B C D
Analysis of Variance for yield, using Adjusted SS for Tests
Source      DF    Seq SS   Adj SS  Adj MS      F      P
mine         2    36.167   36.167  18.083   4.09  0.076
thiobaci     3    78.250   78.250  26.083   5.91  0.032
Error        6    26.500   26.500   4.417
Total       11   140.917

Unusual Observations for yield

Obs     yield      Fit    SE Fit   Residual   St Resid
6     26.0000   29.1667   1.4860   -3.1667    -2.13R

R denotes an observation with a large standardized residual.
```

```
Tukey 95.0% Simultaneous Confidence Intervals
Response Variable yield
All Pairwise Comparisons among Levels of thiobaci

thiobaci = A subtracted from:

thiobaci   Lower    Center   Upper      ------+---------+---------+---------
B         -0.945    5.000   10.945                        (------*------)
C         -7.945   -2.000    3.945         (------*------)
D         -4.612    1.333    7.279            (------*------)
                                           ------+---------+---------+--------
                                              -7.0        0.0        7.0

thiobaci = B subtracted from:

thiobaci   Lower    Center   Upper      -------+--------+---------+---------
C         -12.95   -7.000   -1.055      (------*------)
D          -9.61   -3.667    2.279         (------*------)
                                         ------+---------+---------+--------
                                            -7.0        0.0        7.0

thiobaci = C subtracted from:

thiobaci   Lower    Center   Upper      ------+---------+---------+---------
D          -2.612    3.333    9.279                  (------*------)
                                         ------+---------+---------+--------
                                            -7.0        0.0        7.0
```

Tukey Simultaneous Tests
Response Variable yield
All Pairwise Comparisons among Levels of thiobaci
thiobaci = A subtracted from:

Level thiobaci	Difference of Means	SE of Difference	T-Value	Adjusted P-Value
B	5.000	1.716	2.914	0.0954
C	-2.000	1.716	-1.166	0.6673
D	1.333	1.716	0.777	0.8623

thiobaci = B subtracted from:

Level thiobaci	Difference of Means	SE of Difference	T-Value	Adjusted P-Value
C	-7.000	1.716	-4.079	0.0251
D	-3.667	1.716	-2.137	0.2428

thiobaci = C subtracted from:

Level thiobaci	Difference of Means	SE of Difference	T-Value	Adjusted P-Value
D	3.333	1.716	1.943	0.3045

Although Ohin knows there is a difference in the copper content of ore from the three mines, there is only weak evidence of this from such a small experiment. Nevertheless, taking mines as blocks has substantially increased the precision of the comparison of the thiobacilli.

△ **Review of Case 6.5**

The estimated mean yields for the three types of thiobacillus are 25, 30, 23 and 26 for A, B, C and D, respectively. There is reasonable evidence that B is better than C, and weak evidence that it is better than A. No other differences are statistically significant, but differences of this size would be of substantial economic benefit. Ohin decides to run another replicate of the experiment. If the results substantiate C being worse, he will then run two more comparisons of the three thiobacillus types A, B and D.

7

Factorial experiments with factors at two levels

In Chapter 6 we investigated the effects of a single factor on the response. Our examples included: the effect of membrane material on burst strength of filters; the effect of root-stock on four-year girth of apple trees; and the effect of crop rotation regime on the yield of wheat. Although each of these experiments could be described with a single factor, we used each experiment to compare several categories of its factor. There were 13 varieties of root-stock, for example.

In this chapter we consider the effect of several factors on a response, but each factor will be tested at only two levels, which we conventionally refer to as low (L) and high (H). Such experiments are called *two-level factorial experiments*. If a factor is categorical, we must restrict our experiment to the investigation of just two categories. If the factor is a continuous variable, we implicitly assume an approximate linear relationship between it and the response, as the factor ranges from low to high. In some cases, we may claim that this is a reasonable approximation, if the range of the factor is relatively short, but we discuss how to overcome this limitation in Chapter 9.

Case 7.2 is about manufacturing aluminium wheels, and we investigate the effects of six factors on the porosity of the wheels. A wheel must be rejected if its porosity is above a specified level. There are two presses in the factory, and the choice of press is a categorical factor that we investigate. One of the presses is arbitrarily associated with the low level and the other with the high level. We also investigate the effect of the aluminium temperature, which is a continuous variable. We can set the low temperature at 620°C and the high temperature at 700°C.

You might think that a good experimental design would be to investigate each factor with all the others held constant. There are two reasons why it is not. The first is that it is inefficient. The second, which can be much more serious, is that we may miss interaction effects. An *interaction effect* is present when the effect of one factor on the response depends on the level of another factor.

We shall explain the meaning of an interaction effect with a worked example. But we have a problem. This book is about how to *design and analyse your experiment with MINITAB*. We are using the professional version of MINITAB 14 (Pro), and this includes automatic procedures for the design of experiments. However, the Student version of MINITAB 14 does not include these procedures.

So we shall show you first how to design a two-level factorial experiment by using a code-generation procedure that is available in the Student version. Then we shall show you how to design the same experiment using the automatic procedures in Pro.

For our first example, suppose we are growing tomatoes and wish to investigate the effect of two factors, water and fertiliser, on the yield. We have four possible factor combinations: low water and low fertiliser; low water and high fertiliser; high water and low fertiliser; high water and high fertiliser. This is a 2×2 or 2^2 experiment, because there are two factors, each with two levels. The design and analysis of any factorial experiments with factors at two levels is greatly simplified if we code the low and high level of variables as -1 and $+1$ respectively, and this is standard practice.

7.1 Design of experiments by code generation

In MINITAB select **Calc > Make Patterned Data > Simple Set of Numbers**. In the box at the top of the window enter a name for the first variable, **Fertiliser**. We require a sequence L H L H, so enter **-1** in the **From first value** box, **+1** in the **To last value** box, and **2** in the **In steps of** box. We need to **List each value 1 times** and **List the whole sequence 2 times** (see Figure 7.1).

For the second variable, **Water**, we require the sequence L L H H, so enter **From first value -1 To last value 1, In steps of 2. List each value 2 times** and **List the whole sequence 1 times**. In the MINITAB worksheet you will now see the generated design (the first two columns of Figure 7.2).

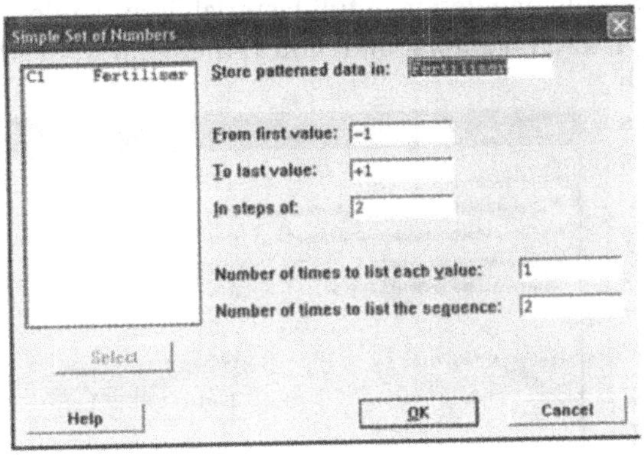

Figure 7.1

Worksheet 1 ***

↓	C1 Fertiliser	C2 Water	C3 Yield	C4
1	-1	-1	25	
2	1	-1	0	
3	-1	1	40	
4	1	1	50	
5				

Figure 7.2

Create a third column. Head it **Yield** and enter values **25, 0, 40, 50**.

We can now create an interactions plot using **Graph > Scatterplot;** select **With Connect and Groups**. Enter **Yield** as the Y variable and **Water** as the X variable and **Fertiliser** as the categorical variable for grouping (Figure 7.3). The interactions plot appears as in Figure 7.4. You will see in the interactions plot that the effect of water on yield (the increase in yield corresponding to the change of water from low to high value) is much greater for the high level of fertiliser (the dashed line) than for the low level of fertiliser

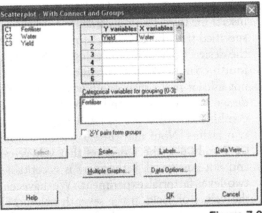

Figure 7.3

(the solid line). It seems that a high level of fertiliser increases the yield provided we have a high
level of water, but kills the plants if it is applied
with a low level of water. We say that the amount
of fertiliser and the amount of water interact.

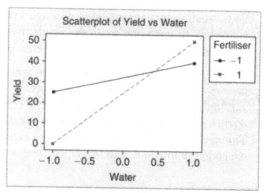

You can get another perspective on the interaction if you plot yield against fertiliser with water
as the categorical variable for the grouping.

Figure 7.4

7.2 Design of experiments with MINITAB's automatic procedure

If you have MINITAB 14 Pro you can design this 2^2 experiment using MINITAB's automatic procedure and follow this with entry of yield values and then an interactions plot.

Begin with **Stat > DOE > Factorial > Create Factorial Design**. Choose **2-level factorial (default generators)** and specify 2 factors. Click on **Designs**. A window shows all the designs that can be produced
using the information entered so far. In this case, only one design, a full factorial, is available but
you must select it before the next step. Click on **Factors** and enter the names of the
factors (A and B) as **Fertiliser** and **Water**, each with
low and high values of **-1** and **+1**. Choose **Options**
and remove the tick next to **Randomize runs**. This
is because, in this case, we do not want to randomise the order of the runs although generally
we should. Our reason here is to make it easy
to explain what we mean by *standard order*
(**StdOrder** in Figure 7.6).

MINITAB's generated design is shown in Figure
7.6. The first column is **StdOrder**, which we shall
explain in due course. The second is **RunOrder**
which, since we switched off **Randomize runs**, is
the same as **StdOrder**. MINITAB automatically
inserts the third column **CenterPt**, even though we
specified that there should be no centre points in
the design. So here it is meaningless. Similarly, the
fourth column, **Blocks**, is redundant since we did
not ask for more than one block of runs. The 2^2
design is in columns 5 and 6, expressed as all four
combinations of low (-1) and high $(+1)$ of the
two factors. Note that these columns are headed
C5-T and **C6-T**: the T denotes that values are text
and not numerical. Either way is acceptable in a
two-level factorial experiment. We have entered
the values of **Yield** in the seventh column. These
values must be numerical.

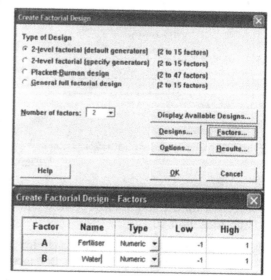

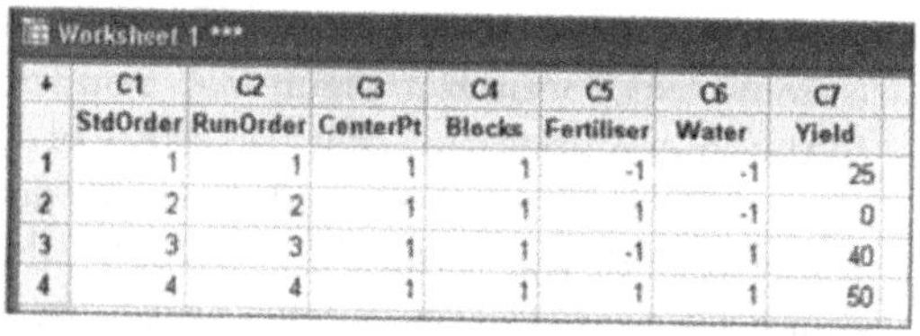

Figure 7.5

+	C1	C2	C3	C4	C5	C6	C7
	StdOrder	RunOrder	CenterPt	Blocks	Fertiliser	Water	Yield
1	1	1	1	1	-1	-1	25
2	2	2	1	1	1	-1	0
3	3	3	1	1	-1	1	40
4	4	4	1	1	1	1	50

Figure 7.6

To obtain the interaction plot follow **Stat > DOE > Factorial > Factorial Plots**. Choose **Interaction Plot** and click **Setup**. In the **Setup** window choose **Yield** as the response variable and **Fertiliser** and **Water** as factors to be included in the plot (Figure 7.7). The interaction plot appears as in Figure 7.8, which is similar to Figure 7.4, showing again that the amount of fertiliser and the amount of water interact: there is an *interaction*.

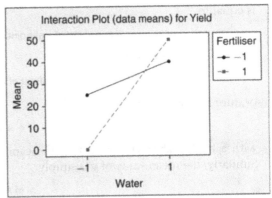

Figure 7.8

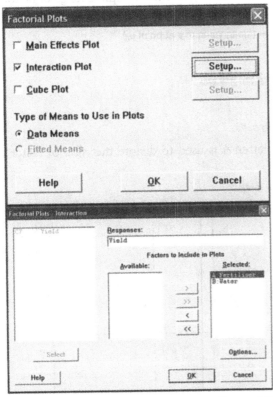

Figure 7.7

In our wheel manufacturing case below, another factor is injection pressure. It is quite possible that the effect of injection pressure depends on the temperature of the aluminium. If we test every possible combination of factors at low and high levels we can investigate all interactions. The number of runs in such an experiment with m factors will be 2^m, and we call this a single replicate of a full factorial experiment with m factors at two levels. If m is small, 2 or 3 say, we may choose to replicate the experiment. If m is large the number of runs may be excessive and we can use the methods described in Chapter 8 to provide a more practical design.

7.3 Standard order and notation

When there is good reason to believe that over the range of values of a control variable the response variable (y) is related to the control variable (x) by a linear function of the form:

$$y = a + bx + \text{(error)}$$

where a and b are coefficients to be estimated, then a and b can be estimated with the greatest precision if all observations are divided equally between the two ends of the range of x. In this equation, the effect on y of a change in x of one unit is represented by the coefficient b, which is the slope of the line.

Another way to represent the relationship between x and y is achieved by using a different notation. In this notation, the control variables (the x) are called factors and are represented by capital letters: $A, B, C, \ldots$. The range of a factor is specified by the two ends of the range: the high and the low values of the factor. These are represented by lower-case letters with suffices. For example, in the single-factor experiment, the high and low values of factor A would be a_1 and a_0, respectively. This lower-case notation is also used to represent the observed values of the response variable at the corresponding observation points.

The estimated effect of factor A on the response variable (y) over the complete range of factor A is equal to:

(mean value of y at point a_1) − (mean value of y at point a_0)

which can be expressed as

$$\text{effect of } A = \bar{y}\,(a_1) - \bar{y}\,(a_0)$$

or, more briefly,

$$A = a_1 - a_0$$

with a further abbreviation in that the capital letter A is used to denote the *effect* of factor A. Similarly, the mean value of y is simply

$$M = (a_1 + a_0)/2$$

Now consider an experiment with two factors, A and B, which can be represented as two variables in a plane with the response variable y along a third dimension perpendicular to the plane (coming out of the paper). If A and B are continuous then y defines a response surface. The design is the chosen values for A and B, and is shown in Figure 7.9. The high and low values of B are b_1 and b_0. If observations of y are made only at points defined by the extreme ranges of the two factors, there are four points which can be denoted by the combinations of letters as: a_0b_0, a_1b_0, a_0b_1, and a_1b_1. The notation can be abbreviated further by writing a and b for a_1 and b_1, and (1) for a_0 and b_0 The four design points are then represented as: (1), a, b, ab. The symbol (1) denotes the observation point at which all the factors are at their low levels. The point a is where factor A is at its high level but factor B is at its low level. The point ab is where both factors are at their high levels. The rule is that the high and low levels of factors are represented by the presence or absence, respectively, of lower-case letters.

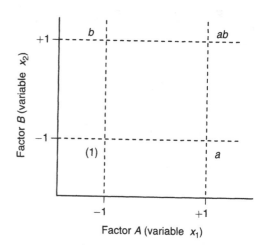

Figure 7.9

Now look at the order of the variable combinations, using the coding −1 and +1 for low and high respectively, and compare them with the corresponding order using the alphabetic notation in Table 7.1. The estimated main effect of A is the mean response of the two runs for which A is at the high level ($X_1 = +1$) minus the mean response for the two runs when A is at the low level ($X_1 = -1$):

$$A = (a + ab)/2 - (1 + b)/2$$

X_1	X_2	AB notation
−1	−1	(1)
+1	−1	a
−1	+1	b
−1	+1	ab

Table 7.1

Similarly, the estimated main effect of B is:

$$B = (b + ab)/2 - (1 + a)/2$$

The estimated two-factor interaction AB is the difference between the estimated effect of A when B is high and the estimated effect of A when B is low:

$$AB = (ab - b)/2 - (a - 1)/2 = (ab + 1)/2 - (a + b)/2$$

AB can equally well be expressed as the difference between the estimated effect of B when A is high and the estimated effect of B when A is low. There are two reasons to note that the interaction can also be expressed as the mean of the runs for which A and B are at the same level $(X_1X_2 = +1)$ minus the mean of the runs for which A and B are at different levels $(X_1X_2 = -1)$. The first is that it demonstrates that the order of A and B is arbitrary, and the second is that it easily generalises to more than two factors.

For the tomato growing experiment, the estimated main effect of fertiliser is

$$(0 + 50)/2 - (25 + 40)/2 = -7.5$$

The estimated main effect of water is

$$(40 + 50)/2 - (25 + 0)/2 = 32.5$$

Finally, the estimated interaction is

$$(50 + 25)/2 - (40 + 0)/2 = 17.5$$

If there is a substantial interaction, the main effects on their own may be meaningless. We cannot give a recommendation about using fertiliser unless we know how much water the plants will be given.

We can introduce a third factor (C) representing a third control variable (X_3) into our experiment. Copy the block of four rows for our two factor experiment in Table 7.1 below the original four rows and insert an X_3 column. In the X_3 column, enter -1 in the first four rows and $+1$ in the second four rows. You can now generate the alphabetic sequence easily. To introduce the letter c just multiply the sequence above it by c. The table becomes Table 7.2.

The main effect of C, for example, is the mean of the four runs with X_3 equal to $+1$ minus the mean of the four runs with X_3 equal to -1. In terms of letters:

$$C = (c + ac + bc + abc)/4 - (1 + a + b + ab)/4$$

The two-factor interaction AB, for example, is the mean of the four runs with the product X_1X_2 equal to $+1$ minus the mean of the four runs with the product X_1X_2 equal to -1. In terms of letters:

$$AB = (abc + ab + c + 1)/4 - (a + b + ac + bc)/4$$

The three-factor interaction ABC is the mean of the four runs with the product $X_1X_2X_3$ equal to $+1$ minus the mean of the four runs with $X_1X_2X_3$ equal to -1. In terms of letters:

$$ABC = (abc + a + b + c)/4 - (ab + ac + bc + 1)/4$$

ABC is the difference between the estimated two-factor AB interaction with C high and the estimated AB interaction with C low. It is also the difference of the BC interaction at the two levels of A and so on.

X_1	X_2	X_3	AB notation
-1	-1	-1	(1)
$+1$	-1	-1	a
-1	$+1$	-1	b
$+1$	$+1$	-1	ab
-1	-1	$+1$	c
$+1$	-1	$+1$	ac
-1	$+1$	$+1$	bc
$+1$	$+1$	$+1$	abc

Table 7.2

You can repeat the procedure to introduce a fourth factor (D). This gives you 16 rows. For a fifth factor (E), repeat with the letter e and you will have 32 rows. This is *standard order*.

In summary:

- The low level is coded -1 and the high level is coded $+1$.
- Lower-case letters ($a, b, c, \ldots$) represent both points in the two-level factorial design, with presence of the letter representing the high level; and observed values of the response variable at those points.
- Capital letters ($A, B, C, \ldots$) represent both names of factors and effects of those factors.
- Combinations of capital letters ($AB, AC, ABC, \ldots$) represent interaction effects.
- Main effects are estimated by the mean of runs with the letter present ($+1$) less the mean of the runs with the letter absent (-1).
- Interaction effects are estimated by the mean of runs with the letter combination corresponding to a product of $+1$ less the mean of the runs with a letter combination corresponding to a product of -1.

This notation is conventional in discussion of factorial experiments and is used in MINITAB. It is particularly useful in the development of fractional two-level experiments, which we shall discuss in Chapter 8.

7.4 Regression analysis of a factorial experiment

The MINITAB regression procedure is a convenient way to analyse factorial experiments. A regression model for our tomato experiment is

$$\text{yield}(i) = \text{OM} + b_1 x_1(i) + b_2 x_2(i) + b_3 x_1(i) x_2(i) + \text{error}(i)$$

where i represents observations 1, 2, 3, and 4 in the corresponding rows of Table 7.2. OM is the overall mean. $x_1(i)$ represents fertiliser and is -1 or $+1$ according to Table 7.2. $x_2(i)$ represents water. Do not confuse the unknown constant coefficients b_1, b_2, b_3 with levels of variable B. The coefficients in regression models are customarily denoted by bs. The main effect of fertiliser corresponds to x_1 changing from -1 to $+1$. The estimated effect is twice the estimated regression coefficient, and the estimated regression coefficient is half the estimated effect. If we rename the fertiliser column **x1** and the water column **x2** and calculate the product of **x1** and **x2** in another column called **x1x2** we can fit the regression using the MINITAB command **Stat > Regression > Regression**, with **yield** in the **Response box** and **x1 x2 x1x2** in the **Predictors** box. The first few lines of the output give the regression equation:

```
Regression Analysis: yield versus x1, x2, x1x2

The regression equation is
yield = 28.8 - 3.75 x1 +16.3 x2 +8.75 x1x2
```

You can check that these estimated coefficients are half the estimated effects. You should also notice that in the regression equation the random errors in the model have been replaced by their expected values of zero, which is our best estimate of the value of an unknown error.

▽ Case 7.1 (UoE)

This case is based on Singh *et al.* (1994).

Crispin Chartreuse is continuing his study of advertising techniques, and is investigating the effect of message spacing on people's recall of television commercials. The response variable was whether people remember the brand of product advertised and the claims made for it. He decided to investigate three factors, each at a high and low level: the number of commercials between the replicates of the test commercial, age, and time after seeing the commercial before recall was tested. The number of intervening commercials is 1 (L) or 4 (H), the age is 20–25 (L) or 62–83 (H), and the time until the recall test is five minutes (L) or one day (H).

A single replicate of the 2^3 factorial experiment has eight runs. He used **Stat > DOE > Factorial > Create Factorial Design**. We denote L as -1 and H as $+1$.

Crispin used some ingenuity in measuring the response variable. He placed a advertisement around the university, and in the local paper, asking for volunteers in either of the age ranges to take part in an experiment on memory recall. He had a good response, and randomly selected 120 people in the low age group to be randomly assigned to the four combinations of number of intervening commercials and time until recall: LL, LH, HL, and HH. He did the same for 120 people in the high age group. We shall explain why Crispin thought 30 people in a group for each combination of factors was a sufficient number at the end of the case. He decided to augment the basic design by setting up two control groups of people who saw the commercial only once. This would allow him to test whether seeing the commercial twice improves recall. The remaining younger and remaining older volunteers formed these two control groups.

Crispin prepared three video tapes, each of which included seven commercials within a news programme.

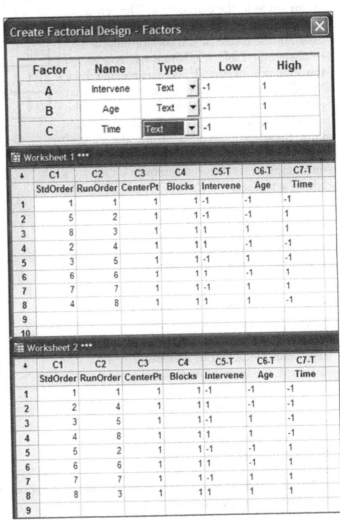

Figure 7.10

The test commercial, X, was for a vegetarian alternative to a hamburger and had relatively high information content. The sequence for the low number of intervening commercials ran: C1, C2, C3, C4, X, C5, X. The sequence for the high number of intervening commercials ran: C4, X, C2, C3, C5, X. Notice that X followed the same other commercials in both cases. The tape for the control group had the sequence: C4, C6, C1, C2, C3, C5, X. Crispin did not tell participants that the questions would be about the commercials. The recall score, for each participant, was one for remembering the brand name and one extra point for each of the eight product claims made in the commercial. The responses were averages of the 30 scores, or rather more than 30 in the case of the control groups. The results are in Table 7.3, where we have sorted the columns of the 2^3 design generated by MINITAB into a standard order. This is the first eight rows of the worksheet. The results for the control groups are in rows 9–12. The fourth column (**repetition**) is coded –1 for the control groups who saw the commercial only once, and +1 for the groups who saw it twice.

row	intervene	age	time	repetition	recall
1	−1	−1	−1	1	1.98
2	1	−1	−1	1	1.45
3	−1	1	−1	1	1.2
4	1	1	−1	1	0.78
5	−1	−1	1	1	1.17
6	1	−1	1	1	1.38
7	−1	1	1	1	0.32
8	1	1	1	1	0.83
9		−1	−1	−1	1.37
10		−1	−1	−1	0.52
11		1	−1	−1	0.85
12		1	−1	−1	0.24

Table 7.3

We visualise the first eight results in a cube plot (Figure 7.11) in which each result is attached to the corner of the cube representing the 2^3 design.

We could use ANOVA to analyse factorial designs, but it is easier to interpret the results if we fit a regression model. Also, it is easier to adapt the regression model to extended designs, such as the inclusion of the control groups who saw the commercial only once. The analyses are equivalent. We start by fitting a regression model to all 12 recall scores with the objective of determining whether seeing the commercial twice improves recall. We disregard spacing and omit **Intervene**. The regression model is:

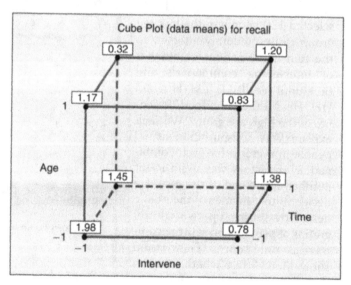

Figure 7.11

$$\text{recall}(i) = OM + b_1 \times \text{age}(i) + b_2 \times \text{time}(i) + b_3 \times \text{repetition}(i) + \text{error}(i)$$

where i represents the observations 1, . . ., 12 in the corresponding rows. The values of age(i), time(i) and repetition(i) are -1 or $+1$. MINITAB fits the model and gives us the estimates of OM, b_1, b_2 and b_3. The estimate of b_1 is one-half of the difference between the mean of the six observations when age is $+1$ and the mean of the six observations when age is -1. The half arises because the difference, which is the effect, is $2b_1$. The estimate of b_2 can be obtained in the same way. Since the four control recall scores have all four age time combinations the estimate of b_3 is the mean of the eight observations with repetition less the mean of the four without. The estimates are obtained in this straightforward manner because the predictor factors (predictors): age, time, and repetition are uncorrelated. In the context of the design of experiments, uncorrelated predictors are usually referred to as orthogonal, which may help emphasise that they are chosen by the experimenter rather than sampled observations. However, the estimate of the overall mean is not the mean of the 12 observations because there are eight observations with repetition and only four without repetition. The residuals are the estimates of the errors. Thus:

$$\text{residual}(i) = \text{recall}(i) - (\text{est.OM} + \text{est.}b_1 \times \text{age}(i) + \text{est.}b_2 \times \text{time}(i) + \text{est.}b_3 \times \text{repetition}(i))$$

In any regression that includes the overall mean, the sum of the residuals must equal zero. The estimates (est.) of the parameters are the values which minimise the sum of squared residuals. This is always the case, whether or not the predictor variables are uncorrelated, and is known as the principle of least squares. It was first published in 1805 by Adrien-Marie Legendre in his work on the estimation of comet orbits. For regression analysis in MINITAB use **Stat > Regression > Regression**. Put **recall** in the **Response** box and **age time repetition** in the **Predictors** box. This gives the following result:

Regression Analysis: recall versus age, time, repetition

The regression equation is
recall = 0.835 − 0.304 age − 0.214 time + 0.304 repetition

Predictor	Coef	SE Coef	T	P
Constant	0.8350	0.1246	6.70	0.000
age	−0.3042	0.1017	−2.99	0.017
time	−0.2138	0.1246	−1.72	0.125
repetition	0.3038	0.1246	2.44	0.041

S = 0.352397 R-Sq = 65.5% R-Sq(adj) = 52.6%

The estimated model, known as a *regression equation*, is followed by a table which gives the estimates of the coefficients followed by their estimated standard error (**SE Coef**), a t-ratio of the coefficient to its standard error and a P-value for testing the individual hypothesis that the coefficient in the corresponding population is zero, with a two-sided alternative. We have evidence, at the 5% level, that recall decreases with age, decreases with time elapsed before questioned, and increases if the advert is repeated because both the P-values are less than 0.05. We can construct 95% confidence intervals for the parameters given by **Coef** $\pm t \times$ **SE Coef**, where t is the upper 0.025 value of a t-distribution with degrees of freedom equal to the residual degrees of freedom. The residual degrees of freedom are the number of data less the number of parameters estimated from the data: $12 - 4 = 8$ (don't forget to count the constant). If the residual degrees of freedom are reasonably large, more than 10 say, t is roughly 2.0. The next line gives S which is the estimate of the standard deviation of the errors. The estimate of the variance of the errors is the sum of the squared residuals divided by their degrees of freedom, and S is the square root of this.

The extra information that MINITAB provides, the analysis of variance of the regression, helps to understand this:

```
Analysis of Variance

Source            DF      SS      MS      F      P
Regression         3   1.8892  0.6297   5.07   0.030
Residual Error     8   0.9935  0.1242
Total             11   2.8826

Source            DF   Seq SS
age                1   1.1102
time               1   0.0408
repetition         1   0.7381
```

The value of S is a good guide for fitting regression models. Generally, the smaller it is, the better the fit of the regression model to the data. However, we may decide that a simpler model with a slightly larger value of S is preferable to a model with a large number of terms. Also, we may decide to retain only those predictors that are statistically significant. There is no correct model; they are all approximations and we wish to select the best one for our purposes.

R-sq or R^2 is the *coefficient of determination* and can be thought of as the percentage proportion of the variance of the recalls that is explained, or described, by the model. It is calculated as:

$$\frac{\text{Total SS} - \text{Residual SS}}{\text{Total SS}} \times 100$$

where **Total SS** is the sum of the squared differences between recalls and their mean and **Residual SS** is the sum of squared residuals. The numerical values of these sums of squares are given in the following ANOVA table, and the numerator is referred to as **Regression SS** (so $R^2 = 1.8892/2.8826 = 0.655$). The limitation of R^2 is that it must increase if you add another predictor variable, because the sum of squared residuals can only decrease, whereas S can increase because of the loss of a degree of freedom. The adjusted R^2 (**R-sq(adj)**) allows for this and increases or decreases according to whether S decreases or increases.

The F-ratio and associated P-value test a hypothesis that b_1, b_2 and b_3 are simultaneously equal to zero: essentially a hypothesis that the model has no predictive power. We can certainly reject this overall null hypothesis. In particular, we do now have evidence that seeing the commercial twice improves recall and can go on to investigate the effect of intervening commercials.

We now fit the model

$$\text{recall}(i) = \text{OM} + b_1 \times \text{intervene}(i) + b_2 \times \text{age}(i) + b_3 \times \text{time}(i) + \text{error}(i)$$

for i from 1, ..., 8. The output from MINITAB is as follows:

```
The regression equation is
recall = 1.14 - 0.029 intervene - 0.356 age - 0.214 time

8 cases used 4 cases contain missing values

Predictor       Coef    SE Coef      T      P
Constant      1.1388    0.1082   10.53   0.000
interven     -0.0287    0.1082   -0.27   0.804
```

```
age           -0.3562   0.1082  -3.29  0.030
time          -0.2138   0.1082  -1.98  0.119

S = 0.3060 R-Sq = 78.7% R-Sq(adj) = 62.8%

Analysis of Variance

Source            DF       SS       MS      F      P
Regression         3  1.38744  0.46248   4.94  0.078
Residual Error     4  0.37445  0.09361
Total              7  1.76189

Source      DF   Seq SS
interven     1  0.00661
age          1  1.01531
time         1  0.36551
```

Crispin was rather disappointed with this as there is no hint that the number of intervening commercials has any effect at all. A colleague suggested that he might try including interaction terms using products of the predictors. First, Crispin must calculate columns with the products that represent the interactions. He could use the **Calc** routine but it is quicker to use the session window:

$$Let\ c7 = c2*c3$$

$$Let\ c8 = c2*c4$$

$$Let\ c9 = c3*c4$$

Label the new columns **inven*age, inven*time age*time**. Now use the regression routine with **Predictors: intervene, age, time, inven*age, inven*time, age*time** to obtain the following:

```
The regression equation is
recall = 1.14 - 0.0288 intervene - 0.356 age - 0.214 time + 0.0512
inven*age + 0.209 inven*time + 0.0062 age*time

8 cases used 4 cases contain missing values

Predictor       Coef   SE Coef       T      P
Constant     1.13875   0.02375   47.95  0.013
interven    -0.02875   0.02375   -1.21  0.440
age         -0.35625   0.02375  -15.00  0.042
time        -0.21375   0.02375   -9.00  0.070
inven*ag     0.05125   0.02375    2.16  0.276
inven*ti     0.20875   0.02375    8.79  0.072
age*time     0.00625   0.02375    0.26  0.836

S = 0.06718 R-Sq = 99.7% R-Sq(adj) = 98.2%

Analysis of Variance

Source            DF       SS       MS      F      P
Regression         6  1.75738  0.29290  64.91  0.095
Residual Error     1  0.00451  0.00451
Total              7  1.76189
```

```
Source      DF   Seq SS
interven     1   0.00661
age          1   1.01531
time         1   0.36551
inven*ag     1   0.02101
inven*ti     1   0.34861
age*time     1   0.00031
```

The interaction between the number of intervening adverts and time has a high t-ratio but its statistical significance is marginal because there is only one degree of freedom. For this reason, it would have been better if Crispin had run the experiment as 30 replicates with individual responses, rather than a single experiment with averages of 30 as the measured recall, although estimates would be the same. Given that we only have the latter, a pragmatic approach is to remove, at least, the age and time interaction, which is far from statistically significant, from the model. But when analysing a designed experiment, it is usually advisable to retain the main effect terms, whether or not they are statistically significant, if their interactions appear in the model. There are exceptional cases. For example, steel may corrode in an atmosphere that is humid and contains sulphur dioxide, but it is the interaction between these factors that causes the corrosion, not either factor on its own.

The interpretation of the main effects is that recall is lower for the older age group and lower if there is a day's delay before asking the questions. The number of intervening commercials has a slight, but not statistically significant, reduction effect when averaged over the short and long time before questioning. However, the positive interaction between time and number of intervening commercials must be taken into account. It suggests that a greater number of intervening commercials increases recall after one day, whereas a smaller number increases recall after five minutes. For example, consider expected recalls for someone in the younger age group. Then **age** is –1, and if there is only one intervening commercial **interven** is –1 and, if we ask after one day, time is **+1**. So the prediction is:

$$1.1388 - 0.0288 \times (-1) - 0.356 \times (-1) - 0.214 \times (+1) + 0.0512 \times (-1 \times -1) + 0.209 + (-1 \times +1) = 1.152$$

We can easily make predictions in MINITAB by asking for the fits. Fits are predictions for the values of predictors used in the experiment. For the younger respondents:

Intervening ads	Time	
	5 mins (−1)	1 day (+1)
1 (−1)	2.00	1.15
4 (+1)	1.42	1.41

For the older respondents:

Intervening ads	Time	
	5 mins (−1)	1 day (+1)
1 (−1)	1.18	1.15
4 (+1)	0.81	0.80

Interpretation of the two tables suggests that a three-way interaction should be added to the model because the two-way interaction is apparent only for the younger respondents.

We need to be careful about ascribing age effects in this study to the population as a whole because participants were volunteers rather than a random sample from a well-defined population. UoE psychology students are likely to be over represented in the younger group. Also, the experiment considered recall of a commercial, and it would have been interesting to compare this with recall of the news items. It is possible that the older group would perform better than the young group on this.

We shall now explain how Crispin decided on a sample size of 30. He knows from his experience of the advertising industry that young participants are unlikely to score more than 5.0 on a recall test taken directly after seeing commercials and that an average of 2.5 under these circumstances is typical. If the scores are approximately normally distributed, 95% would lie within two standard deviations of the mean. Thus he thinks that an estimated standard deviation of $(5 - 2.5)/2 = 1.25$ would be realistic for the sample size calculations. He decided that he wanted a power of about 0.7 of detecting a difference of 0.4 in population means at the two factor levels.

Power and Sample Size

2-Level Factorial Design

Sigma = 1.25 Alpha = 0.05

Factors: 3 Base Design: 3, 8
Blocks: none

Center Points Per Block	Effect	Reps	Target Power	Actual Power
0	0.4	31	0.7000	0.7087

The MINITAB calculation gives 31 replicates. He then checked a sample of size 30 which corresponds to a power of 0.6944. If he had randomised participants to 30 replicates of the 2^3 design, subject to the age distributions, he would have had $240 - 7$ degrees of freedom for error in a model which includes the constant, three main effects, and three two-factor interactions. The model would be for individual responses rather than for averages of 30 responses, but the estimated coefficients would be identical. Their standard errors (**SE Coef**) might be slightly different. If he has the original data he could even do this retrospectively, but the practical conclusions would be unlikely to change.

△ **Review of Case 7.1**

There is evidence that recall improves if people see a commercial twice. The estimated increase is by a factor of approximately 2. This is impressive, and it may justify the increased cost, especially for a launch of a new product. If the commercial is to be repeated in a sequence of commercials during a break of the TV programme, for a younger audience it is better to separate them with three other commercials than just one other commercial. The converse applies to an older generation.

▽ **Case 7.2 (SeaDragon)**

A division of SeaDragon makes aluminium wheels for the automotive sector. Its main customers are well-known car manufacturers who are continually pressing them to deliver better quality in shorter series at increasingly low prices. The quality supplied is good, but specifications are achieved at the cost of reprocessing (recasting) 20–25% of the wheels that are produced. This increases costs significantly and reduces productivity considerably, leading quite often to delays in delivery with the corresponding complaints and occasional economic penalties.

One year ago, Alinject became part of the multinational group SeaDragon, and the new owners gave a clear indication that unless the operation costs were reduced drastically the plant would be closed. The costs of reprocessing were estimated at €850,000 each year. Joe Debossa, the production manager, had recently attended a presentation on the uses and benefits of industrial statistics. He decided to apply these methods to reduce the need to reprocess wheels.

The first step was to prepare a detailed flow chart for the process. Figure 7.12 is a simplified version.

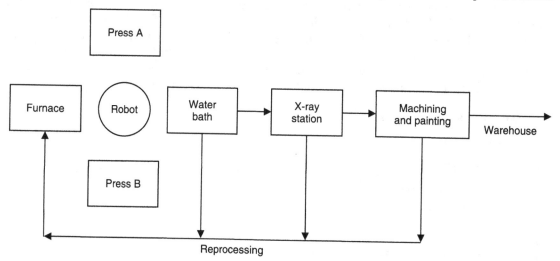

Figure 7.12

The process begins with the melting of the aluminium, which remains in a furnace at a temperature of about 650°C. A robot moves a measure of aluminium to the moulds that are in two presses close to the furnace. When the lower part of the mould is full, the upper part is lowered on top to seal it off and pressure is applied. The mould is cooled on the exterior so that the aluminium solidifies. The mould opens automatically, and a worker uses a crane to extract the wheel and quench it in water. If he notices any defects, such as in-filling, attachments and adhered materials, he returns the wheel to the furnace for remelting. Otherwise, he sends it to the X-ray station, where the porosity is examined. The X-ray machine examines the wheel from various angles and classifies the wheel's porosity from 0 to 10 (0 would be a perfect wheel and 10 would be a 'Swiss cheese'). If the score exceeds 3.75 the wheel is reprocessed, otherwise it proceeds to the machining and painting stations, where other defects may be detected.

Joe assembled a table of frequencies of different types of defect and, using MINITAB, produced a Pareto chart.

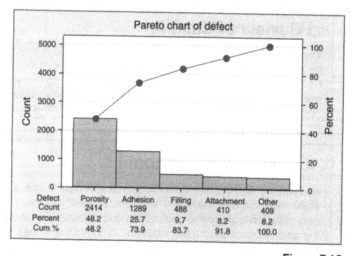

Defect	Porosity	Adhesion	Filling	Attachment	Other
Count	2414	1289	488	410	409
Percent	48.2	25.7	9.7	8.2	8.2
Cum %	48.2	73.9	83.7	91.8	100.0

Figure 7.13

In a MINITAB worksheet, enter a column of defect names and a columns of counts. Choose **Stat > Quality Tools > Pareto Chart** and click on the **Chart defects table** radio button. In **Labels in**, enter a column of defect names. In **Frequencies**, enter a column of counts. The Pareto chart in Figure 7.13 shows the causes of reprocessing during a typical month.

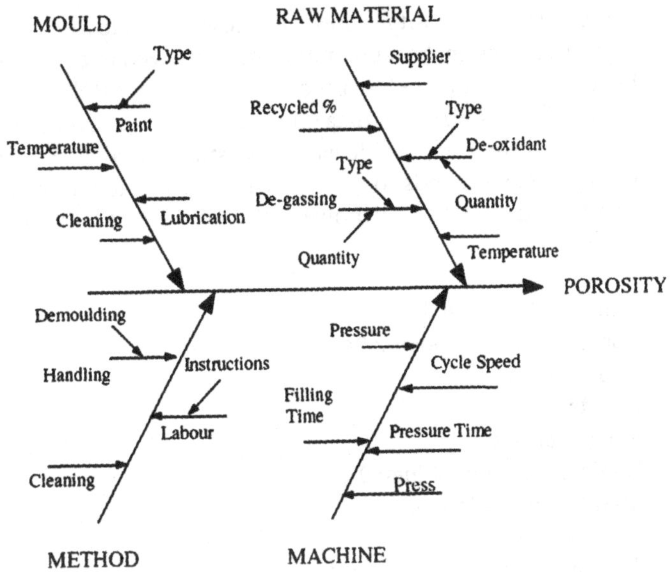

It was clear to Joe that to reduce reprocessing he should concentrate on reducing the number of defects due to porosity and adhered material. As a start, he asked Wyanet White to do an experiment aimed at reducing the porosity of the wheels using the reading provided by the X-ray machine as the response (Y). Her first step was to prepare a cause–effect diagram as a guide to establish the controllable factors that could affect porosity (X variables). Her diagram is shown in Figure 7.14.

After confronting opinions and analysing the existing data, poor and unreliable in the opinion of some of the engineers, Wyanet concluded that there were six factors that could most affect the appearance of pores in the wheels, and the next step was to confirm and quantify their relationship using a designed experiment.

Figure 7.14

Factors (X variables)	Range	
Press	Press A	Press B
Aluminium temperature (°C)	620	700
Injection pressure (kPa)	800	1000
Paint mould	Type A	Type B
Recycling (%)	0	20
Degassing	No	Yes

Table 7.4

The possible variation range of each factor was limited by the manufacturing conditions and by the technical knowledge and experience of the working team. Table 7.4 shows the team's conclusions.

The current operating settings, as described in the company's Quality Assurance Manual, yield an average porosity index of 3.6. The settings are in Table 7.5.

So there will be six factors in the experiment. A full factorial experiment will need 2^6 (64) runs of the process. Wyanet is prepared to do this many runs provided the factors remain within the limits the working team drew up. The 64 wheels produced can be added to inventory

Factors (X variables)	Range
Press	Both
Aluminium temperature (°C)	640
Injection pressure (kPa)	900
Paint mould	Type A
Recycling (%)	10
Degassing	Yes

Table 7.5

provided their porosity scores are below 3.75. However, she is not keen to run more than one replicate of a 2^6 design, because of the time involved in changing the process settings after each wheel has been moulded. But, before proceeding, she uses MINITAB to check the power of her proposed experiment. She has plenty of data on porosity at the current operating conditions. The mean is estimated as 3.6 and the standard deviation is estimated as 0.19. If she assumes porosity is normally distributed this corresponds to 21% of wheels being above the specified limit, and having to be recycled. She tries the MINITAB power calculation **Stat > Power and Sample Size > 2-Level Factorial** and enters:

Number of factors: 6
Number of corner points: 64
Replicates: 1
Effects:
Power value: 0.8
Number of center points: 0
Standard deviation: 0.19

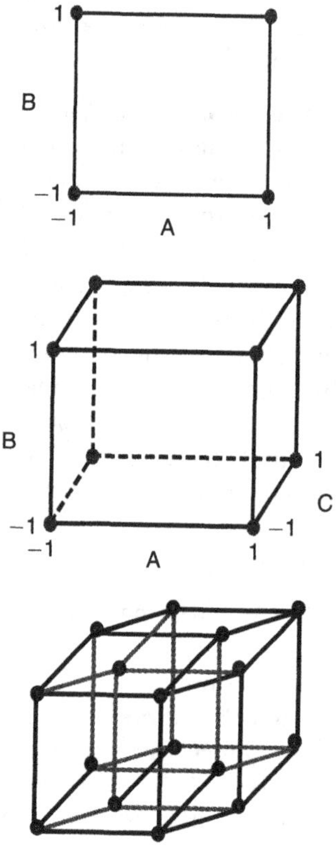

The number of corner points is just the number of runs. It is easiest to explain why they are called corner points in the context of a 2^2 design. If we call the two factors X_1 and X_2 and plot the design points in axes representing X_1 and X_2 they lie at corners of a square. They are called corner points because they are at the corners of a square. For a 2^3 design the design points are at the corners of a cube, and you can imagine the principle applying in higher dimensions. Figure 7.15 shows the sequence: the four points of a 2^2 design (square), the eight points of a 2^3 design (cube), the 16 points of a 2^4 design (four-dimensional cube or tesseract).

Wyanet clicks on OK, but receives an error message telling her that there are no degrees of freedom for error. MINITAB is assuming that she is fitting the *full regression model*:

- 1 constant
- 6 main effects
- 15 two factor interaction terms
- 20 three-factor interaction terms
- 15 four-factor interaction terms
- 6 five-factor interaction terms
- 1 six-factor interaction term

Figure 7.15

The design is orthogonal because any two from the 64 predictors in this list have a correlation of zero and are hence orthogonal, *provided each variable is coded to have values that are balanced symmetrically about zero*. The standard coding is $(-1, +1)$ as illustrated in Figure 7.9. A similar result holds for any 2^m design. The main practical advantage of an orthogonal design is that an estimate of any coefficient of a control variable remains the same whichever other control variables are included in the model. This makes the interpretation of the experiment and the choice of a suitable model easier.

In practice, we usually assume that interactions involving four or more factors are negligible. The interpretation of a three-factor interaction is that the two-factor interaction depends on the level of the third factor, so we usually ignore three-factor interactions as well.

Wyanet is going to fit only a constant, six main effects and 15 two-factor interactions. She is omitting 42 terms from the model. She follows: **Stat > Power and Sample Size > 2-Level Factorial Design**. She wants to know what effect can be detected with a power of 0.8. She clicks on **Design** and enters 42 terms omitted from the model. Her results are as follows:

```
2-Level Factorial Design

Alpha = 0.05   Assumed standard deviation = 0.19

Factors:   6    Base Design: 6, 64
Blocks: none

Number of terms omitted from model: 42

Center        Total
Points  Reps  Runs  Power   Effect
   0     1     64    0.8   0.136216
```

Thus, if there is a difference of 0.136 or more between two levels of a factor, there is a 0.8 probability of rejecting an overall null hypothesis, of no difference, at the 5% level, with a two-sided alternative, if there really is such a difference. Wyanet thinks that the power is quite sufficient for her investigation, as she is hoping to find more substantial differences than this. It makes little difference to the power calculation if she includes three-factor interaction terms in her model and specifies 22 terms omitted. The degrees of freedom for the residuals are equal to the number of terms omitted and affect the power calculation only through the t-value.

She decides to take the following values for the factors in her experiment:

Factor	Low	High
Press	PressA	PressB
Aluminium temperature	635	685
Pressure	840	960
Paint mould	Type A	Type B
Recycling	0%	20%
Degassing	no	yes

She has not used the full extent of the feasible operating ranges for the temperature and pressure in this experiment.

MINITAB provides a convenient worksheet, including a random order of runs, for the design. She selects **Stat > DOE > Factorial > Create Factorial Design**. She clicks the radio button for **2-level factorial (default generators)** and then the **Designs** control button. In the **Designs** window she selects **Full factorial** and clicks **OK**. Back in the **Create Factorial Design** window she clicks OK again. MINITAB produces the design. She runs her experiment and obtains the results in Table 7.6. The data set is also available in www.greenfieldresearch.co.uk/doe/data.htm.

Press	Paint	Degas	Temp	Pressure	Recycle	Porosity
−1	−1	−1	−1	−1	−1	5.13
−1	−1	1	−1	−1	−1	3.76
−1	1	−1	−1	−1	−1	5.28
−1	1	1	−1	−1	−1	3.97
1	−1	−1	−1	−1	−1	5.41
1	−1	1	−1	−1	−1	4.04
1	1	−1	−1	−1	−1	5.45
1	1	1	−1	−1	−1	3.8
−1	−1	−1	−1	−1	1	5.45
−1	−1	1	−1	−1	1	3.88
−1	1	1	−1	−1	1	3.58
−1	1	−1	−1	−1	1	5.35
1	−1	−1	−1	−1	1	5.64
1	−1	1	−1	−1	1	3.68
1	1	−1	−1	−1	1	5.56
1	1	1	−1	−1	1	3.63
−1	−1	−1	−1	1	−1	3.61
−1	−1	1	−1	1	−1	3.43
−1	1	−1	−1	1	−1	3.97
−1	1	1	−1	1	−1	3.91
1	−1	−1	−1	1	−1	3.63
1	−1	1	−1	1	−1	3.69
1	1	−1	−1	1	−1	3.9
1	1	1	−1	1	−1	3.64
−1	−1	−1	−1	1	1	3.72
−1	−1	1	−1	1	1	3.63
−1	1	−1	−1	1	1	3.86

Table 7.6

Press	Paint	Degas	Temp	Pressure	Recycle	Porosity
−1	1	1	−1	1	1	3.72
1	−1	−1	−1	1	1	3.45
1	−1	1	−1	1	1	4.08
1	1	−1	−1	1	1	3.65
1	1	1	−1	1	1	3.71
−1	−1	−1	1	−1	−1	4.65
−1	−1	1	1	−1	−1	2.93
−1	1	−1	1	−1	−1	4.91
−1	1	1	1	−1	−1	3.1
1	−1	−1	1	−1	−1	4.32
1	−1	1	1	−1	−1	2.96
1	1	−1	1	−1	−1	4.56
1	1	1	1	−1	−1	3.1
−1	−1	−1	1	−1	1	4.19
−1	−1	1	1	−1	1	2.5
−1	1	−1	1	−1	1	4.09
−1	1	1	1	−1	1	3.01
1	−1	−1	1	−1	1	4.39
1	−1	1	1	−1	1	3.08
1	1	−1	1	−1	1	4.66
1	1	1	1	−1	1	2.94
−1	−1	−1	1	1	−1	3
−1	−1	1	1	1	−1	2.66
−1	1	−1	1	1	−1	2.95
−1	1	1	1	1	−1	3.02
1	−1	−1	1	1	−1	3.05
1	−1	1	1	1	−1	2.83

Table 7.6   (Continued)

Press	Paint	Degas	Temp	Pressure	Recycle	Porosity
1	1	−1	1	1	−1	2.95
1	1	1	1	1	−1	2.83
−1	−1	−1	1	1	1	2.88
−1	−1	1	1	1	1	3.01
−1	1	−1	1	1	1	3.22
−1	1	1	1	1	1	3.33
1	−1	−1	1	1	1	3.37
1	−1	1	1	1	1	2.79
1	1	−1	1	1	1	3.27
1	1	1	1	1	1	3.28

Table 7.6   (Continued)

She starts by fitting only the main effects using the MINITAB regression procedure:
Stat > Regression > Regression. The results are:

Regression Analysis: Porosity versus press, paint, . . .

The regression equation is
Porosity = 3.77 + 0.0256 press + 0.0525 paint − 0.406 degas − 0.397
Temp − 0.390 Pressure + 0.0025 Recycle

Predictor	Coef	SE Coef	T	P
Constant	3.76625	0.05577	67.54	0.000
press	0.02563	0.05577	0.46	0.648
paint	0.05250	0.05577	0.94	0.350
degas	−0.40625	0.05577	−7.28	0.000
Temp	−0.39656	0.05577	−7.11	0.000
Pressure	−0.39000	0.05577	−6.99	0.000
Recycle	0.00250	0.05577	0.04	0.964

S = 0.446130   R-Sq = 72.9%   R-Sq(adj) = 70.1%

Analysis of Variance

Source	DF	SS	MS	F	P
Regression	6	30.5805	5.0967	25.61	0.000
Residual Error	57	11.3448	0.1990		
Total	63	41.9253			

Interpretation seems easy. Temperature, pressure and degassing have dramatic effects, and the other variables do not. Given the negative signs of the coefficients she should use: high temperature, high pressure and degassing. She was relieved that there was no evidence that recycling rejected wheels in the melt increased porosity.

She was about to write her report when she noticed that the estimated standard deviation of the errors (**S** = 0.446) was more than twice as high as the standard deviation of porosity on the current settings. She could think of at least two explanations. The first was that the variability increased if the plant was operated away from the current settings. She would need to run the process for a few days at different settings to check this. The second explanation was that two-factor interaction terms were important and should be included in the model. She can test this immediately using MINITAB. If you have only the Student version, you can use the regression procedure but this involves calculating 15 columns for the two-factor interactions. With Pro you can select: **Stat > DOE > Factorial > Analyze Factorial Design**. Click on the control button **Terms**. In the **Terms** window, select only main effects and first order (two-factor) interactions. At the top of this window there is an instruction **Include terms in the model up through order**. . . . Enter **2**.

Factorial Fit: Porosity versus press, paint, . . .

Estimated Effects and Coefficients for Porosity (coded units)

Term	Effect	Coef	SE Coef	T	P
Constant		3.7663	0.02439	154.44	0.000
press	0.0512	0.0256	0.02439	1.05	0.299
paint	0.1050	0.0525	0.02439	2.15	0.037
degas	−0.8125	−0.4062	0.02439	−16.66	0.000
Temp	−0.7931	−0.3966	0.02439	−16.26	0.000
Pressure	−0.7800	−0.3900	0.02439	−15.99	0.000
Recycle	0.0050	0.0025	0.02439	0.10	0.919
press*paint	−0.0725	−0.0363	0.02439	−1.49	0.145
press*degas	−0.0112	−0.0056	0.02439	−0.23	0.819
press*Temp	0.0069	0.0034	0.02439	0.14	0.889
press*Pressure	−0.0388	−0.0194	0.02439	−0.79	0.431
press*Recycle	0.0588	0.0294	0.02439	1.20	0.235
paint*degas	−0.0038	−0.0019	0.02439	−0.08	0.939
paint*Temp	0.0581	0.0291	0.02439	1.19	0.240
paint*Pressure	0.0438	0.0219	0.02439	0.90	0.375
paint*Recycle	−0.0350	−0.0175	0.02439	−0.72	0.477
degas*Temp	−0.0056	−0.0028	0.02439	−0.12	0.909
degas*Pressure	0.7550	0.3775	0.02439	15.48	0.000
degas*Recycle	0.0062	0.0031	0.02439	0.13	0.899
Temp*Pressure	0.0956	0.0478	0.02439	1.96	0.057
Temp*Recycle	0.0069	0.0034	0.02439	0.14	0.889
Pressure*Recycle	0.1138	0.0569	0.02439	2.33	0.025

S = 0.195092   R-Sq = 96.19%   R-Sq(adj) = 94.28%

Analysis of Variance for Porosity (coded units)

Source	DF	Seq SS	Adj SS	Adj MS	F	P
Main Effects	6	30.580	30.580	5.09675	133.91	0.000
2-Way Interactions	15	9.746	9.746	0.64975	17.07	0.000
Residual Error	42	1.599	1.599	0.03806		
Total	63	41.925				

```
Unusual Observations for Defects

Obs  StdOrder  Defects      Fit   SE Fit   Residual  St Resid
 29        29  3.45000  3.84813  0.11438  −0.39812     −2.52R
 43        43  4.09000  4.44125  0.11438  −0.35125     −2.22R
```

R denotes an observation with a large standardized residual.

The **Effect** (column 2) is twice the regression coefficient. This is because the regression coefficient is multiplied by −1 for low and +1 for high and the estimated effect, which is the difference between low and high, is twice the coefficient. The estimates of the main effects are unchanged by the addition of the interactions because the design is orthogonal. The residual mean square is 0.03806, so the estimated standard deviation of the errors is its square root, which equals 0.195. This closely agrees with the value estimated from current production. Several two-factor interactions are statistically significant, but one dominates and is of considerable practical importance. The coefficient of the pressure–degassing interaction is +0.3375, about as large in absolute value as the negative coefficients of the pressure and degassing main effects. The upshot is that there is no advantage to having both a high temperature and degassing. Both incur costs, and either one will produce much the same effect as both.

For her report, Wyanet decides to include a regression model which includes only the dominant effects: temperature, pressure, degassing, and the interaction of the last two.

Regression Analysis: Defects versus degas, Temp, Pressure, Pressure*Deg

```
The regression equation is
Porosity = 3.77 − 0.406 degas − 0.397 Temp − 0.390 Pressure
           + 0.378 Pressure*Degas

Predictor          Coef  SE Coef        T      P
Constant        3.76625  0.02544   148.06  0.000
degas          −0.40625  0.02544   −15.97  0.000
Temp           −0.39656  0.02544   −15.59  0.000
Pressure       −0.39000  0.02544   −15.33  0.000
Pressure*Degas  0.37750  0.02544    14.84  0.000

S = 0.203497   R-Sq = 94.2%   R-Sq(adj) = 93.8%
```

The estimated standard deviation of the errors is only slightly increased from the more complicated model, 0.203 rather than 0.195. The higher temperature is clearly desirable. Wyanet calculates predictions for the high temperature with all four combinations of pressure and degassing being high or low.

Pressure	Degas	Porosity
−1	−1	4.54
−1	1	2.98
1	−1	3.01
1	1	2.95

Since degassing is more expensive than raising the pressure Wyanet thinks that the best choice is high pressure with no degassing. As well as the point estimate of the mean porosity for these settings, 3.01, MINITAB gives a 95% confidence interval (CI) for this mean and a 95% prediction interval (PI) for the porosity of an individual wheel.

Predicted Values for New Observations

```
New Obs    Fit    SE Fit       95.0% CI            95.0% PI
1       3.0084   0.0569   (2.8946, 3.1223)   (2.5856, 3.4312)
```

Values of Predictors for New Observations

```
New Obs    Temp    Pressure    degas    prss*deg
1          1.00      1.00      -1.00      -1.00
```

From the PI we anticipate only 2.5% of wheels with a porosity of more than 3.43: well below the upper specification limit of 3.75.

To conclude her report Wyanet compares current and recommended operating conditions. Since including recycled wheels in the melt has no discernible effect she recommends any amount up to 20%. If her recommendations (Table 7.7) are as good as she hopes, there will not be much material available for recycling. She sends her report to her boss, Joe, and to the CEO of Alinject, Olivia Orange. We shall follow Olivia's response in Chapters 8 and 9.

Factors	Current	Recommended
Press	both	both
Paint Mould	Type A	both
Aluminium temperature	640	685
Mould pressure	900	960
Recycling	10%	Any up to 20%
Degassing	yes	no

Table 7.7

8

Fractional factorial experiments with factors at two levels

8.1 2^{m-p} experiments

At the end of Chapter 7 Wyanet sent her report with the analysis of her 64-run experiment, and her recommendations for changes in process operating conditions to Olivia Orange. Olivia was impressed, but said she would like a confirmatory follow-up experiment, with temperature and pressure at the ends of their feasible ranges. Wyanet was unenthusiastic about repeating the work and wondered whether it was essential to use an experiment with 64 runs. In this chapter we shall discuss her options, but first we look at a simpler case.

The general principle is to use a fraction of the full design which still leads to estimates of main effects and low order interactions (two or three factors). Fractional factorial experiments are widely used.

▽ **Case 8.1 (AgroPharm)**

Geoff Gold's project was to investigate the effects of cooking on niacin retention in vegetables. He wanted to look at a wide variety of vegetables so he needed to use designs with as few runs as possible for detecting important effects. He thought temperature, sieve size, and cooking time might affect niacin retention.

Geoff chose asparagus for his first experiment. If he ran a full factorial with each of these three factors at low (L) and high (H) levels his experiment would have 2^3 (=8) runs. For a pilot study he decided to try a *half fraction* of this design with only four runs. The half fraction is written 2^{3-1}. Although such a design is extremely parsimonious, it does demonstrate most of the general principles involved in constructing a fractional factorial experiment. To begin with, we will refrain from using the MINITAB DOE procedure and instead rely on MINITAB **Calc**. We shall also, as it is the usual convention, refer to the three factors by capital letters: *A* (temperature), *B* (sieve size), and *C* (time).

To construct the design we start with the full factorial. In MINITAB standard order the three columns will be: -1 $+1$ repeated four times; -1 -1 $+1$ $+1$ repeated twice, and -1 -1 -1 -1 $+1$ $+1$ $+1$ $+1$ once only (Table 8.1). These

A	B	C
−1	−1	−1
+1	−1	−1
−1	+1	−1
+1	+1	−1
−1	−1	+1
+1	−1	+1
−1	+1	+1
+1	+1	+1

Table 8.1

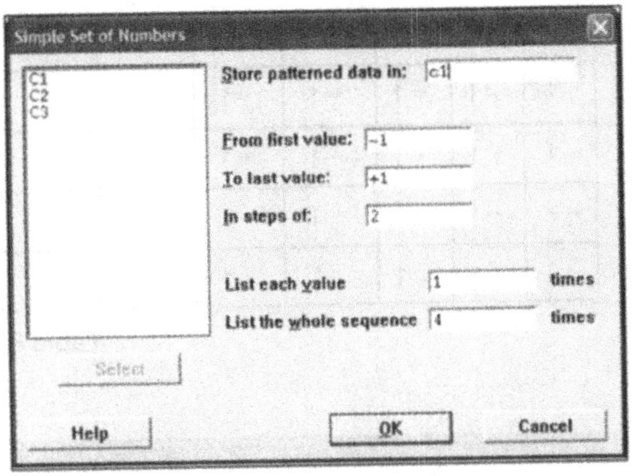

Figure 8.1

	1st	2nd	3rd
Store patterned data in	C1	C2	C3
From first value	−1	−1	−1
To last value	+1	+1	+1
In steps of	2	2	2
List each value * times	1	2	4
List the whole sequence * times	4	2	1

Table 8.2

columns are generated by: **Calc > Make Patterned Data > Simple Set of Numbers** (see Figure 8.1) which you use three times (Table 8.2).

Instead of using the windows interface, you can enter commands in the session window. If you prefer to do this, switch to the session window and click on **Editor > Enable Commands**. Now enter:

MTB > Set c1

DATA > 4(−1 : 1 / 2)1

DATA > End.

MTB > Set c2

DATA > 2(−1 : 1 / 2)2

DATA > End.

MTB > Set c3

DATA > 1(−1 : 1 / 2)4

DATA > End.

Again, following convention, we shall label these columns as **A**, **B** and **C** rather than **x1**, **x2**, and **x3**. With this convention **A** is used as both the name of the variable and the effect of the variable: it represents both the variable **x1**, if we fit a regression model, and the estimate of the main effect of **A**, which is the difference between the average of the responses for which the entry in column **A** is +1 and the average of the responses for which the entry in column **A** is −1. We construct the three two-factor and one three-factor interactions (Table 8.3) using **Calc** and specifying the products. These columns are labelled **AB**, **AC**, **BC**, and **ABC** respectively.

At this stage we can see that the three-factor interaction effect **ABC** is estimated by the average

row	A	B	C	AB	AC	BC	ABC
1	−1	−1	−1	1	1	1	−1
2	1	−1	−1	−1	−1	1	1
3	−1	1	−1	−1	1	−1	1
4	1	1	−1	1	−1	−1	−1
5	−1	−1	1	1	−1	−1	1
6	1	−1	1	−1	1	−1	−1
7	−1	1	1	−1	−1	1	−1
8	1	1	1	1	1	1	1

Table 8.3

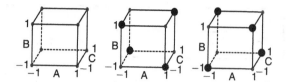

Figure 8.2

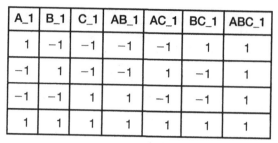

A_1	B_1	C_1	AB_1	AC_1	BC_1	ABC_1
1	−1	−1	−1	−1	1	1
−1	1	−1	−1	1	−1	1
−1	−1	1	1	−1	−1	1
1	1	1	1	1	1	1

Table 8.4

of shaded rows 2, 3, 5 and 8 less the average of rows 1, 4, 6 and 7. Suppose we assume this three-factor interaction is negligible and forgo its estimation. Then we can consider either the four runs corresponding to the +1, or the four runs corresponding to the −1 in the **ABC** column. We choose the former because MINITAB adopts this alternative. We could equally well select the other four runs; some other design programs do this.

We can select the corresponding rows and copy the columns using **Data > Copy > Columns to Columns** (Figure 8.3). We can see what these look like when represented as the points of a cube as in Chapter 7. The first cube on the left in Figure 8.2 shows the full factorial design. The middle cube shows the four points as specified in Table 8.4. The third cube shows the alternative set of four points.

We now have four rows which form a half fraction of the full factorial design (Table 8.4). You can check that all the entries in **ABC** are +1. If you look carefully, you will see that some of the columns are duplicates. In fact, **A** is the same as **BC**, **B** is the same as **AC**, and **C** is the same as **AB**. It follows that our estimate of the main effect of **A** will be identical to our estimate of the interaction **BC**, whatever the data happen to be. We say that **A** *is aliased with* or *is an alias of* **BC**. If we are prepared to assume the interaction is small compared to the main effect we can estimate the latter. The alias relationship follows from some simple algebra:

$$AA = A^2 = 1$$

because both $-1 \times -1 = 1$ and $+1 \times +1 = 1$. We choose the half fraction for which

$$ABC = 1$$

Multiply both sides by A

$$AABC = A(ABC) = A1 = A$$

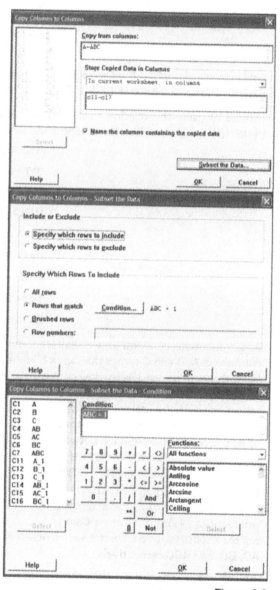

Figure 8.3

But

$$AA = 1$$

so

$$BC = A$$

Similarly, we have the other two alias relationships. We may as well remove the redundant columns AB_1, AC_1, BC_1 and ABC_1 from our worksheet.

A	B	C	Design	Niacin
1	−1	−1	a	83
−1	1	−1	b	85
−1	−1	1	c	68
1	1	1	abc	71

Table 8.5

Geoff performed this small experiment, and the four niacin measurements (ppm) are shown in Table 8.5.

We have added a column into this table to show the design represented in lower-case letters as we described in Chapter 7, deleted the original columns representing the full factorial, and renamed X1 as A and so on. The MINITAB procedure Stat > Regression > Regression yields the results:

```
Regression Analysis: niacin versus A, B, C

The regression equation is

niacin = 76.8 + 0.250 A + 1.25 B − 7.25 C

Predictor       Coef  SE Coef  T  P
Constant     76.7500        *  *  *
A             0.250000       *  *  *
B             1.25000        *  *  *
C            −7.25000        *  *  *

S = *
```

There are no degrees of freedom for error because the number of coefficients is the same as the number of data and we are bound to have an exact fit. The factor C has the dominant effect. If we fit it on its own we do at least have two degrees of freedom for the error.

```
Regression Analysis: niacin versus C

The regression equation is

niacin = 76.8 − 7.25 C

Predictor      Coef   SE Coef      T        P
Constant    76.7500   0.9014   85.15   0.000
C           −7.2500   0.9014   −8.04   0.015

S = 1.80278   R-Sq = 97.0%   R-Sq(adj) = 95.5%
```

Although we should be wary of a regression obtained by selecting the dominant factor, it is as expected. The longer the cooking time, the lower the niacin content.

Note that the main effect of C is twice its coefficient ($-7.25 \times 2 = -14.5$) because when C is −1 niacin is estimated as 76.8 + 7.25 and when C is +1 niacin is estimated as 76.8 − 7.25. That is, we estimate that the increase in cooking time from its low level to its high level will reduce the niacin content of the asparagus by 14.5 ppm, which is nearly 20%.

△ Review of Case 8.1

Geoff has estimated that the effect of the longer cooking time rather than the shorter cooking time would be to reduce the niacin content of asparagus by about 20%. But, with only four runs, this estimate is not precise. A reduction of niacin with cooking time was expected, and the objective of the experiments is to obtain reasonably precise estimates of the effects of all three factors on different vegetables. He therefore decides to proceed using the full design, rather than a half replicate.

8.1.1 MINITAB's design of experiments routine

Geoff Gold designed the half factorial (2^{3-1}) experiment by following **Calc > Make Patterned Data > Simple Set** and **Data > Copy > Columns to Columns** (Figure 8.2). We described this procedure to explain the way in which a half of a full factorial can be selected. We assume that a high-order interaction (in this case the three-factor interaction ABC) has negligible effect. This enables us to divide the full design into two parts so that we can select one part.

StdOrder	RunOrder	CenterPt	Blocks	A	B	C
1	1	1	1	−1	−1	1
2	2	1	1	1	−1	−1
3	3	1	1	−1	1	−1
4	4	1	1	1	1	1

Table 8.6

MINITAB has a routine that automatically divides a full factorial into two parts, or four parts, or 8, or 16 and so on, depending on the number of factors in the full factorial. We shall demonstrate this routine with the simplest possible case: a half factorial (2^{3-1}) experiment. The routine starts as described in Chapter 7: **Stat > DOE > Factorial > Create Factorial Design**. In the window (Figure 7.5) choose **2-level factorial (default generators)** and specify three factors. Click on **Designs** and choose **1/2 fraction**, then click on **OK**. Choose **Options** and remove the tick next to **Randomize runs**. This is because, in this case, we do not want to randomise the order of the runs although generally we should. The design generated by MINITAB is in Table 8.6.

The first column is **StdOrder**. The second is **RunOrder** which, since we switched off **Randomize runs**, is the same as **StdOrder**. MINITAB automatically inserts the third column **CenterPt**, even though we specified that there should be no centre points in the design. So here it is meaningless. Similarly, the fourth column, **Blocks**, is redundant since we did not ask for more than one block of runs. The 2^{3-1} design is in columns 5 to 7. It is the same as in Table 8.5 except that the rows are in a different order.

In the session window of MINITAB, you will see:

Fractional Factorial Design

```
Factors:  3  Base Design:    3, 4  Resolution:  III
Runs:     4  Replicates:        1  Fraction:    1/2
```

```
Blocks:    1 Center pts (total):     0
```

* NOTE * Some main effects are confounded with two-way interactions.

Design Generators: C = AB

Alias Structure

```
I + ABC
A + BC
B + AC
C + AB
```

This records that there are three factors, the experiment has four runs, there is only one block, there is only one replicate of each run; there are no centre points; and the design is a half fraction. We shall explain the meaning of *resolution* in the discussion of Case 8.2 where we shall also explain the statement about **Design Generators**.

Alias Structure lists the aliasing between main effects and interactions, as we described earlier. **A + BC** means that **A** is aliased with **BC**; we estimate the coefficient of **(A + BC)** but, since we assume that the interaction **BC** is negligible, we assign the total estimate to the main effect **A**. Likewise, **B** is aliased with **AC** and **C** is aliased with **AB**. However, **I + ABC** does not mean that ABC is aliased with a strange factor labelled with the letter I; in fact I is the roman symbol for 'one', which is useful in the algebra of two-level factorials. The statement **I + ABC** means that the full factorial has been split into two parts by reference to the high-order interaction **ABC**; **ABC** is aliased with the overall mean.

☐ Case 8.2 (SeaDragon)

Wyanet decides to look at her options: she selects **Stat > DOE > Factorial > Create Factorial Designs** and clicks on **Display Available Designs**. With six factors MINITAB offers: Full with 64 runs; Resolution VI with 32 runs; Resolution IV with 16 runs; and Resolution III with 8 runs. These are the full factorial 2^6, half fraction 2^{6-1}, quarter fraction 2^{6-2} and eighth fraction 2^{6-3}, respectively.

The definition of *resolution* is that a design is of resolution R, conventionally written as a roman numeral, if no p-factor interaction is aliased with another effect containing less than $R - p$ factors.

Generally, we want to use fractional designs that have the highest possible resolution consistent with the degree of fractionation required. Resolution III, IV, and V designs are particularly important.

Resolution III	No main effects are aliased with any other main effect, but main effects are aliased with two-factor interactions and two-factor interactions are aliased with each other. We can make plausible estimates of the main effects only if it is reasonable to suppose that the effects of two-factor interactions are relatively small.
Resolution IV	No main effects are aliased with any other main effect or two-factor interactions, but two-factor interactions are aliased with each other. We can detect the presence of two-factor interactions but we cannot identify them uniquely.
Resolution V	No main effects or two-factor interactions are aliased with any other main effect or two-factor interactions, but two-factor interactions are aliased with three-factor interactions.

Wyanet decides to consider in more detail the quarter fraction, of resolution IV, with 16 runs. The design is constructed from the full factorial by setting two high-order interactions equal to 1. You might think that: **ABCDEF = 1** and **ABCDE = 1** would be good choices. However, if **ABCDEF = 1** and **ABCDE = 1** then **ABCDEFABCDE = 1** and since the left hand side reduces to **F** we have **F = 1**. All the runs would be with **F** at the high level and we would not be able to estimate its main effect. The best we can do is to choose two four-factor interactions, with the smallest number of letters in common, which is 2, to set equal to 1. The MINITAB default is to choose **ABCE = 1** and **BCDF = 1** and hence their product **ADEF = 1**, but you can make your own choice if you prefer. The three aliases of the main effect of **A**, for example, are found by:

$$A = AABCE = BCE; A = ABCDF;$$

$$A = A(ABCEBCDF) = A(AEDF) = EDF$$

An example of the three aliases of a two-factor interaction is the aliases of **BC**:

$$BC = BCABCE = AE; BC = BCBCDF = DF; BC(AEDF) = ABCDEF$$

You can obtain the 16 runs by using **Calc** in a similar way to Case 8.1. Generate six columns, headed **A**, **B**, **C**, **D**, **E**, **F**, by **Calc > Make Patterned Data > Simple Set of Numbers**, which you use six times (Table 8.7).

Use **Calc > Calculator** to create two new columns, **ABCE** and **BCDF**, as the products of four corresponding columns in each case.

	1st	2nd	3rd	4th	5th	6th
Store patterned data in	A	B	C	D	E	F
From first value	−1	−1	−1	−1	−1	−1
To last value	+1	+1	+1	+1	+1	+1
In steps of	2	2	2	2	2	2
List each value * times	1	2	4	8	16	32
List the whole sequence * times	32	16	8	4	2	1

Table 8.7

Copy columns applying the condition **ABCE = 1** **And**

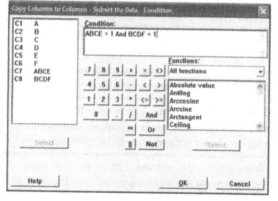

Figure 8.4

BCDF = 1, as in Figure 8.4. This produces the design as in Table 8.8. We have added a column to show the number of runs (16, which is 2^{6-2} or a quarter of 2^6). In another column we have listed the design in the alphabetic code. This is useful to check that the design is balanced: each letter appears eight times, indicating that each factor is present at its high level eight times and at its low level eight times.

However, if you have the Pro version of MINITAB, follow **Stat > DOE > Factorial > Create Factorial Design**. In the display of available designs, see that a design with six factors and 16 runs is available. In the **Designs** windows,

Run	A_1	B_1	C_1	D_1	E_1	F_1	Design
1	−1	−1	−1	−1	−1	−1	(1)
2	−1	+1	+1	−1	−1	−1	bc
3	+1	+1	−1	1	−1	−1	abd
4	+1	−1	+1	1	−1	−1	acd
5	+1	−1	−1	−1	1	−1	ae
6	+1	+1	1	−1	1	−1	abce
7	−1	+1	−1	1	1	−1	bde
8	−1	−1	1	1	1	−1	cde
9	+1	+1	−1	−1	−1	1	abf
10	+1	−1	1	−1	−1	1	acf
11	−1	−1	−1	1	−1	1	df
12	−1	+1	1	1	−1	1	bcdf
13	−1	+1	−1	−1	1	1	bef
14	−1	−1	1	−1	1	1	cef
15	+1	−1	−1	1	1	1	adef
16	+1	+1	1	1	1	1	abcdef

Table 8.8

choose **1/4 fraction**. You will be presented with the same design as in Table 8.8 but more information will be in the sessions pane:

Fractional Factorial Design

```
Factors:  6  Base Design:      6, 16  Resolution:  IV
Runs:     16  Replicates:          1  Fraction:    1/4
Blocks:    1  Center pts (total):  0
```

Design Generators: E = ABC,   F = BCD

Alias Structure

I + ABCE + ADEF + BCDF

A + BCE + DEF + ABCDF
B + ACE + CDF + ABDEF
C + ABE + BDF + ACDEF
D + AEF + BCF + ABCDE
E + ABC + ADF + BCDEF
F + ADE + BCD + ABCEF
AB + CE + ACDF + BDEF
AC + BE + ABDF + CDEF

```
AD + EF + ABCF + BCDE
AE + BC + DF + ABCDEF
AF + DE + ABCD + BCEF
BD + CF + ABEF + ACDE
BF + CD + ABDE + ACEF
ABD + ACF + BEF + CDE
ABF + ACD + BDE + CEF
```

I + ABCE + ADEF + BCDF states that the three design generators are all aliased with each other and with the overall mean.

The MINITAB output is expressed slightly differently from our explanation, but it is easy to show they are equivalent. We expressed the design generators as **ABCE = 1** and **BCDF = 1**. Multiply both sides of the first by **E** and the second by **F** to obtain the MINITAB expressions. We demonstrated that **A = BCE = DEF = ABCDF**. MINITAB emphasises that these are aliases by writing **A + BCE + DEF + ABCDF**. This means that that we can interpret the corresponding estimated effect as only the main effect of **A** when we assume three-factor and higher interactions are negligible. We say that the aliases **A**, **BCE**, **DEF**, and **ABCDF** are *confounded*, or *aliased*, because we cannot distinguish between them.

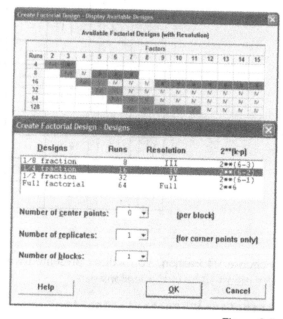

Figure 8.5

Wyanet's original full factorial experiment had found that the dominant interaction was pressure with degassing, but there was also some evidence of a pressure and recycle interaction and a temperature and pressure interaction. With this evidence, she decides that when she fits a model to the data she will include only three interactions: pressure × degassing, temperature × pressure, and pressure × recycle.

She now checks the power of the design. The full model would have 16 terms: the constant, six main effects, seven confounded two-factor interactions, and two confounded three-factor interactions. She intends fitting a regression with a constant, six main effects, and three interactions (**BC**, **CD**, **CE**): 10 terms. Therefore there are six terms missing from the model. She uses the same estimate of the standard deviation (0.19) as in Case 7.2 and wonders what the power would be if she tried to detect an effect of 1.5 standard deviations (0.285). She uses the MINITAB power calculation **Stat > Power and Sample Size > 2-Level Factorial Design** and enters:

Number of factors: 6

Number of corner points 16

Replicates: 1

Effect: 0.285

Number of center points: 0

Standard deviation: 0.19

She clicks on **Design** and specifies six Terms omitted from the model. MINITAB returns the following:

Power and Sample Size

2-Level Factorial Design

Alpha = 0.05   Assumed standard deviation = 0.19

Factors: 6 Base Design: 6, 16
Blocks: none

Number of terms omitted from model: 6

Center Points	Effect	Reps	Total Runs	Power
0	0.285	1	16	0.706712

She has a probability of only 0.71 of detecting an effect if it is as large as 1.5 standard deviations. She wonders how the power would increase if she were to replicate the design, so she enters the same effect but with two replicates:

Power and Sample Size

2-Level Factorial Design

Alpha = 0.05   Assumed standard deviation = 0.19

Factors: 6 Base Design: 6, 16
Blocks: none

Number of terms omitted from model: 6

Center Points	Effect	Reps	Total Runs	Power
0	0.285	2	32	0.981741

So, with two replicates, she has a very high probability (0.98) of detecting such an effect.

A half fraction is an alternative to two replicates of a quarter fraction, and has almost the same power. The slight discrepancy comes from the former having only 10 degrees of freedom if all possible interactions are fitted. The advantage of the half replicate is that the two-factor interactions are not aliased. The advantage of two replicates of a quarter fraction is that we estimate the standard deviation of the errors on 16 degrees of freedom without assuming that our model accounts for all the variability because we are replicating runs at the same factor settings. For a first experiment we would recommend the half fraction, but the latter has some attraction for a follow-up study if we think that only a few interactions are likely to be important.

Wyanet is tempted to run two replicates of the quarter fraction by making two wheels for each of the 16 factor combinations. This would require only the same number of process changes as a single replicate; the process changes that take the time and add cost to the experiment. On reflection, she decides against this because the assumption that the errors are independent would be seriously compromised. For example, one source of error is that the temperature of aluminium differs from its target value. If this is so it will be the same for two consecutive wheels. Ideally, she should complete one replicate before starting the second, and randomise the order within replicates. But if it took a long time for the aluminium temperature to stabilise it might be better to have a set of runs at the low temperature and then a set of runs at the high temperature because the carry-over effects of thermal inertia are of more concern than errors about the target.

Wyanet decides to go ahead with one replicate of a quarter factorial. Once she has the results she will contemplate a second replicate. The low and high levels for the six factors are now at the ends of the possible ranges recommended by the working team (Table 8.9).

We said earlier that the quarter fraction design is constructed from the full factorial by setting two high-order interactions equal to 1. The MINITAB default is to choose: **ABCD = 1** and **BCDF = 1**, but you can make your own choice if you prefer. Wyanet decided to use **ABEF = 1** and **ACDF = 1**. She followed the same procedure as described in Table 8.7 and Figure 8.4 and then converted the low (−1) and high (+1) values according to Table 8.9. This led to the design in Table 8.10 to which she has added a column of results. Notice that she has used

Factor	L (−1)	H (+1)
Press	PressA	PressB
Temperature	620	700
Pressure	800	1000
Recycle	0%	20%
Degassing	No	Yes
Paintmould	Type A	Type B

Table 8.9

Factor>>	A	B	C	D	E	F	Response
Run	Press	Paint	Temp	Pressure	Recycle	Degas	Porosity
1	PressA	TypeA	620	800	0	No	6.9
2	PressB	TypeB	700	800	0	No	5.81
3	PressB	TypeB	620	1000	0	No	4.25
4	PressA	TypeA	700	1000	0	No	3.17
5	PressA	TypeB	620	800	20	No	6.87
6	PressB	TypeA	700	800	20	No	5.33
7	PressB	TypeA	620	1000	20	No	4.41
8	PressA	TypeB	700	1000	20	No	2.59
9	PressB	TypeA	620	800	0	Yes	4.83
10	PressA	TypeB	700	800	0	Yes	3.72
11	PressA	TypeB	620	1000	0	Yes	4.78
12	PressB	TypeA	700	1000	0	Yes	3.34
13	PressB	TypeB	620	800	20	Yes	4.53
14	PressA	TypeA	700	800	20	Yes	3.99
15	PressA	TypeA	620	1000	20	Yes	4.79
16	PressB	TypeB	700	1000	20	Yes	3.88

Table 8.10

the actual values of the variables, rather than their -1 and $+1$ coding, for company records. We have added a row at the top to label the factors alphabetically since we shall refer to these labels during the analysis.

She first fits all the main effects and the three two-factor interactions. She has a choice for the analysis: either use the standard regression analysis or use the analysis that MINITAB provides for factorial experiments. The latter has several advantages. One is that it will deliver the regression results for both coded data (using −1, +1) and uncoded data (such as **Temp = 620**). A second advantage is that you can specify interactions in the model without having to calculate them as you would for the standard regression. A third advantage is that many extras can be delivered such as effects and interaction charts, plots of variables against residuals and experimental order, normal plots, tables of observed, predicted and residual values and a table of aliased effects. So she decides to use the analysis for factorial experiments.

Wyanet had saved the data of Table 8.10 (except the top row) in an Excel spreadsheet. This is available at www.greenfieldresearch.co.uk/doe/data.htm. She copied and pasted this into a new MINITAB worksheet. Then she followed: **Stat > DOE > Factorial > Analyze Factorial Design**. MINITAB doesn't recognise the table as a designed experiment so asks for some details (Figure 8.6).

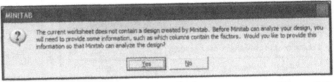

Figure 8.6

She is then able to follow the sequence shown in Figure 8.7. To leave the screen shown in Figure 8.7a, click on **Low/High**. For the text variables, you have to type in the name of the low and high level as it is in the column of your worksheet. To leave this screen, click on **OK** twice. After specifying the response, click on **Terms**. You should now see the screen in Figure 8.7d. Click **OK** and return to the screen of Figure 8.7c when you select **Graphs**. The screen of Figure 8.7e should appear. Complete this as shown.

The analysis results are as follows:

Factorial Fit: Porosity versus Press, Paint, . . .

Estimated Effects and Coefficients for Porosity (coded units)

Term	Effect	Coef	SE Coef	T	P
Constant		4.5744	0.08068	56.70	0.000
Press	−0.0538	−0.0269	0.08068	−0.33	0.750
Paint	−0.0413	−0.0206	0.08068	−0.26	0.807
Temp	−1.1913	−0.5956	0.08068	−7.38	0.000
Pressure	−1.3463	−0.6731	0.08068	−8.34	0.000
Recycle	−0.0512	−0.0256	0.08068	−0.32	0.762
Degas	−0.6838	−0.3419	0.08068	−4.24	0.005
Temp*Pressure	−0.1212	−0.0606	0.08068	−0.75	0.481
Pressure*Recycle	0.0837	0.0419	0.08068	0.52	0.622
Pressure*Degas	1.2763	0.6381	0.08068	7.91	0.000

S = 0.322720   R-Sq = 97.17%   R-Sq(adj) = 92.92

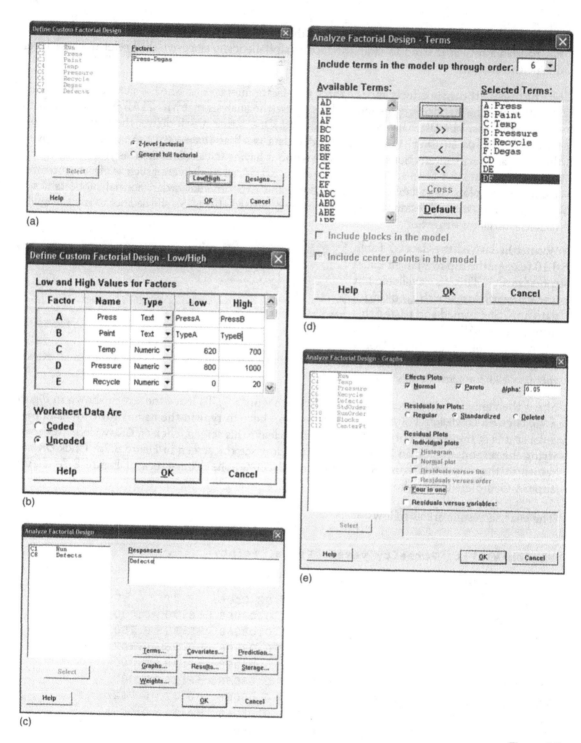

Figure 8.7

Estimated Coefficients for Porosity
using data in uncoded units

Term	Coef
Constant	11.8600
Press	−0.0268750
Paint	−0.0206250
Temp	−0.0012500
Pressure	0.0028531
Recycle	−0.0402500
Degas	−6.08500
Temp*Pressure	−1.51563E−05
Pressure*Recycle	4.18750E−05
Pressure*Degas	0.00638125

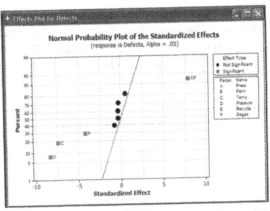

Figure 8.8

There is rather a lot in this output to be able to grasp it easily. The first list shows the analysis for the coded data in which every factor was coded −1, +1. The coefficients are those that are fitted to a regression model. The effect of a factor on the response (porosity) is the estimated change in the response caused by a change in the factor from its low to its high level: two units. The coefficient is the effect of changing the factor by one unit so it is half of the effect. The second list shows the regression coefficients for the uncoded, or raw, data.

The results are easier to interpret if you refer to the charts supplied with the analysis. If, in Figure 8.7e, you had checked the box for **Normal** effects plots, you would get the chart in Figure 8.8. This is a normal probability plot for the effects estimated from the coded data. If none of the control variables, or their interactions, affects the response the estimated effects would be scattered about the straight line. If a point is far removed from this line it indicates a significant effect. Here it shows that temperature, pressure, degassing and the pressure × degassing interaction are significant. The P-column in the listed results support this (any value of 0.05 or less suggests that the effect is significant). So Wyanet fits a simpler model, using again the sequence of Figure 8.7 but she selects only the terms **C, D, F, DF** (Figure 8.9).

The normal probability plot for the effects estimated from the coded data (Figure 8.10) supports this choice. The Pareto chart and residual plots (not shown) are straightforward to interpret.

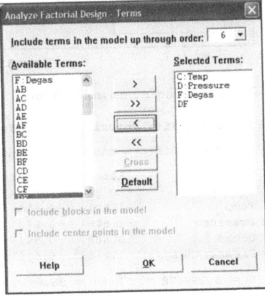

Figure 8.9

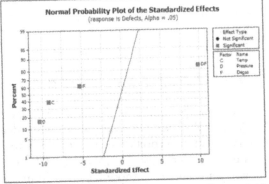

Figure 8.10

Further plots are available to help with the interpretation of the analysis. Follow: **Stat > DOE > Factorial > Factorial Plots** and tick **Main Effects Plot** and **Interaction Plot**. For the latter, click on the **Setup** button, then **Options** and check **Draw full interactions plot matrix**. The charts are in Figures 8.11 and 8.12. These clearly show the strong effects of the three main factors and the one interaction included in the analysis.

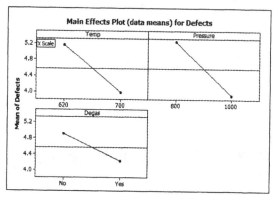

Figure 8.11

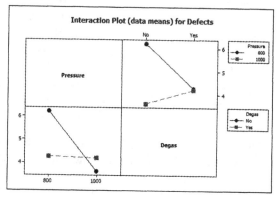

Figure 8.12

The listing of the factorial analysis follows:

Factorial Fit: Porosity versus Temp, Pressure, Degas

Estimated Effects and Coefficients for Porosity (coded units)

Term	Effect	Coef	SE Coef	T	P
Constant		4.5744	0.06487	70.52	0.000
Temp	−1.1913	−0.5956	0.06487	−9.18	0.000
Pressure	−1.3462	−0.6731	0.06487	−10.38	0.000
Degas	−0.6837	−0.3419	0.06487	−5.27	0.000
Pressure*Degas	1.2763	0.6381	0.06487	9.84	0.000

S = 0.259478   R-Sq = 96.64%   R-Sq(adj) = 95.42%

Estimated Coefficients for Porosity using data in uncoded units

Term	Coef
Constant	20.4603
Temp	−0.0148906
Pressure	−0.00673125
Degas	−6.08500
Pressure*Degas	0.00638125

The model predicts lowest porosity when temperature and pressure are high. It also seems to predict that the lowest porosity occurs when degassing is used. But look at the interaction effect: this changes the interpretation. The lowest porosity occurs when the pressure is high and no degassing is used. Under these conditions the CI for the mean porosity and PI for the porosity of a single wheel are similar to those obtained from the full factorial. Wyanet had also asked for predicted values (**Prediction** from Figure 8.7c) and for these to be stored. The results appear as extra columns in the MINITAB worksheet, from which we show two rows in Table 8.11, chosen because temperature and pressure are high and degassing is low.

The estimated standard deviation of porosity is 0.26. The upper limit for porosity is 3.75. The estimated mean under the process settings of Table 8.11 is 3.01. We can refer to the standard normal distribution (mean = 0, standard deviation = 1) to estimate the proportion of wheels exceeding

Run	Temp	Pressure	Degas	Porosity	PFit1	PSEFit1	CLimLo1	CLimHi1	PLimLo1	PLimHi1
5	700	1000	No	2.59	3.01	0.15	2.69	3.33	2.36	3.66
6	700	1000	No	3.17	3.01	0.15	2.69	3.33	2.36	3.66

Table 8.11

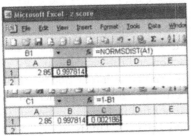

Figure 8.13

the upper limit. The standardised distance of the upper limit from the mean is $(3.75 - 3.01)/0.26 = 2.85$. The probability that the standardised value of porosity exceeds 2.85 is 0.002. You can check this value in a normal probability table or in MINITAB, or you can use MS Excel as in Figure 8.13. Enter 2.85 in cell **A1**; select cell **B1** and apply the Excel function **NORMSDIST(A1)** which returns the area under the standard normal distribution up to the value in **A1**; select cell **C1** and set its value equal to **1.0 – B1**. This is the area under the distribution to the right of the value in **A1**. The final result is 0.002. Alternatively, you can avoid the explicit standardisation by using NORMDIST. Specify the mean as **3.01**, standard deviation as **0.26**, and set **A1** as **3.75**.

This figure is subject to considerable uncertainty because it is based on several assumptions: the estimates of mean and standard deviation are assumed to precisely represent the mean and standard deviation of the underlying population and the underlying distribution is assumed to be normal.

Wyanet is keen to run the process at these settings, but Olivia raises another doubt. It is possible that some intermediate values of temperature and pressure might give even lower porosity. We follow this up in Chapter 9.

△ Review of Case 8.2

Wyanet's experiment was restricted to high and low values of six control variables: temperature, pressure, degassing, press, paint mould, and percentage recycled aluminium. She recommends running the process at the high temperature of 700 and high pressure of 1000 without degassing. Under these conditions she predicts a mean porosity of 3.01 and that 95% of wheels will have porosity between 2.36 and 3.66. Since the upper limit for porosity is 3.75 these operating conditions should lead to a dramatic reduction in the proportion of wheels that are recycled. Wyanet considers the choice of press and paint moulds, and the inclusion of recycled wheels in the melt, are irrelevant for porosity.

Despite Wyanet's success, Olivia has asked for an investigation of the plant performance that includes mid-point values of the continuous variables temperature and pressure. Olivia's reasoning is that the feasible range of these control variables would have been set around values that the plant designer thought to be best and that she expects a curved response with a minimum corresponding to control variables being set near their mid-point.

8.2 Confounding

In some applications we want to run a full factorial experiment, but our experimental units are arranged in blocks, and there are not enough units in a block to perform the whole experiment. The blocks might be days, or batches of aluminium ingots, or fields in farms. For example, in the full factorial experiment on the porosity of wheels it is possible that a batch of aluminium is sufficient for only 40 wheels. In this case, the idea, known as *confounding*, is to perform half the experiment, 32 wheels, on one batch of aluminium and the other half of the experiment on a second batch. The runs that make up the halves of the experiment are not, however, left to chance. Each half of the experiment corresponds to a half fraction. One has the generator **ABCDEF = 1** the other has the generator **ABCDEF = –1**. The difference between the mean porosity for the wheels made from the two batches is an estimate of the six-factor interaction and the difference between batches. If there is a substantial difference between the means we cannot say whether it is due to a six-factor interaction, or a difference in batches, or a combination of both. The six-factor interaction and the batch difference are aliased and said to be confounded. We might reasonably assume the six-factor interaction is negligible and attribute the difference to variation between batches, but this conclusion relies on the assumption. There are variations on confounding. If we replicated our full factorial experiment we might confound batches with a five-factor interaction in the second replicate. This is known as *partial confounding* and it becomes more useful if we have at least four blocks and are confounding on a three-factor interaction and two-factor interactions.

▽ Case 8.3 (UoE)

Greta Green has been asked to advise a local education board about the design of an experiment to evaluate the effects of three factors on children's progress in learning Mandarin as a second language.

A child's progress will be a score based on a combination of a written and an oral test taken at the end of the first year of study. The factors are: use of phonetic English spelling (A); use of TV programmes on Chinese culture (B); use of audio tapes (C). Each factor was either not present (L) or present (H). Children in the same state as UoE start learning Mandarin in Year 4, and Greta has recruited a group of eight specialist language teachers who are keen to take part in the experiment. Her first thought was that each teacher should try all eight factor combinations with their class. But, after discussions with the teachers, she accepted that randomly assigning children in a Year 4 class to one of eight factor combinations, which the team called teaching strategies, was impractical. The teachers thought random assignment of a class between four strategies might be feasible, but they would prefer to halve their classes between just two strategies. The response will be the average of the test scores for the children in a group (a half or a quarter of a class) taught by one teacher using one strategy. It is essential that children within a class are assigned to the groups at random. Greta designed a blocked experiment to suit this strategy by following: **Stat > DOE > Factorial > Create Factorial Design**. She specifies 3 factors. Clicking on **Design**, she selects **Full factorial**, with no centre points, 4 replicates and 8 blocks.

The three-factor interaction is confounded with the differences between blocks 1 and 2, 3 and 4, 5 and 6, and 7 and 8. Table 8.12 summarises the design delivered by MINITAB. Each pair of blocks contains the full 2^3 design. The block numbers correspond to teachers, and factor combinations correspond to teaching strategies. The errors in the model represent the differences in scores obtained by children in the four groups within a block, if they were to be taught with the same strategy. The block effects include any differences between teachers and differences between the average innate ability of classes to learn Mandarin; the team referred to these as teacher−class effects.

	A	B	C		A	B	C		A	B	C		A	B	C
1	1	1	−1	3	1	1	−1	5	−1	1	1	7	1	−1	1
1	−1	1	1	3	−1	1	1	5	1	−1	1	7	1	1	−1
1	−1	−1	−1	3	−1	−1	−1	5	−1	−1	−1	7	−1	−1	−1
1	1	−1	1	3	1	−1	1	5	1	1	−1	7	−1	1	1
2	1	−1	−1	4	1	1	1	6	1	1	1	8	−1	1	−1
2	1	1	1	4	−1	−1	1	6	1	−1	−1	8	1	1	1
2	−1	1	−1	4	−1	1	−1	6	−1	−1	1	8	1	−1	−1
2	−1	−1	1	4	1	−1	−1	6	−1	1	−1	8	−1	−1	1

Table 8.12

MINITAB summarises the design, in the session window, as follows:

Full Factorial Design

Factors: 3   Base Design: 3, 8   Resolution with blocks: IV
Runs: 32   Replicates: 4
Blocks: 8   Center pts (total):   0

Block Generators: ABC, replicates

Alias Structure
I
Blk = ABC
A
B
C
AB
AC
BC

If each teacher is restricted to two strategies, Greta can replicate the design only twice, but the class sizes are twice the size so the standard deviation of the response will be reduced by a factor of $\sqrt{2}$. The real limitation is that she now has quarter fractions of the full factorial dispersed over four blocks and the two-factor interactions are confounded with block effects. The block effects are differences between teacher–class combinations and could be substantial. Also, there may be important two-factor interactions among the three factors that comprise the eight teaching strategies. For example, the benefit of the TV programmes on Chinese culture may be enhanced if the children use the audio tapes as well.

The first four columns of Table 8.13 are the design delivered by MINITAB with the following sequence of commands: **Stat > DOE > Factorial > Create Factorial Design**. The number of factors is specified as 3, and the design is a full factorial with no centre points, 2 replicates and 8 blocks. Under **Options, Randomize runs** is unchecked. The column labelled **Teacher** was labelled **Blocks** by MINITAB but Greta has relabelled it. MINITAB summarises the design, in the session window, as follows:

Full Factorial Design

```
Factors:   3  Base Design:       3, 8  Resolution with blocks: III
Runs:     16  Replicates:          2
Blocks:    8  Center pts (total):  0
```

* NOTE * Blocks are confounded with two-way interactions.

Block Generators: AB, AC, replicates

Alias Structure

```
I

Blk1 = AB
Blk2 = AC
Blk3 = BC

A
B
C
ABC
```

Teacher	A	B	C	score
1	1	−1	−1	49
1	−1	1	1	57
2	1	1	−1	48
2	−1	−1	1	61
3	−1	1	−1	55
3	1	−1	1	64
4	−1	−1	−1	65
4	1	1	1	70
5	1	−1	−1	43
5	−1	1	1	85
6	1	1	−1	56
6	−1	−1	1	52
7	−1	1	−1	61
7	1	−1	1	63
8	−1	−1	−1	61
8	1	1	1	77

Table 8.13

You need to be careful about interpreting the alias structure from this list. The three lines **Blk1 = AB, ... , Blk3 = BC** warn us that the three two-factor interactions are aliased with three distinct differences between the blocks. We can find the detail of the differences from Table 8.13. For example the **AB** interaction is aliased with the difference between the means of **blocks 2, 4, 6** and **8**, and **blocks 1, 3, 5,** and **7**. This difference between means is an example of a contrast. A contrast is any linear combination of observed responses such that the sum of the coefficients defining the contrast is zero.

After some discussion, Greta and her team of teachers decided to proceed with the design shown in Table 8.13. They agreed that there might be some interesting interactions but the practical issues associated with dividing classes into four groups were too great a drawback. They thought that the schools from which the volunteers teachers came were similar in catchment area and reputation, and hoped the teacher–class effects might be relatively small. They also noted that interactions were associated with groups of teachers rather than any particular pair. For example, the **BC** interaction, which corresponds to the combined effect of TV programmes on Chinese culture and the audio tapes, is confounded with the comparison between teachers **1**, **4**, **5** and **8** and teachers **2**, **3**, **6** and **7**. The results for the experiment, average score for the half class, are in the final columns of Table 8.13.

If you have Pro, copy Table 8.13 from www.greenfieldresearch.co.uk/doe/data.htm into a blank MINITAB spreadsheet. Select **Stat > DOE > Factorial > Analyze Factorial Design**. Choose **score** as the response. In the **Terms** window choose **Include terms in the model through order 2**, then tick **Include blocks in the model**. The default analysis excludes the confounded blocks and two-factor interactions. Greta removed all the interactions from the model. Remember you can include only seven of the eight blocks as the eighth is confounded with the overall mean. MINITAB fits blocks 1–7, omitting block 8, although any choice of seven from eight can be fitted.

Only the main effect of **C** was statistically significant at the 10% level (P = 0.065). Since none of the block effects was statistically significant she repeated the analysis without block effects, without the three-factor interaction, and with two-factor interactions. So she dropped the indicator variables and fitted a regression with only the main effects and two-factor interactions. The results are:

Factorial Fit: score versus A, B, C

Estimated Effects and Coefficients for score (coded units)

Term	Effect	Coef	SE Coef	T	P
Constant		60.438	2.030	29.77	0.000
A	−3.375	−1.687	2.030	−0.83	0.427
B	6.375	3.188	2.030	1.57	0.151
C	11.375	5.687	2.030	2.80	0.021
A*B	1.625	0.812	2.030	0.40	0.698
A*C	8.125	4.063	2.030	2.00	0.076
B*C	5.875	2.938	2.030	1.45	0.182

S = 8.12105   R-Sq = 65.73%   R-Sq(adj) = 42.88%

Analysis of Variance for score (coded units)

Source	DF	Seq SS	Adj SS	Adj MS	F	P
Main Effects	3	725.69	725.69	241.90	3.67	0.056
2-Way Interactions	3	412.69	412.69	137.56	2.09	0.172
Residual Error	9	593.56	593.56	65.95		
Lack of Fit	1	60.06	60.06	60.06	0.90	0.370
Pure Error	8	533.50	533.50	66.69		
Total	15	1731.94				

If you have Student you can obtain the same results using **Stat > Regression**. You need to use **Calc > Make Indicator Variables** for the blocks, and to calculate the products of columns for the interactions **AB**, **AC**, and **BC**. Before you make indicator variables for the blocks label columns **Block1**, . . . , **Block 8**, and specify these columns for the indicator variables. When you specify the predictors in the regression specify just seven of these eight columns, **Block1-Block7** for example. Then the coefficients of **Block1** to **Block7** estimate the differences between them and **Block 8**. The rationale for doing this is explained in detail in the next section.

We need to be a little careful over interpretation of the P-values. The assumption that the errors in the model are independent requires us to assume also that there are no teacher effects. Although we have no strong evidence of teacher–class effects, we should not take this to imply there are none. There almost certainly will be some, but we might argue that they are relatively small and take the P-values as an approximate guide. If we do this, we can claim slightly stronger evidence (P = 0.021) that factor C, the audio tapes, helps the children to learn. There is some weak evidence of a positive interaction between use of phonetic English spelling and the audio tapes, but as the coefficient of the use of phonetic English spelling is negative, albeit not significantly so, we would wish to research further before making any recommendations about using phonetic English spelling. The estimated coefficients of factor B, the use of TV programmes on Chinese culture, and its interaction with C are positive and, taken together, predict as great an increase in average score as use of audio tapes.

She estimated that the use of TV programmes on Chinese culture, in conjunction with the audio tapes, further improves test scores but she is less confident about this finding. Nevertheless, she does recommend the use of these TV programmes because children enjoyed them and she thought they had a valuable educational content beyond learning the language.

Greta cannot make any recommendations about the use of phonetic English spelling from this experiment. She will perform a follow-up experiment next year. All children will be taught with audio tapes and the TV programmes. Each class will be randomly divided into two groups. One group will be taught with the use of phonetic English spelling and the other group will be taught without the use of phonetic English spelling. This should lead to a precise assessment of the effect of using phonetic English spelling. There is a possibility that children in the no phonetic spelling group will pick up phonetic spelling from the others. Greta doesn't expect this to be widespread, although she wouldn't wish to discourage children talking about their classes. She will ask children whether they have talked to friends about phonetic spelling, and if so, whether they would have preferred to have been taught using it. She could include knowledge of phonetic spelling from other sources as a concomitant variable in the analysis.

△ Review of Case 8.3

Greta is reasonably confident that the use of audio tapes increases children's test scores in the Mandarin language after one year's tuition during school Year 4. Her best estimate of this increase is about 20% from a mean score of 60, but this estimate is subject to considerable uncertainty.

8.2.1 Coded variables and indicator variables

Coded variables and indicator variables need some detailed explanation. You met a limited form of this in Chapter 7 where two levels of a factor could be coded in a single variable with values of [0, 1] or, preferably, in that context, [−1, +1]. Such coding is suitable for both quantitative and qualitative variables, and this is one of the great advantages of two-level factorials. The blocks in Case 8.3, and some variables in later chapters, need coding into more than two levels. If a factor has three levels it can be coded into a single variable, such as [0, 1, 2] or [−1, 0, 1], only if the factor is quantitative and if the levels are equally spaced, such as temperature = [150°C, 200°C, 250°C]. If the factor is qualitative or if it is quantitative with unequally spaced levels, MINITAB has a procedure for generating a separate variable for each of the levels. These variables are called indicator variables. We shall illustrate this with a very simple example: a qualitative factor with three levels. With this example, we shall highlight a potential problem that the procedure creates, a problem that first influences the number of extra columns in a worksheet and secondly affects the regression analysis. We shall show how to avoid this problem.

Our example is of an experiment with only one control variable or factor. A screen is placed in front of an observer who has three keys to press, marked R, G and B. The screen is flooded with a colour, red, green or blue, and the time from the appearance of the colour to the pressing of the correct key is recorded. So we might have results, perhaps in a random order, as in Table 8.14. We start by heading three blank columns **Blue**, **Green** and **Red** to represent the three levels. Note that the column headings must be entered in alphabetical order: the order in which MINITAB will process the text

Colour	Time (seconds)
R	0.4
B	0.6
G	0.5
R	0.35
B	0.65
G	0.45

Table 8.14

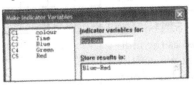

Figure 8.14

values. We then follow **Calc > Make Indicator Variables** (Figure 8.14). Enter **colour** as the indicator variable and **Blue**, **Green** and **Red** as the columns for the results (Table 8.15). Now we can use **Stat > Regression** to fit the model Time = mean +

$a_1 \times$ Red + $a_2 \times$ Blue + $a_3 \times$ Green + error

The regression analysis yields:

Regression Analysis: Time versus Blue, Green, Red

```
* Red is highly correlated with other X
variables
* Red has been removed from the equation.

The regression equation is
Time = 0.375 + 0.250 Blue + 0.100 Green
```

Colour	Time	Blue	Green	Red
R	0.4	0	0	1
B	0.6	1	0	0
G	0.5	0	1	0
R	0.35	0	0	1
B	0.65	1	0	0
G	0.45	0	1	0

Table 8.15

So what has happened? The table values do not suggest that there is any correlation between the three variables but we check and confirm this with **Stat > Basic Statistics > Correlation**:

```
        Blue    Green
Green  -0.500
Red    -0.500  -0.500
```

MINITAB is really telling us, but without much detail, that there is something wrong with the data for regression analysis. The full data that are used in the regression analysis are not shown. There is one more indicator variable. It is as if the model to be fitted was

$$y = \text{mean} \times x_0 + a_1 x_1 + a_2 x_2 + a_3 x_3$$

in which the values of x_1, x_2 and x_3 are the values of **Blue**, **Green** and **Red** in Table 8.16, but $x_0 = 1$ in every case. The data can be seen as in Table 8.16. This table reveals the problem. In every row $x_0 = x_1 + x_2 + x_3$ and the four variables are said to be *linearly dependent*. Whenever this happens, regression analysis is stymied: it won't work. MINITAB helps by throwing away one of the variables and then offers a cryptic explanation: **Red is highly correlated with other X variables.** In fact, **Red** is highly correlated with a linear combination of other X variables; specifically, the correlation between **Red** and the sum of **Blue** and **Green** is -1.

Let us look again at the regression fitted by MINITAB:

$$\text{Time} = 0.375 + 0.250\ \text{Blue} + 0.100\ \text{Green}$$

X_0	X_1	X_2	X_3
1	0	0	1
1	1	0	0
1	0	1	0
1	0	0	1
1	1	0	0
1	0	1	0

Table 8.16

If the colour is blue, with **Blue** equal to 1.00, then the predicted time is 0.625. The coefficient of the variable **Blue** is an estimate of the difference between time with red and time with blue. Similarly, the coefficient of the variable **Green** represents the difference between time with red and time with green. The predicted time, if the colour is green, is 0.475. The predicted time if the colour is red is 0.375. MINITAB selected **Red** as the indicator variable to be left out, but this is an arbitrary choice and we get the same predictions whichever indicator variable we choose to leave out. Rather than elicit warning messages from MINITAB, and accepting its selection, it is better to make our own choice of indicator variable to omit. You use **Calc > Make Indicator Variables**, as before, but delete one column.

Another advantage of deleting a column is to save tabular space, especially when there are interactions. For example, if there are two factors, one with five levels and another with four levels, the number of indicator variables reduces from 9 to 7 and the number of extra columns for interactions reduces from 20 to 12. This improves the efficiency of the regression analysis, avoids checks for linear dependency and rejection of variables. The benefits are even more striking when there are many more factors. You use MINITAB's **Calc > Make Indicator Variables** but then you must delete columns.

Choosing which one of the indicator variables to delete for a particular categorical variable is arbitrary, inasmuch as the predicted values are unaffected. However, if the numbers of cases in each category differ you will get the most precise estimates of differences if you make comparisons with the largest category, by deleting its indicator variable. In other applications, there may be a standard category against which we wish to make comparisons. Alternatively, after a first analysis, you may wish to make comparisons against the category with the highest, or lowest, predicted response. Then you can reanalyse the data with a different indicator variable removed.

8.2.2 Summary of terms

1. Any fraction of a two-level factorial, other than a full design, implies that it may not be able, from observed data, to distinguish between some effects. For example, it may not be possible to distinguish between a main effect and some interactions; or it may not be possible to distinguish between some interactions. Any effects than cannot be distinguished are said to be *aliased*. They may also be said to be *confounded*. The two words are synonyms, but *confounded* is generally used when we cannot distinguish between a block effect, or replicate effect, and a factor effect. *Aliased* is usually used when we cannot distinguish factor effects, including interactions.

2. A *contrast* is any linear combination of observed responses, such that the sum of the coefficients defining the contrast equals zero. An effect is a particular example of a contrast.

3. An effect that is aliased with the overall mean is called a *defining contrast* because it defines how the experiment is split and what other effects are aliased. It is also called a *generator* because it is used to generate the design.

4. The design procedure is as follows:

   (a) State the model that is to be fitted as a set of effects, including interactions, *that must not be aliased*, with interactions of fewer than some specified number of terms. This is called *the requirements set*. It is based on prior knowledge such as theory or data from a previous experiment or from previous analysis. For example, in Case 8.2, Wyanet used her first analysis to construct a requirements set comprising three main effects and one interaction.

   (b) Construct an aliasing matrix such that *no* elements of the requirements set are aliased. You must be able to estimate all elements of the requirements set, so not one element should be aliased with any other.

   (c) Use the defining contrasts of that aliasing matrix to construct a fractional two-level factorial.

Step (b) is difficult. It can be done, as in Cases 8.1 and 8.2, by experimenting with defining contrasts, but this can be very time-consuming and tedious. Unfortunately, MINITAB does not allow the user to specify a requirements set before asking for a fractional design. There are computer packages that do have this feature. One of these is WinDEX (see www.greenfieldresearch.co.uk).

9

Response surfaces

9.1 Curved predictor response relations

In Case 2.2 Gerard Grey investigated the effect of the concentration of an additive on the drying time of a varnish. If you look back to Figure 2.17 you will see that an assumption of a linear relationship over the range 0–10 mg/g would be seriously misleading. A linear approximation over shorter ranges such as 0–5, 5–10, or even 3–7 might do, but extrapolation beyond those ranges would be extremely unreliable and would lead to physically impossible results in some cases. In Case 9.1 we consider some work done before Gerard took over.

▽ **Case 9.1 (AgroPharm)**

Before he ran his experiment, Gerard read a report of an earlier investigation by a colleague, Steve Slate. AgroPharm had been routinely using 10 mg/g of additive in the varnish mix. Steve, who was a new recruit at the time, asked why this was done. He had been told that it had always been done that way, and further investigation was discouraged. Nevertheless, he decided to run a small bench-top experiment over eight days.

On four of the days he timed the drying time for varnish with no additive and on the other four days he timed the drying time for varnish with 10 mg/g of additive. He randomised the order of testing in the hope of breaking up any systematic patterns in temperature or humidity. The drying times in minutes are given in Table 9.1. Steve used

$$\text{coded additive} = (\text{additive} - 5.0)/5.0$$

so that no additive is coded as −1 and an additive of 10 mg/g is coded as +1.

Additive	Time	Additive	Time
−1	12	1	12
−1	13	1	10
−1	13	1	11
−1	12	1	9

Table 9.1

He fitted a regression using the MINITAB procedure **Stat > Regression > Regression** with the following results:

additive	time
−1	12
−1	13
−1	13
−1	12
0	7
0	10
0	8
0	5
1	12
1	10
1	11
1	9

Table 9.2

```
Regression Analysis: time versus additive

The regression equation is
time=11.5-1.00 additive

Predictor      Coef    SE Coef      T       P
Constant     11.5000    0.3536    32.53   0.000
additive     -1.0000    0.3536    -2.83   0.030

S = 1     R-Sq = 57.1%     R-Sq(adj)  = 50.0%
```

The effect of the expensive additive was statistically significant, but Steve was not impressed by the average reduction of 2 minutes' drying time. Fortunately, he thought of trying some experiments in the middle of the range at 5 mg/g. With our coding, the middle of the range, 5.0, becomes zero. These results are shown in Table 9.2.

He plotted the results using **Stat > Regression > Fitted Line Plot**, with time in the **Response (Y)** box, **additive in the Predictor (X)** box, and clicking the radio button for the **Quadratic** regression model. This instruction tells MINITAB to plot the data and add a curved line corresponding to a quadratic function. Figure 9.1 shows the resulting graph and fitted quadratic. Steve noticed that zero, corresponding to 5 mg/g, seemed better than both −1 and 1, corresponding to 0 and 10 mg/g.

Although the fitted line plot routine does produce the fitted equation and the R^2 value, it does not produce the full table of test values that the regression routine produces, so he used the latter for further analysis.

He first created a column of values of a new variable, additive squared (**add**2**), using MINITAB's calculator. Then he fitted a regression including the squared (quadratic) term with the following result:

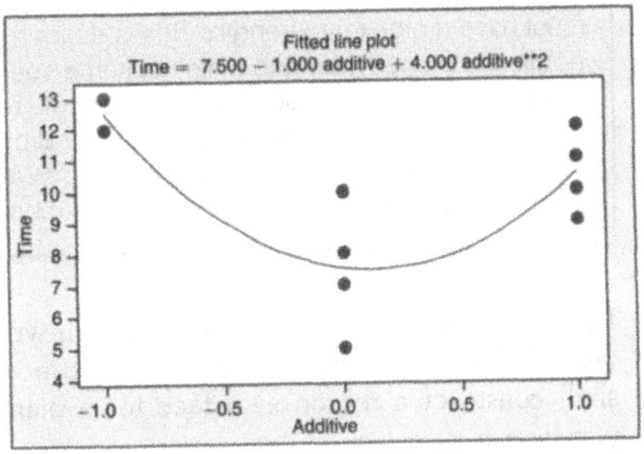

Figure 9.1

```
Regression Analysis: time versus
additive, add**2

The regression equation is
time = 7.50-1.00 additive + 4.00 add**2

Predictor      Coef    SE Coef      T        P
Constant     7.5000    0.7265    10.32    0.000
additive    -1.0000    0.5137    -1.95    0.083
add**2       4.0000    0.8898     4.50    0.001

S = 1.45297    R-Sq = 72.7%    R-Sq(adj) = 66.7%
```

Steve had found the minimum time predicted by the fitted regression equation. He differentiated time with respect to additive and set the derivative equal to zero. The predicted minimum time is 7.4375 minutes with a coded value for the additive of 0.125, corresponding to a real value of 5.625 mg/g. Steve told the plant manager of his findings and, since then, the company has been using 5.6 mg/g. However, the works manager decided to review the process and asked Gerard to investigate.

Gerard fitted a quadratic regression to the data of Case 2.2, shown in Figure 2.17. He estimated a minimum time of 7.5 with 5.4 mg/g additive: remarkably close to Steve's results.

△ Review of Case 9.1

Gerard recommends an addition of 5.4 mg additive to 1 g of varnish to minimise the drying time. The estimated drying time is then 7.5 minutes, compared with an estimated 12.5 minutes without additive.

If we have only one predictor variable the response curve can be drawn on an ordinary graph. If we have two predictor variables we have a response surface, and if we wish to represent this graphically we need a three-dimensional representation. Our next case involves two predictor variables.

▽ Case 9.2 (AgroPharm)

AgroPharm provides a rubber material to be used for padding dashboards in motor vehicles. It must have a high tear strength. This is defined as the resistance to growth of a nick when tension is applied to a specimen with a nick of some specified length. Other desirable characteristics are high tensile strength, elongation hardness, dimple recovery, and stiffness. Here we concentrate on the tear strength (y, kN/m) and how it is affected by the percentage of compound A in the rubber and the percentage of compound B in the resin. We take x_1 and x_2 to represent the percentages, of compounds A and B respectively, coded to be in the range $[-1, 1]$.

Before discussing the experimental design we will show you some typical forms of response surface. We use MINITAB's 3D wire frame procedure, although other options are available. First, we shall construct a response surface for a plane which, in general, has an equation of the form $y = b_0 + b_1 x_1 + b_2 x_2$.

You may be more used to using variables x, y and z for describing surfaces in three dimensions. The reason for preferring x_1, x_2 and y for describing our response surfaces is that, in general, we will have more than two control variables. If there are more than two control variables, the algebra follows the same principles, but we can no longer visualise the shape in three dimensions.

MINITAB has a routine that enables you to store a function, **z**, of two variables **x** and **y** that you can use to generate data needed to plot the surface. This is in a macro file called **Userfunc.mac**.

1. Start a word processor or text editor. **Notepad** is the easiest to use because you can access it with **Tools** in MINITAB.

2. Open the file **Userfunc.mac** The file is in the Macros folder of your main MINITAB directory.
3. Back up the file by saving it as a different name, such as **Userfunc.bak**. Save the file as plain text (ASCII file) and specify **Save as Type: All files** (otherwise you'll end up with **Userfunc.bak.txt**).
4. In **Userfunc.mac** you can add your own function by following the instructions at the top of the file. In the macro you will assign your function a number between 1 and 1000. However, you just need to copy the precise syntax we give you for the following examples.
5. Save **Userfunc.mac** as a plain text file. Again, remember to **Save as type: All Files.**

For this example, store the function, number 10, $40 - 10*x 1 + 5*x 2$ by adding to the code (15 lines up from the end of the file before **endif**) the following two lines:

elseif nf = 10

let z = 40 − 10*x + 5*y

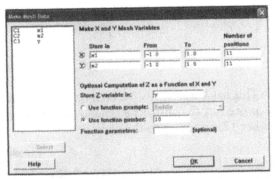

Figure 9.2

Move to the MINITAB worksheet, head three columns as **x1**, **x2** and **y** and follow: **Calc > Make Mesh Data**. Under **Store in X** and **Store in Y** enter variables **x1** and **x2**. In **Use function number**, enter **10**. Enter ranges from **−1.0** to **1.0** with **11** positions (Figure 9.2). This will generate data in the three columns.

Chart the data by following **Graph > 3D Surface Plot** and selecting **Wireframe**. Specify **y** as the **Z** variable, **x2** as the **Y** variable and **x1** as the **X** variable (Figure 9.3). Click on **Labels** and enter **y = 40 + 5 x1 − 10 x2** as the title. Choose fonts for all the labels. The graph is shown in Figure 9.4

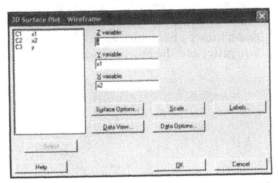

Figure 9.3

Follow the same procedure to construct a wireframe surface plot in which the response is a linear combination of **x1**, **x2**, and their product which represents an interaction. In **Userfunc.mac**, store the function, number 11, by adding to the code (after your two lines for function nf = 10):

elseif nf = 11

let z = 40 − 10*x + 5*y + 15*x*y

The graph is shown in Figure 9.5. Note that there are no curves in the Wireframe (every line in the wire frame is straight because it is parallel to one or other axis) but there are twists. The twists represent the interaction.

A general quadratic surface has the form

$$y = b_0 + b_1x_1 + b_2x_2 + b_3x_1^2 + b_4x_1x_2 + b_5x_2^2$$

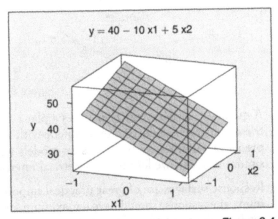

Figure 9.4

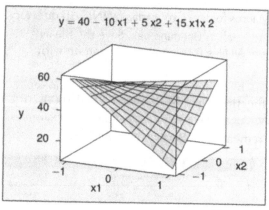

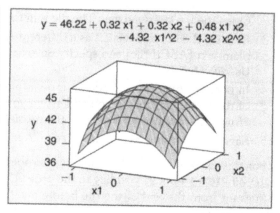

Figure 9.5

Figure 9.6

This has linear terms, an interaction term and two quadratic terms (x_1^2 and x_2^2). Follow the same procedure to construct a general quadratic surface plot. In **Userfunc.mac**, store the function, number 12, by adding to the code:

elseif nf = 12

let z = 46.22 + 0.32*x + 0.32*y + 0.48*x*y − 4.32*x1**2 − 4.32*x2**2

The graph is in Figure 9.6. Note that with negative coefficients for the quadratic terms the graph shows a maximum value of the response surface. With positive values for the quadratic terms, the graph shows a minimum value of the response surface (Figure 9.7). If one of the quadratic terms is negative and the other is positive the response surface is a saddle (Figure 9.8).

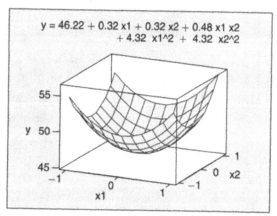

y = 50 − 10 x1^2 + 10 x2^2

Figure 9.7

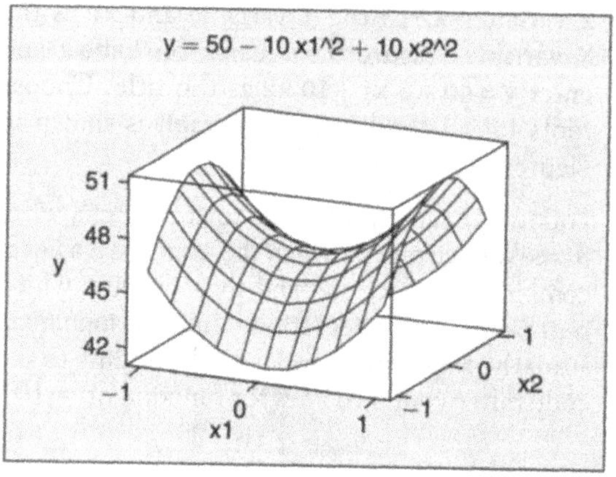

Figure 9.8

A quadratic surface, which includes a plane as a special case, is usually good enough to visualise the results of an industrial experiment with two predictor variables. If we have more than two predictor variables we fit regression models with squared terms and two-factor interactions in a similar way, but we lose the geometrical interpretation.

Response surfaces are of great practical importance because, if a process is running near its optimum, the response should have a maximum, or minimum, within the design region for the process settings. We want to find the process settings that give the optimum. It is common to run these experiments as part of routine production, in which case there is no need to specify a particular number of replicates in advance.

▽ Case 9.2 (AgroPharm) continued

Ever since the 'Slate incident', which became widely known, as Steve was not a particularly modest young man, AgroPharm management has been keen on optimisation. This has contributed towards the company's success. Myfanwy Maroon has been asked to investigate the process for producing the rubber material. There are two process control variables (x_1) and (x_2), and she aims to maximise the tear strength. The process engineer recommends that the percentage of A in the rubber should be within the range [0, 20] and that the percentage of B in the resin should be in the range [10, 30]. We code these ranges as $[-1, 1]$. Myfanwy wants to allow for a possible maximum strength within these ranges, although she realises that the greatest strength might be at a boundary point.

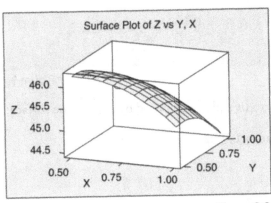

Figure 9.9

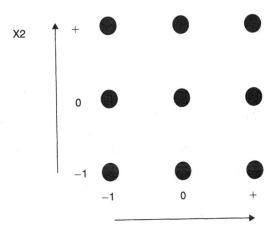

Figure 9.10

x_1	x_2	Strength
−1	−1	45.5
0	−1	45.1
1	−1	42.1
−1	0	44
0	0	45.9
1	0	40.3
−1	1	42.8
0	1	42.1
1	1	41

Table 9.3

For example, look again at Figure 9.6. Values of both x_1 and x_2 might be constrained to be greater than 0.5. This would create a boundary such that the greatest theoretical value of y could not be within the defined ranges of x_1 and x_2. Thus, the greatest practical value of y must be some point on the boundary (see Figure 9.9).

A quadratic surface will identify either situation if there are sufficient data to fit the coefficients reasonably precisely. She therefore needs at least three points in the ranges of **x1** and **x2**. To begin with she tries a single replicate of a 3 × 3 factorial design with **x1** and at **x2** all possible combinations of **−1, 0** and **1**, as in Figure 9.10. She performs the nine runs in a random order, and the results are in Table 9.3.

A useful first step is to look at various graphs of the data. Selecting **Graph > 3D Scatterplot** and clicking on **Simple** gives the graph in

Figure 9.11. We then look at a surface interpolated through them, obtained by following **Graph > 3D Surface Plot**, and clicking on **Wireframe**; under **Surface Options**, we select **Custom**, and enter **X-Mesh Number 10, Y-Mesh Number 10**. The resulting graph is shown in Figure 9.12. These provide a guide to the general shape of the surface and the variation about it.

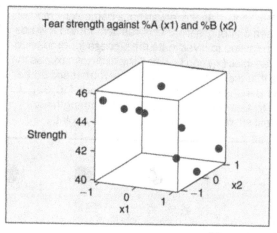

Figure 9.11

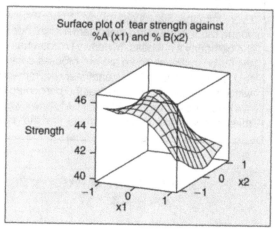

Figure 9.12

Myfanwy uses MINITAB's regression procedure to fit a plane, followed by a quadratic surface:

Regression Analysis: strength versus x1, x2

```
The regression equation is
strength = 43.2 - 1.48 x1 - 1.13 x2
```

Predictor	Coef	SE Coef	T	P
Constant	43.2000	0.4665	92.60	0.000
x1	−1.4833	0.5713	−2.60	0.041
x2	−1.1333	0.5713	−1.98	0.095

```
S = 1.39950      R-Sq = 64.0%      R-Sq(adj) = 52.0%
```

Regression Analysis: strength versus x1, x2, x1*x2, x1*x1, x2*x2

```
The regression equation is
strength = 44.6 - 1.48 x1 - 1.13 x2 + 0.400 x1*x2 - 1.75 x1*x1 - 0.300
x2*x2
```

Predictor	Coef	SE Coef	T	P
Constant	44.5667	0.9435	47.24	0.000
x1	−1.4833	0.5168	−2.87	0.064
x2	−1.1333	0.5168	−2.19	0.116
x1*x2	0.4000	0.6329	0.63	0.572
x1*x1	−1.7500	0.8950	−1.96	0.146
x2*x2	−0.3000	0.8950	−0.34	0.760

```
S = 1.26579      R-Sq = 85.3%      R-Sq(adj) = 60.8%
```

Myfanwy was able to predict the best operating point by using the Solver procedure in Microsoft's Excel. Open Excel and set

$$A1 = 0$$

$$B1 = 0$$

$$C1 = 44.6 - 1.48*A1 - 1.13*B1 + 0.4*A1*B1 - 1.75*A1*A1 - 0.3*B1*B1$$

Note that **Max** is the default. Now click on **C1** and **Tools > Solver**. Click on **Guess** to get **A1** and **B1** into **By Changing Cells**. Add the constraints **A1 >= –1** and and **B1 >= –1**. Then click on **Solve**. This predicted the maximum tear strength of **45.9 with x1** $= -0.54$ and **x2** $= -1.0$.

The results are equivocal. Inclusion of the quadratic terms has resulted in a somewhat reduced estimate of the standard deviation of the errors, but the overall model is not statistically significant (in the analysis of variance, not shown in the text). This is partly because there are so few degrees of freedom for error $(9 - 6 = 3)$. She decides to replicate the experiment. That is, she does the same set of nine runs again and obtains another nine values of tear strength as in Table 9.4. She repeats the regression analysis for a quadratic fit, using all 18 observations. These 18 data are available in two tables at www.greenfieldresearch.co.uk/doe/data.htm. The analysis indicates that the two main effects and x1*x1 are strong but that the interaction and x2*x2 are weak. The fit of 83.4% is quite good but there is some unexplained variation, so Myfanwy may consider other factors to include in future experiments.

x_1	x_2	Strength
−1	−1	46.6
0	−1	48.1
1	−1	40.7
−1	0	42.3
0	0	46
1	0	38.7
−1	1	43.2
0	1	42.8
1	1	38.7

Table 9.4

Regression Analysis: strength versus x1, x2, x1*x2, x1*x1, x2*x2

```
The regression equation is
strength = 44.8 - 1.91 x1 - 1.46 x2 + 0.375 x1*x2 - 2.84 x1*x1 + 0.358
x2*x2

Predictor        Coef      SE Coef        T          P
Constant       44.7611      0.6883     65.03      0.000
x1             -1.9083      0.3770     -5.06      0.000
x2             -1.4583      0.3770     -3.87      0.002
x1*x2           0.3750      0.4618      0.81      0.433
x1*x1          -2.8417      0.6530     -4.35      0.001
x2*x2           0.3583      0.6530      0.55      0.593

S = 1.30604     R-Sq = 83.4%     R-Sq(adj)  = 76.5%
```

Myfanwy again used Excel Solver to predict the best operating point. This predicted the highest tear strength of **47.0** with **x1** = **–0.40** and **x2** = **–1.0**. She used **MINITAB Stat > Regression > Regression**; clicking on **Options**, she entered –.4 –1.0 0.4 0.16 1.0 (the values of x1, x2, x1*x2, x1*x1, x2*x2) in the **Prediction intervals for new observations** box to get a 95% confidence interval for the mean tear strength as [45.6, 48.5] and 95% prediction limits for tear strength of an individual test piece as [43.8, 50.2]. Both the confidence interval and prediction limits allow for sampling error in the parameter estimates. She states her recommendations in terms of the physical units.

9.2 Design of response surface experiments: central composite designs

The idea of three-level factorial designs, which allow the inclusion of quadratic terms in a regression, can be extended to any number of continuous factors. But the size of a full factorial experiment is usually excessive if the number of factors exceeds 3. If there are n factors, then the number of design points in a full three-level factorial is 3^n. Two factors have nine points, three factors have 27 points, four factors have 81 points, etc. Fractions are possible, but their design is not straightforward and they are rarely used because there is a far neater solution known as a *central composite design*. We code two values of every factor as –1 and

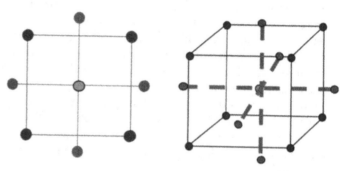

Figure 9.13

+1 so that we start with a two-level factorial design, or a fraction of one, and then we add a *star design*. The mid-value of every factor's range is coded **0**. The point at which every factor is **0** is called the *centre point*. The star design adds a centre point, or possibly several centre points, and two axial points at the end of each of the coordinate axes, typically the same distance from the origin as the corner points. You can see the idea for two and three factors in Figure 9.13. In general, if there are m factors the star design adds $2m$ axial and at least one centre point.

To construct a star design we have to specify the distance of the axial points from the centre and the number of centre points. With finesse, provided by MINITAB, we can achieve an orthogonal design by appropriate choices of these parameters. The MINITAB default takes axial points to be the same distance from the origin as the corner points, mathematically equal to the square root of the number of factors. This choice of distance has some aesthetic appeal but may be well beyond operating limits for the process. Also, the number of centre points may seem excessive. It is not essential to have an orthogonal design but it does have two advantages:

- Estimates of the main effects do not change when quadratic and product terms are added.
- Estimates are made with high efficiency for a given number of runs.

These advantages are not overwhelming. The first is useful if we want to make individual hypothesis tests for each coefficient, but we usually just need to know whether the quadratic model is an improvement on a model with linear terms only. This can be assessed from the estimated standard deviation of the errors. Case 9.3 does use the default MINITAB construction for an orthogonal design. In Case 9.4 we use an improvised approach.

The distance of an axial point from the centre is called alpha (α). It is illustrated in Figure 9.14. The MINITAB design procedure allows you to enter a value for alpha: default, face centred, or custom. There are arguments for using values other than MINITAB's default. It is common, for example, for the controllable intervals of variables to be constrained. Also there are other theoretical criteria by which alpha may be calculated.

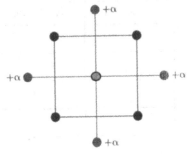

Figure 9.14

Central composite design experiments are often performed using a two-level factorial design, or a fraction of this, first, followed by a star design. Each part is randomised, but the sequence is not random. Thus any change in the response over time will be confounded with the results from the star design, but we should be running a composite design only when we are sure we have a relatively stable process. After all, the purpose of the experiment is to find the optimum process settings for future production. Even so, we can allow for a change in the mean level by using an indicator variable for the time difference between the original factorial design and the follow-up star design. The limitation is that this does not allow for any interaction between time and the control factors.

▽ Case 9.3 (AgroPharm)

This case is based on Wu and Hamada (2000) and Morris *et al.* (1997). Ranitidine hydrochloride is an ingredient of a drug that is widely prescribed for the treatment of stomach ulcers. It is separated from a mixture of other compounds in a buffer solution by electrophoresis, and Petra Purple has been asked to improve the separation process.

Petra's team has found that three factors appear to affect the chromatographic exponential function (CEF) which is an indicator of the separation achieved at the end of the process, and which they wish to minimise. The factors are: the pH of the buffer solution; the voltage used in the

Factor	Lowest permissible	Highest permissible
A pH	2	9
B voltage (kV)	9.9	30.1
C alpha-CD(mM)	0	10

Table 9.5

electrophoresis; and the concentration of a particular component (alpha-CD) of the buffer solution. The ranges of these factors, and their allocation to letters, are shown in Table 9.5.

We generate the design in MINITAB using **Stat > DOE > Response Surface > Create Response Surface Design**, with **3** as the **Number of factors** and accepting the default **Designs**. This produces the design in Table 9.6. We have sorted it into standard order so that you can see the structure of the additional points. MINITAB's default value for alpha, for this experiment, is 1.68. The lowest permissible value of each variable corresponds to -1.68 on the coded scale, the highest permissible value corresponds to 1.68, and we take the mid-points of the ranges as zero. So, for voltage for

StdOrder	RunOrder	A	B	C	CEF	ln(CEF)
1	4	−1	−1	−1	17.3	2.85
2	10	1	−1	−1	45.5	3.82
3	11	−1	1	−1	10.3	2.33
4	3	1	1	−1	11757.1	9.37
5	16	−1	−1	1	16.9	2.83
6	8	1	−1	1	25.4	3.23
7	18	−1	1	1	31697.2	10.36
8	5	1	1	1	12039.2	9.40
9	12	−1.68	0.00	0.00	16548.7	9.71
10	20	1.68	0.00	0.00	26351.8	10.18
11	15	0.00	−1.68	0.00	11.1	2.41
12	2	0.00	1.68	0.00	6.7	1.90
13	17	0.00	0.00	−1.68	7.5	2.01
14	7	0.00	0.00	1.68	6.3	1.84
15	19	0	0	0	9.9	2.29
16	13	0	0	0	9.6	2.26
17	6	0	0	0	8.9	2.18
18	9	0	0	0	8.8	2.17
19	14	0	0	0	8	2.08
20	1	0	0	0	8.1	2.09

Table 9.6

example, 9.9 kV is coded as −1.68 and 30.1 kV is coded as 1.68, and 0 in coded units corresponds to 20 kV.

The experiment was run in the run order generated by MINITAB and the results have been added to Table 9.6. The CEF measurements are on a scale that allows for some extraordinarily high values, presumably because of the exponentiation, which will dominate the regression fit. Petra decides to use its natural logarithm as the response. She added these values to Table 9.6. The aim is to minimise **ln(CEF)** because if **ln(CEF)** decreases then **CEF** decreases too. Petra uses **Stat > DOE > Response Surface > Analyze Response Surface Design**. You can find the data on www.greenfieldresearch.co.uk/doe/data.htm. When you are asked to specify **Low/High** you should declare the Worksheet data to be coded units. When you are asked for the **Response** declare **Analyze** data using **coded units**. If you specified the data as **uncoded** and analyse as **coded**, MINITAB will code the minimum and maximum values of the range of the control variables at **–1** and **1** respectively. Specifying the data as **uncoded** and analysing as **uncoded** gives the same result as specifying both as **coded**. Specifying as **coded** and analysing as **uncoded** makes no sense as MINITAB cannot infer the coding you used, although you do get output based on **uncoded 0** and **1** being **0** and the upper value of the **coded** variable respectively!

So Petra uses **Stat > DOE > Response Surface > Analyze Response Surface Design** to regress **ln(CEF)** on:

- the main effects only;
- the main effects and the three two-factor interactions;
- the main effects, two-factor interactions and three squared terms;
- the main effects, two-factor interactions and three squared terms with the three-factor interaction added.

She clicks on the **Terms** button. In the **Terms** window she is able to follow the sequence for analyses as shown in Figure 9.15. The estimated standard deviations of the errors are 3.30, 3.46, 1.85, and 1.75 respectively. The standard deviation of the response, **ln(CEF)**, values is 3.32 if no regression model is fitted. This is the estimated standard deviation of the errors in the model:

$$\text{ln(CEF)} = \text{OM} + \text{error}$$

where OM is the overall mean. Such a model, without any regression predictors, is sometimes called the *null model*. In this case, the estimated standard devotion of the errors in the models without the squared terms is either almost the same as, or larger than, the standard deviation of the ln(CEF) values. We have to include quadratic (squared) terms if we are to present a plausible model.

Follow **Stat > DOE > Response Surface > Analyze Response Surface Design**. Specify terms for full quadratic with **ln(CEF)** as the response. Click on **Graphs** and ask for a normal plot of residuals. The plot of residuals is in Figure 9.15. The results of the regression analysis, presented in the session window, are as follows:

Response Surface Regression: ln(CEF) versus A, B, C

The analysis was done using coded units.

Estimated Regression Coefficients for ln(CEF)

Term	Coef	SE Coef	T	P
Constant	2.15582	0.7559	2.852	0.017
A	0.60339	0.5015	1.203	0.257
B	1.30867	0.5015	2.609	0.026
C	0.52458	0.5015	1.046	0.320
A*A	2.89308	0.4882	5.926	0.000
B*B	0.13890	0.4882	0.284	0.782
C*C	0.05758	0.4882	0.118	0.908
A*B	0.58875	0.6553	0.898	0.390
A*C	−1.07125	0.6553	−1.635	0.133
B*C	1.08375	0.6553	1.654	0.129

```
S = 1.853    R-Sq = 83.6%       R-Sq(adj) = 68.8%
```

Analysis of Variance for ln(CEF)

Source	DF	Seq SS	Adj SS	Adj MS	F	P
Regression	9	175.097	175.097	19.4552	5.66	0.006
Linear	3	32.119	32.119	10.7064	3.12	0.075
Square	3	121.628	121.628	40.5427	11.80	0.001
Interaction	3	21.350	21.350	7.1166	2.07	0.168
Residual Error	10	34.353	34.353	3.4353		
Lack-of-Fit	5	34.316	34.316	6.8633	935.48	0.000
Pure Error	5	0.037	0.037	0.0073		
Total	19	209.450				

The normal score plot for the residuals (Figure 9.15) is not so much to check that they are normal as to assess whether there are any outlying observations. These would indicate that further investigation is required.

There are no clear outliers from this experiment. If the three-factor interaction is included its coefficient, −0.93, with estimated standard error 0.62, is not statistically significant (P = 0.17). All pairs of predictors in the model, including ABC, are orthogonal, or almost so, as we can check in MINITAB by generating a correlations matrix for all the terms in the model. You will have noted that MINITAB automatically generates the quadratic terms for regression analysis but it does not save them as extra columns in the worksheet. Thus, to obtain a correlations matrix, you need to create the extra columns and use MINITAB's calculator. Use **State > Basic Statistics > Correlation** then choose the variables.

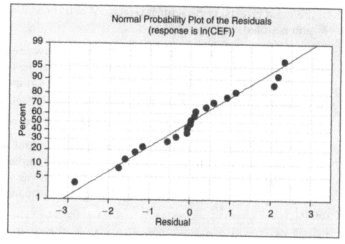

Figure 9.15

Correlations: A, B, C, AB, AC, BC, A^2, B^2, C^2

	A	B	C	AB	AC	BC	A^2	B^2
B	0.000							
C	−0.000	−0.000						
AB	−0.000	−0.000	0.000					
AC	0.000	0.000	−0.000	−0.000				
BC	0.000	0.000	−0.000	−0.000	0.000			
A^2	−0.000	−0.000	0.000	0.000	−0.000	−0.000		
B^2	−0.000	−0.000	0.000	0.000	−0.000	−0.000	−0.090	
C^2	−0.000	0.000	0.000	0.000	−0.000	−0.000	−0.090	−0.090

Note that almost all the correlations are zero and only three are slightly different. This is an expression of orthogonality, overlooking the negligible correlations between the squared terms. If the data had been coded differently, such as (0, 1) or (1, 2) instead of –1, +1, there would have been strong correlations between the main factors and the interactions as well as with the quadratic terms. It is the centring that leads to the zero correlations between main effects and interactions. Zero correlations between main effects and interactions, in a designed experiment, can be achieved with raw values provided they are centred before cross-products are calculated; that is, they are transformed simply by subtracting the mean values of variables.

A nice consequence of orthogonality is that the coefficients of the predictors remain the same regardless of those that are retained in the regression. In particular, if we removed the interaction and squared terms in **B** and **C**, the coefficients of **A**, **B**, **C**, and **A²** would be just the same. Therefore, we do not need to consider whether or not it is advisable to include the terms that are not statistically significant when interpreting the relative importance of the more significant control variables. But, the optimisation does depend on which predictors are included in the model. The positive coefficient of **A²** is the largest, but we usually would not consider a quadratic effect without the linear term **A** unless we had a good physical reason to do so. A quadratic effect without the linear term **A** is unlikely. Taking the two together, and ignoring the relatively small interaction terms, elementary calculus gives a minimum when **A** equals −0.10. The statistically significant positive coefficient of **B** suggests that it should be set at the largest negative value within the range investigated, that is, −1.67. The **BC** interaction has a positive coefficient which, although it is not statistically significant at the 10% level, is the next largest estimate, so we might recommend a positive value for **C** despite its estimated coefficient being positive. For Table 9.7 we used Excel Solver to predict optimum operating conditions to minimise ln(CEF) based on different models.

Model terms	Estimated standard deviation of errors (s)	Estimated values for control variables (A,B,C) to achieve minimum	Estimated minimum
A,B,A^2	1.932	(−0.105, −1.68,0)	0.068
A,B,C,A^2,AC,BC	1.695	(0.208, −1.68,1.68)	−2.201
A,B,C,A^2,AB,AC,BC	1.698	(0.379, −1.68,1.68)	−2.492
A,B,C,A^2,B^2,C^2,AB,AC,BC	1.853	(0.378, −1.68,1.68)	−2.078

Table 9.7

However, Petra has some reservations about the model and hence about recommending values for the factors far from zero. The **Response Surface Regression** breaks down the **Residual Error** with ten degrees of freedom into **Lack-of-Fit** on five degrees of freedom and **Pure Error** on five degrees of freedom. The **Pure Error** is estimated from the six replicate runs at the centre of the design. The degrees of freedom for the estimates are fairly small, but the variance ratio of 936 with a P-value of 0.000 cannot be ignored. The pure error mean square is 0.0073, so the estimate of the standard deviation of the response, at the centre of the design, is the square root of this, 0.086, very much smaller than the value of *S* from the regression, 1.855. It may be that the full quadratic model is inadequate, but a more likely explanation is that the process standard deviation increases if we move away from the centre. Even so, the estimated decrease in the response from 2.2 at the centre to less than −2.0 seems promising. Petra decides to follow up this experiment with several runs of the process with A at −0.10, B at −1.68 and C at 0 and several runs with A at 0.38, B at −1.68 and C at 1.68, before writing her report. She realises that a small increase in the process mean could be counter-productive if it were accompanied by an increase in the standard deviation.

△ Review of Case 9.3

Petra has clear evidence that pH affects the separation and that any substantial deviations from the centre of its range, 5.5, will increase CEF. She also has evidence that CEF increases with increasing voltage over the range 9.9–30.1 kV. At this stage she does not think she has identified a suitable model from which to make clear recommendations of optimum operating conditions, particularly as variability may increase if the process is operated away from the centre conditions. She will investigate performance over six runs at two sets of process conditions: the first is pH 5.17, voltage 9.9 kV and alpha-CD at 5 mM; the second is pH 6.74, voltage 9.9 kV and alpha-CD at 10 mM. She will then consider a sequence of hill climbing experiments (Chapter 10).

▽ Case 9.4 (SeaDragon)

Wyanet had recommended that the wheel production process should be run with the pressure and temperature at the high end of the ranges she had investigated, and be run without degas (Case 8.2). This advice had been followed, and the quality of the wheels has improved, though not quite as much as she had hoped. Olivia continues to point out that Wyanet still has not investigated the process performance with temperature and pressure in the middle of their ranges. So, Wyanet decides she had better investigate the possibility that a model including quadratic terms would better describe the wheel production process. Three of the six variables are categorical and have only two levels. She can change the other three, which include the temperature and pressure, over a continuous range.

She decides to augment her original factorial experiment with the three continuous variables at the centres of their ranges. She will do this for all 2^3 combinations of the other factors. This may provide evidence of a quadratic effect, but she will not be able to attribute it to a particular factor because the squares of the numbers in the columns for pressure, temperature and recycling are identical and the three quadratic effects are confounded. To tell them apart she must add some star points. Her low and high levels for temperature in the original experiment were 635 and 685, although the team had thought a range of 620–700 was feasible. In terms of the coded factor, she can take axial points for temperature a distance $40/25 = 1.6$ from the centre. Her low and high levels for pressure were 840 and 960, although the team had thought a range of 800–1000 was feasible. She can take axial points a distance as far as $100/60 = 1.67$, in coded units, from the centre. She used the full range for the percentage recycled in the original experiment, but as there was no evidence this had any effect she does not investigate any more values for this variable and leaves it at its mid-point of 10%, or 0 in coded units. She will replicate the four star points for a half fraction of the 2^3 design for the three categorical variables: press, paint mould, and degas. You can see (Table 9.8) that the first eight runs are the full 2^3 design for the three categorical variables with temperature and pressure at their mid-values. The next eight runs have the half 2^3 replicated with temperature at the extreme ends of its axis and with pressure at its mid-value. The next eight runs have the half 2^3 replicated with pressure at the extreme ends of its axis and with temperature at its mid-value. The full table of results is on www.greenfieldresearch.co.uk/doe/data.htm where you will find the Excel workbook for Case 9.4. In that workbook, there are two spreadsheets: one showing the data for the full design; the other showing the augmenting experiment (Table 9.8 in rows 65–88 of the full design). She has entered the porosities in the last column.

In MINITAB, she used the calculator to create columns containing interactions and quadratic terms. Note that in these she abbreviated variable names such as **Pure** for 'Pressure'.

Run	Press	Paint	Degas	Temp	Pressure	Recycle	Porosity
1	−1	−1	−1	0	0	0	3.41
2	1	−1	−1	0	0	0	3.65
3	−1	1	−1	0	0	0	3.44
4	1	1	−1	0	0	0	3.79
5	−1	−1	1	0	0	0	2.85
6	1	−1	1	0	0	0	2.76
7	−1	1	1	0	0	0	2.47
8	1	1	1	0	0	0	2.68
9	−1	−1	1	−1.6	0	0	4.87
10	1	1	1	−1.6	0	0	4.88
11	1	−1	−1	−1.6	0	0	5.46
12	−1	1	−1	−1.6	0	0	5.49
13	−1	−1	1	1.6	0	0	3.75
14	1	1	1	1.6	0	0	3.65
15	1	−1	−1	1.6	0	0	4.51
16	−1	1	−1	1.6	0	0	4.26
17	−1	−1	1	0	−1.67	0	2.77
18	1	1	1	0	−1.67	0	3.01
19	1	−1	−1	0	−1.67	0	5.13
20	−1	1	−1	0	−1.67	0	5.06
21	−1	−1	1	0	1.67	0	2.85
22	1	1	1	0	1.67	0	2.53
23	1	−1	−1	0	1.67	0	2.42
24	−1	1	−1	0	1.67	0	2.38

Table 9.8

We can calculate the correlations between some of the predictor variables in MINITAB:

```
Results for: WHEELS.MTW

Correlations: Temp, Pressure, Recycle, degas, Temp^2, Pure^2, recycle^2,
Temp*Pu

              Temp Pressure Recycle   degas Temp^2 Pure^2 recycle^ Temp*Pur
Pressure    0.000
Recycle     0.000    0.000
degas       0.000    0.000   0.000
Temp^2     -0.000    0.000   0.000   0.000
Pure^2      0.000    0.000   0.000   0.000 -0.492
recycle^   -0.000   -0.000   0.000  -0.000  0.103  0.046
Temp*Pur    0.000    0.000   0.000   0.000  0.000  0.000    0.000
Pure*deg    0.000    0.000   0.000   0.000  0.000  0.000    0.000   0.000

Cell Contents: Pearson correlation
```

The design is not completely orthogonal because of the correlation between the square of temperature and the square of pressure. Although we have not included all 15 interaction terms, they are all uncorrelated with the main effects and the squared terms.

Here we digress for a moment to explain the important distinction between correlated predictor variables and an interaction between predictor variables. The former is a consequence of the experimental design, and has nothing to do with the response. The latter refers to the effect on the response of one predictor depending on the value of the other. Interaction is modelled by including the product term in the regression. In particular, we do this with our orthogonal 2^n designs. So, uncorrelated predictor variables can, and often do, interact.

If our predictors are correlated we will still obtain unbiased estimates of the model parameters, whether or not the model includes an interaction, provided we include all the predictors in the model. This is because if predictors are correlated, then estimates of coefficients will depend on which other predictors are included in the model. It is neater, and more efficient, to use an orthogonal, or nearly orthogonal design, if we can do so and hence avoid correlation between the predictors. However, a correlation between predictor variables does not imply that they necessarily interact.

Wyanet uses MINITAB for various regression analyses and finds strong evidence of a quadratic effect with temperature. She wonders if the mean of the process has changed, so she adds a variable coded as 0 for the runs in the original design and 1 for the runs in the star design. This is not significant (P = 0.85) so there is no evidence that the mean of the process has changed. She omits **press, paint** and **recycle** because there is no evidence they affect the porosity directly. Also, with the possible exception of recycle with pressure, there is no evidence that they interact with the other variables. If all 15 interactions are included, for which we suggest you use **Stat > DOE > Responses Surface > Analyze Response Surface Design**, the coefficient of the pressure recycle interaction is 0.057 with a P-value of 0.017. However, as the coefficient of recycle itself is 0.0025 with a P-value of 0.914, and none of its other interactions is significant, and a large number of P-values are being calculated, with the associated caveat of multiple comparisons, Wyanet felt justified in omitting **recycle** from the model, for the moment at least. If she eventually recommends running the process at high pressure, she will do further experiments with recycled material. So she obtained the following model:

```
Regression Analysis: Porosity versus degas, Temp, . . .

The regression equation is
Porosity = 3.14 - 0.408 degas - 0.386 Temp - 0.401 Pressure +
           0.376 Pure*deg
         + 0.0478 Temp*Pure + 0.578 Temp^2 + 0.0500 Pure^2
```

Predictor	Coef	SE Coef	T	P
Constant	3.13582	0.06118	51.25	0.000
degas	−0.40830	0.01983	−20.59	0.000
Temp	−0.38622	0.02023	−19.09	0.000
Pressure	−0.40121	0.02002	−20.04	0.000
Pure*deg	0.37647	0.02002	18.81	0.000
Temp*Pur	0.04781	0.02325	2.06	0.043
Temp^2	0.57814	0.03594	16.08	0.000
Pure^2	0.05001	0.03314	1.51	0.135

S = 0.1860   R-Sq = 95.9%   R-Sq(adj) = 95.5%

There is no doubt about the statistical significance of the dominant terms, and it is usual to include all quadratic terms if some are significant when there is a large number of degrees of freedom for the error. There are no outliers in the normal plot of the residuals (Figure 9.16)

Wyanet decides to display the relationship by plotting contours of porosity against temperature and pressure for the two levels of degas. MINITAB makes this easy. But, if you have Pro, unless you have already analysed the experiment, the contour/surface plots button will be greyed out. Use **Stat > DOE > Response Surface > Contour/Surface Plots**. Select both **Contour** and **Surface** plots and click on **Setup** for each of them to choose **Display plots using Coded units**.

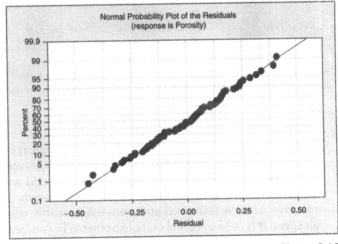

Figure 9.16

If you are using **Regression**, first use **Calc > Make Mesh Data** for **temperature** and **pressure**. Remember to **Store in** two new columns that you name appropriately. Then calculate a column for each predictor in the model, including a column of length 121 with −1.0 in each cell for **degas**, and run the regression with **Options > Predictions** for new observations. **Store Fits**. **Graph > Contour Plot** Then select followed by **Graph > 3D Surface Plot**, each with **Degas = −1** (Figure 9.17). The contour and surface plots with **degas = 1.0** are in Figure 9.18.

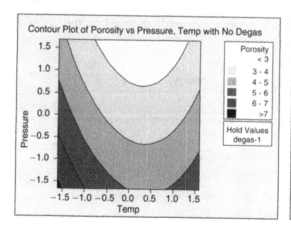

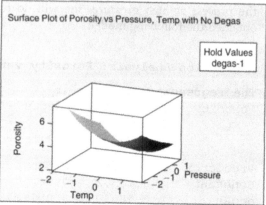

Figure 9.17

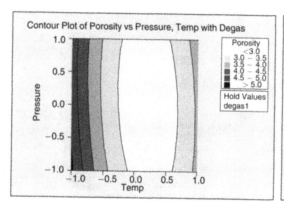

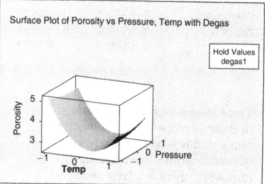

Figure 9.18

Then the conclusion, which she drew after the original factorial experiment, about high pressure and no degas remains unchanged, although she has increased the range of pressures she has investigated and could recommend going above 1, in coded units. So she will recommend no degas and a pressure of 1.67 in coded units. However, she needs to modify her advice about temperature. You can obtain the estimated optimum temperature, when pressure is 1.67, from elementary calculus, and it is 0.26 in coded units. Alternatively you can make a rough assessment from the contour plot, or you could use Solver.

Wyanet has remembered the possible link between increased porosity and use of recycled material when the pressure is high. She thinks the best way to follow this up is to monitor the process during routine production. Given the predicted improvements, they should only ever have a small percentage of recycled material in the melt.

Olivia is pleased, but now has doubts about running at this highest pressure. She is concerned that the process variability may be higher. We shall follow the developments in Chapter 11. The optimisation of the quadratic model in Wyanet's case turned out to be fairly straightforward as she could reduce the problem to considering temperature and pressure at the two levels of a categorical variable, degas, and see what was going on in contour plots. In more complicated cases you can use Excel Solver.

☐ Case 9.5 (SeaDragon)

Mercedes Magenta has been asked to develop an expert system for the control of a rotary kiln for making cement. The meal, which is the technical term for the material fed into the kiln, is a mixture of limestone, clay, sand and iron ore. The main ingredient is the limestone, and the expert system is to maintain the percentage of free lime in the cement product between 1% and 2% while maximising the ratio of feed rate to fuel rate.

After discussions with the plant operators Mercedes identified six control variables: feed rate (A); rotation speed (B); proportion of fuel to oxidant (C); fuel rate (D); speed of fan 1 (E); and speed of fan 2 (F). Small changes in the standard operating conditions would be made to establish their effect on the percentage of free lime in the cement. Some of the experimental runs may lead to cement that is out of specification, but she hopes this product can be blended with cement near the other end of the specification and then sold. The process is continuous and does not respond immediately to changes in the control variables, so two hours will be allowed for the process to reach a steady state after each change. Each experimental run will last for a further two hours during which samples of cement will be taken for X-ray analysis at five-minute intervals. The response (y) is the average of the 24 determinations of the free lime content.

Mercedes thinks the response may depend on quadratic effects and two-factor interactions of control variables as well as the linear terms, and decides to use a central composite design for the experiment. She intends to use a half replicate of a 2^6 factorial design, as that would allow her to identify all two-factor interactions, with at least 13 extra points for the star design. She is not prepared to have the star points more than two scaled units from the origin. She approaches the design with **Stat > DOE > Response Surface > Create Response Surface Design**. MINITAB gives her 54 runs, with custom alpha equal to 2.0, which includes ten centre points. However, she would rather restrict the number of runs to 48, by having only four runs at the centre, so the entire experimental programme would be completed in 12 days. So she specifies the number of centre points in MINITAB again with four custom centre points.

The final design is shown in the first six columns of the table for Case 5.5 in www. greenfieldresearch.co.uk/doe/data.htm. Note that the variables are coded. The design is not precisely orthogonal, because of very small correlations between any two squared terms of -0.04, but this has no practical implications.

Mercedes expects two routinely measured characteristics of the limestone, the first loosely described as its *burnability*, and the second, *water content*, to affect the response. Although these variables are monitored, they cannot be controlled and are called concomitant variables to distinguish them from the control variables. All the control variables and concomitant variables are potential predictors of the response. She abbreviates the names of these two concomitant variables as **burn** and **water**. The results from the experiment (observed values of **burn** and **water** as well as the response variable, *free lime*, which she labels as **y(ppm)**, are in www.greenfield-research.co.uk/doe/data.htm.

If you want to confirm that the design is approximately orthogonal you would need to calculate the 15 two-factor interactions, 24 three-factor interactions, 15 four-factor interactions, 6 five-factor interactions, the six-factor interaction, and the six squared terms. As a partial check, compute values in extra columns (**AD BD BF CC FF**) followed by **Stat > Basic Statistics > Correlation**

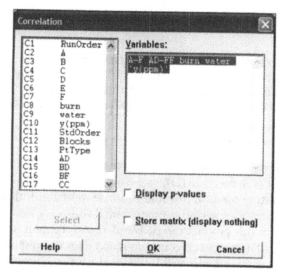

Figure 9.19

(Figure 9.19). There was no need to display P-values since the purpose of the correlations was to check the orthogonality and perhaps to get a feel for the associations between the predictive terms, including the interactions, quadratics and concomitants, and the response variable. So, we have added into this table the correlations with the response variable values observed in the experiment. MINITAB produces the correlations matrix of Table 9.9.

Mercedes performed various regression analyses and found that **burn** and **water** had substantial effects on the free lime content. However, she thought it unlikely that these concomitant variables would have any substantial interactions with the control variables. She decided to keep quadratic and interaction terms involving

	A	B	C	D	E	F	AD	BD	BF	CC	FF	Burn	Water
B	0.00												
C	0.00	0.00											
D	0.00	0.00	0.00										
E	0.00	0.00	0.00	0.00									
F	0.00	0.00	0.00	0.00	0.00								
AD	0.00	0.00	0.00	0.00	0.00	0.00							
BD	0.00	0.00	0.00	0.00	0.00	0.00	0.00						
BF	0.00	0.00	0.00	0.00	0.00	0.00	0.00	0.00					
CC	0.00	0.00	0.00	0.00	0.00	0.00	0.00	0.00	0.00				
FF	0.00	0.00	0.00	0.00	0.00	0.00	0.00	0.00	0.00	−0.04			
Burn	0.18	−0.10	−0.06	0.11	0.05	0.19	−0.08	0.12	0.23	−0.14	−0.12		
Water	0.18	−0.09	0.00	−0.25	−0.02	0.17	0.04	0.18	0.19	0.02	0.05	0.11	
y(ppm)	−0.51	−0.14	0.23	−0.53	0.27	0.25	−0.17	−0.06	0.21	−0.15	−0.11	0.26	0.31

Table 9.9

the control variables, with t-ratios greater than 2.0 in absolute value, in the regression model which is to be incorporated in the expert system. She was able to do this analysis using **Stat > DOE > Response Surface > Analyze Response Surface Design**. In this, the two concomitant variables, **burn** and **water**, need to be declared as **factors**.

```
Response Surface Regression: y(ppm) versus A, B, C, D, E, F,
burn, water
```

The analysis was done using uncoded units.
Estimated Regression Coefficients for y(ppm)

Term	Coef	SE Coef	T	P
Constant	17937.8	126.869	141.389	0.000
A	−1952.8	78.686	−24.818	0.000
B	−260.8	76.499	−3.410	0.002
C	796.0	75.846	10.495	0.000
D	−1591.8	79.206	−20.096	0.000
E	824.5	75.793	10.878	0.000
F	490.4	78.571	6.241	0.000
burn	91.1	7.670	11.881	0.000
water	30.1	3.165	9.498	0.000
C*C	−433.4	87.653	−4.945	0.000
F*F	−326.3	87.488	−3.730	0.001
A*D	−547.2	85.055	−6.433	0.000
B*D	−531.0	87.107	−6.096	0.000
B*F	336.2	89.202	3.769	0.001

S = 478.6   R-Sq = 98.1%   R-Sq(adj) = 97.4%

Analysis of Variance for y(ppm)

Source	DF	Seq SS	Adj SS	Adj MS	F	P
Regression	13	405557943	405557943	31196765	136.21	0.000
Linear	8	375818293	358283827	44785478	195.54	0.000
Square	2	7764925	8264954	4132477	18.04	0.000
Interaction	3	21974726	21974726	7324909	31.98	0.000
Residual Error	34	7787031	7787031	229030		
Total	47	413344974				

Thus the fitted equation, expressing free lime content as ppm, is:

$$4 \text{ (ppm)} = 17{,}938 + 91.1 \times \text{burn} + 30.1 \times \text{water} - 1953A - 261B + 796C - 1592D$$
$$+ 824E + 490F - 547AD - 531BD + 336BF - 433CC - 326FF$$

The specification for free lime is 10,000–20,000 ppm. The estimated standard deviation of the random errors is 479. She thinks it would be prudent to restrict the expert system to a range that leaves six standard deviations at either end before being out of specification, that is, 13,000–17,000, and to aim for the centre of that range. The highest feed-to-fuel ratio can be achieved when feed is at its highest value ($A = 2$ on the coded scale) and fuel at its lowest value ($D = -2$ on the coded scale). The expert system algorithm will also be provided with values of **burn** and **water** for the limestone in the meal, and it will provide suitable values for the remaining control variables: B, C, E and F within the range $[-2, 2]$. It might, for instance, be based on Excel Solver with an objective of minimising the squared difference between the free lime (y) and the centre of its specified range, 15,000 ppm.

10

Hill climbing

We can think of the response surface for our process as a hill, and optimisation as finding the summit. If we are a long way from the summit, we can perform a small two-level factorial experiment, fit a plane as a local approximation, and move in the direction of steepest ascent up the plane. We repeat the experiment and, if the direction of steepest ascent is clearly defined, move in that direction. If there is no clear direction of steepest ascent, our design points may be around the summit. We check this by adding a star design, to make a central composite design, and fit a quadratic surface to the results. The same strategy should locate a minimum, if this is the objective.

There may be more than one stationary point. If this is possible, it is a good idea to start the hill climbing from different points and to compare the results. It is also possible, and perhaps more likely, that there is no stationary point within the ranges of the control variables. If this is the case, hill climbing will find a highest value, or a lowest value, on a boundary.

The hill climbing method is more formally known as *evolutionary operation* and the summit is usually called *the peak*.

▽ Case 10.1 (AgroPharm)

After Petra Purple's success with the ranitidine hydrochloride manufacturing process (Case 9.3), the manager of the penicillin batch fermenter asked her to help him to optimise the process. The biochemistry of the process is well understood, but the yield of penicillin from the specified weights and volumes of raw materials also depends on the rate of stirring and the temperature of the jacket surrounding the reactor. Both variables can be controlled quite accurately, and the specified settings have been 200 rpm and 30°C for as long as anyone can remember. It is safe to operate the process with changes in stir rate and temperature up to at least 10%. Petra proposes experimenting with small changes in these two variables. The plant would continue to produce satisfactory penicillin and the experimental programme should identify optimum settings for stir rate and temperature. If these turned out to be different from the current specification, substantial savings would be made. The manager, Reynard Red, was interested in the possible benefits but remained sceptical about the value of the experiment because he thought this was how the current specified settings had been decided.

Petra suggested that previous experiments might have relied on a one-variable-at-a-time strategy, and explained that this might not have led to the best settings. A possible dependence of yield on stir rate and temperature is represented by the contour diagram in Figure 10.1.

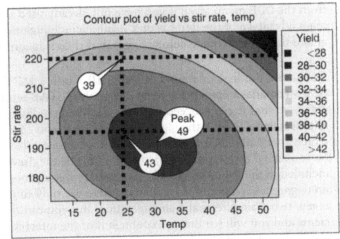

Figure 10.1

This is an example of a response surface, and the general shape shown in the diagram is quite common in the chemical industry. Imagine that this diagram represents the unknown response surface. If we fix the stir rate at 220 rpm and vary the temperature, we will find a highest yield of 39 kg at 24°C. Now suppose we fix the temperature at 24°C and vary the stir rate. The highest yield becomes 43 kg when the stir rate is 195 rpm. This is a long way from the optimum yield of 49 kg. The stir rate and temperature are said to interact; that is, the effect of changing one depends on the value of the other. This is just one possible scenario, and alternatives include the possibility that the present settings already correspond to the highest point or that the response surface is of a quite different shape (multi-peaked or saddle-shaped, for example). The purpose of the proposed experimental programme is to infer the nature of the surface as efficiently as possible, and to use this information to specify optimum operating conditions.

If you fix one variable and investigate the effect of the other, and iterate this procedure, you are constraining your changes to be parallel to the temperature or stir rate axes. You should reach the peak eventually, but it is much more efficient to investigate both variables simultaneously and to then proceed directly towards the peak.

Petra begins with the simplest design that makes changes in both variables, a single replicate of a 2^2 design. Each factor appears as high (coded $+1$) or low (coded -1). This gives four runs, allowing a plane to be fitted with one degree of freedom for error. You may argue that fitting a plane ignores an interaction, but you should think of the plane as a local approximation to the curved response surface. Also, you can check whether an interaction seems to be dominant by comparing coefficients if you fit the saturated model. Petra and an operator did such an experiment and the results, in standard order, are the first four rows of Table 10.1. The data are on www.greenfieldresearch.co.uk/doe/data.htm. Open Case 10.1, where you will find Tables 10.1, 10.2 and 10.3 on three spreadsheets.

The runs were in a random order, which helps to justify the assumption that the random errors in the regression model are independent. The fitted plane, where **y1** is the yield, **S1** is stir rate (units of 10 rpm from 200 rpm) and **T1** is temperature (units of 1 from 30) can be estimated with either the regression procedure or the analysis procedure

S1	T1	y1
−1	−1	43.2
1	−1	44.9
−1	1	43.7
1	1	46.2
−1.4	0	41.9
1.4	0	43.9
0	−1.4	42.1
0	1.4	45.1
0	0	43.6

Table 10.1

from the **DOE** menu, even if you have not already used this menu to design the experiment. The main advantage of the latter is that it can include higher-order terms without your having to calculate the corresponding columns first, but it does incorporate other useful features.

Copy only the first four rows of Table 10.1 into a MINITAB worksheet and then select **Stat > DOE > Factorial > Analyze Factorial Design**. MINITAB sends you a message and asks you for some details so that it can proceed with the analysis (see Figure 8.6). You state the factors, and select **2-level factorial**. Check the high and low values. You could leave the radio button on **Uncoded**, in which case MINITAB will code −1 and 1 to themselves, but it is appropriate to select **Coded**. Now specify the response variable, and check the **Terms**. The interaction **AB** is automatically included in the list of terms to be estimated. If you proceed it fits the saturated model. There are no degrees of freedom for error since there are only four observations from which to estimate a mean, two main effects and an interaction. But you can check the relative magnitude of the coefficients and you will see that the coefficient of the interaction **AB** is the smallest. Remove the interaction **AB**, to obtain the following:

```
Factorial Fit: y1 versus S1, T1

Estimated Effects and Coefficients for y1 (coded units)

Term        Effect      Coef    SE Coef        T       P
Constant             44.5000    0.2000   222.50   0.003
S1          2.1000    1.0500    0.2000     5.25   0.120
T1          0.9000    0.4500    0.2000     2.25   0.266


S = 0.40    R-Sq = 97.03%     R-Sq(adj) = 91.08%
```

The standard error of each of the estimated coefficients is 0.20 but, with only one degree of freedom for error, the 90% confidence intervals are wide and include zero.

As any movement would be away from the specified settings, Petra thought it prudent to augment the experiment with another five runs arranged as a star design to give a composite design. She obtained further data, in a random order, and these are in rows 5–9 of Table 10.1. She fitted a plane to all nine data points. This time Petra no longer had a two-level factorial design, so she needed to use either the regression procedure or **Stat > DOE > Response Surface > Analyze Response Surface Design**. She removed the terms **AA**, **BB**, and **AB** to obtain a plane (no twists and no curves). This is to give Petra an indication of the greatest slope: the direction to move towards the highest point of the response surface.

Note, in Table 10.1, that the variables **S1** and **T1** are in the range −1.4 to +1.4. You need to follow the correct procedure vary carefully, using **Coded** data to produce the right results. The procedure is shown in Figure 10.2.

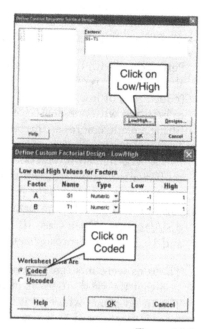

Figure 10.2

Response Surface Regression: y1 versus S1, T1

The analysis was done using coded units.

Estimated Regression Coefficients for y1

Term	Coef	SE Coef	T	P
Constant	43.8444	0.2991	146.603	0.000
S1	0.8838	0.3188	2.772	0.032
T1	0.7576	0.3188	2.376	0.055

S = 0.8972   R-Sq = 69.0%   R-Sq(adj) = 58.6%

Analysis of Variance for y1

Source	DF	Seq SS	Adj SS	Adj MS	F	P
Regression	2	10.7323	10.7323	5.36616	6.67	0.030
Linear	2	10.7323	10.7323	5.36616	6.67	0.030
Residual Error	6	4.8299	4.8299	0.80498		
Total	8	15.5622				

The standard error of each of the coefficients of **S1** and **T1** has actually increased to 0.32 because the reduction due to additional data has been more than offset by what happens to be an increase in the residual standard deviation ($S = 0.897$). The confidence intervals for the coefficients are, nevertheless, narrower because of the increase in degrees of freedom to 6. The two-tailed t-value with 6 degrees of freedom for 90% confidence intervals is 1.943, so the 90% confidence intervals for the coefficients of **S1** and **T1** are $0.884 \pm (1.943 \times 0.319)$ which gives [0.26, 1.50], and $0.756 \pm (1.943 \times 0.319)$, which is [0.14, 1.38].

It is possible to fit a quadratic surface to the data in the composite design, simply by adding the terms **AA**, **BB** and **AB** if you are using **Analyze Response Surface Design**, but this results in an increase in s to 1.17 and a loss of degrees of freedom. Remember the plane is just a local approximation and, on a small scale, we expect a plane to be adequate unless the range of the control variables includes the values which give the peak response.

Petra now feels confident to move the experiment in the direction of steepest ascent. She finds this direction by moving each variable a distance that is proportional to its estimated coefficient. For example, if we increase **S1** by one unit we should increase **T1** by $0.7576/0.8838 = 0.86$ units. An increase in stir rate by one unit (10 rpm) did seem reasonable, and the corresponding increase in temperature was rounded to 0.9°C.

This next set of experiments is centred on a stir rate of $200 + 10 = 210$ rpm and a temperature of $30 + 0.9 = 30.9$°C. Now define **S2** as stir rate (units of 10 rpm from 210 rpm) and **T2** as temperature (units of 1 from 30.9°C). This time, Petra runs two replicates of a 2^2 with a centre point experiment. The results are in Table 10.2. In MINITAB she needs to open a new worksheet.

S2	T2	y2
−1	−1	44.8
1	−1	44.6
−1	1	44.2
1	1	46.1
−1	−1	44.4
1	−1	45.2
−1	1	43.8
1	1	47.5
0	0	45.6
0	0	44.9

Table 10.2

She fits a plane, again using Stat > DOE > Response Surface > Analyze Response Surface Design and removing the terms **AA**, **BB** and **AB**, and she obtains the following:

Response Surface Regression: y2 versus S2, T2

The analysis was done using coded units.

Estimated Regression Coefficients for y2

Term	Coef	SE Coef	T	P
Constant	45.1100	0.2602	173.372	0.000
S2	0.7750	0.2909	2.664	0.032
T2	0.3250	0.2909	1.117	0.301

S = 0.8228   R-Sq = 54.4%   R-Sq(adj) = 41.4%

This is not such a good fit as the initial fitted plane but it is sufficiently convincing for Petra to move the experiment in its direction of steepest ascent. She increases **S2** by one unit, corresponding to 10 rpm, and **T1** by $0.325/0.775 = 0.419$ units, corresponding to 0.42°C. The next set of experiments is centred on a stir rate of $210 + 10\,\text{rpm} = 220\,\text{rpm}$, and a temperature of $30.9 + 0.4 = 31.3°C$. Now define **S3** as stir rate (units of 10 rpm from 220 rpm) and **T3** as temperature (units of 1 from 31.3°C).

Petra starts out with a single replicate of a 2^2 design with a centre point. The five runs are performed in random order, although throughout the description of this case the results are tabled in standard order, and these are shown in the first five data rows of Table 10.3. She fits a plane to these data but the coefficients are not determined with sufficient precision to be useful:

Response Surface Regression: y3 versus S3, T3

The analysis was done using coded units.

Estimated Regression Coefficients for y3

Term	Coef	SE Coef	T	P
Constant	46.7400	0.5373	86.989	0.000
S3	−0.9750	0.6007	−1.623	0.246
T3	−0.7750	0.6007	−1.290	0.326

S = 1.201   R-Sq = 68.2%   R-Sq(adj) = 36.5%

Petra follows this experiment with a star design and the results are the next four lines of Table 10.3. She first fits a plane to all nine data, and then tries a quadratic surface. This time she does not remove the terms **AA**, **BB** and **AB**.

Response Surface Regression: y3 versus S3, T3

The analysis was done using coded units.

Estimated Regression Coefficients for y3

Term	Coef	SE Coef	T	P
Constant	47.9467	1.1578	41.411	0.000
S3	−0.8813	0.4116	−2.141	0.122

S3	T3	y3
−1	−1	47.7
1	−1	46.7
−1	1	47.1
1	1	44.2
0	0	48
−1.4	0	45.9
1.4	0	43.7
0	−1.4	45.9
0	1.4	45.1
−1	−1	46.8
1	−1	46.2
−1	1	47.8
1	1	44.8
0	0	48.2
−1.4	0	48.1
1.4	0	45.8
0	−1.4	47.1
0	1.4	44.2

Table 10.3

```
T3           -0.5328    0.4116    -1.295    0.286
S3*S3        -1.2657    0.6870    -1.842    0.163
T3*T3        -0.9086    0.6870    -1.323    0.278
S3*T3        -0.4750    0.5791    -0.820    0.472

S = 1.158    R-Sq = 77.6%    R-Sq(adj) = 40.2%
```

The estimated standard deviation of the errors (S) is 1.26 for the plane and 1.16 for the quadratic surface. However, even for the quadratic surface the P-values associated with all the coefficients are greater than 0.05, so the coefficients have not been estimated with sufficient precision to infer optimum conditions.

So Petra replicates the central composite design. There is no reason for her not to replicate now she thinks how close she is to the peak, because the process should be operating at near optimum conditions for all the runs and the penicillin produced can all be sold as usual. The results are in the remaining nine lines of Table 10.3.

Petra fits a quadratic model to the 18 data. She uses **Stat > DOE > Response Surface > Analyze Response Surface Design**. She specifies **Coded** data and asks for the fitted values to be stored. They appear in the worksheet in a column headed **FITS1**.

```
Response Surface Regression: y3 versus S3, T3

Estimated Regression Coefficients for y3

Term        Coef      SE Coef      T         P
Constant    48.0706   0.6160       78.037    0.000
S3          -0.8712   0.2190       -3.979    0.002
T3          -0.5480   0.2190       -2.502    0.028
S3*S3       -0.9326   0.3655       -2.552    0.025
T3*T3       -1.0857   0.3655       -2.970    0.012
S3*T3       -0.5375   0.3081
-1.744      0.107

S = 0.8715    R-Sq = 74.3%
R-Sq(adj) = 63.6%
```

Petra now wants a contour plot of the fitted relationship. On the worksheet she used for this analysis, she applies **Stat > DOE > Response Surface > Contour/Surface Plots** She checks **Contour Plot** and clicks on **Setup** in order to accept the default pair of factors. She clicks the radio button for **Coded unit**. This produces the contour plot in Figure 10.3.

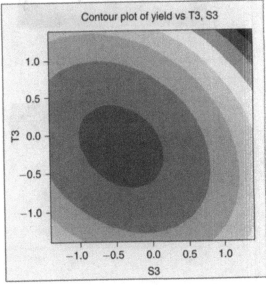

Figure 10.3

If you have the Student version, instead of Pro, you might use **Graph > Contour Plot**. This produces the upper contour plot in Figure 10.4. This looks bad and Petra wouldn't like to show it to the manager of the fermenter. You can obtain a better plot with the method we described in Chapter 9. Open **Userfunc.mac** in Notepad and enter the following code:

```
elseif nf = 21
  if (params = 1) and (num < > 6)
    call writerror 1
  elseif params = 0
    let ktmp.1 = 48.07
    let ktmp.2 = 0.87
    let ktmp.3 = 0.55
    let ktmp.4 = 0.93
    let ktmp.5 = 1.09
    let ktmp.6 = 0.54
  endif
  let z = ktmp.1-ktmp.2*x-ktmp.3*y-ktmp.4*x*x-1.09*y*y-0.54*x*y
```

This is a representation of a general quadratic regression function with two predictors in MINITAB's macro language. If you specify a column for parameters (that is, estimates of the coefficients) it will override the **ktmp** assignments.

Open a new worksheet, head three columns **S3**, **T3** and **Yield** and then follow **Calc > Make Mesh Data.** Specify the ranges of S3 and **T3** as [**–1.4, +1.4**], each with 20 positions. Specify **Yield** as the **Z-variable** and apply **function 21**. This generates an array of data with fine increments, much more suitable for the smooth contour plot on the bottom of Figure 10.4.

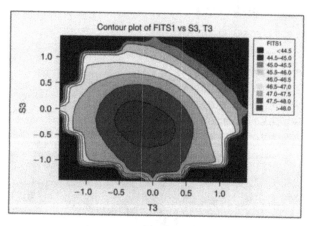

From the contour plot, Petra estimates that the best operating conditions result in a yield greater than 48, when **S3** and **T3** are about -0.4 and -0.2, respectively, in coded units.

The worksheet gives the maximum over the 13×13 grid but the contour plot allows us to interpolate between these points. For more precise values, it is better to use Excel Solver.

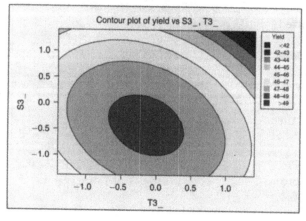

Figure 10.4

If you use Solver, or calculus, with the coefficients to four decimal places, the maximum of 48.30 occurs when **S3** and **T3** are -0.42 and -0.15 respectively, but these estimates are still subject to sampling errors. In physical units this corresponds to a rotation speed of $320 - 0.42 \times 10 = 316\,\text{rpm}$ (to one decimal place), and a temperature of

31.3 − 0.15 = 31.15°C. For her report, she uses MINITAB to calculate a predicted value, a 95% confidence interval for the mean value of all batches made with a stir rate of 316 and temperature 31.15, and a 95% prediction interval for individual batches. She repeats **Stat > DOE > Response Surface > Analyze > Response Surface Design**, clicks on **Prediction** and enters values −0.4 and −0.15 (Figure 10.5). Note that where she entered these values she could have entered the names of the variables (**S3** and **T3**). This would produce a table of predicted values, with confidence intervals, corresponding to all values observed in the experiment.

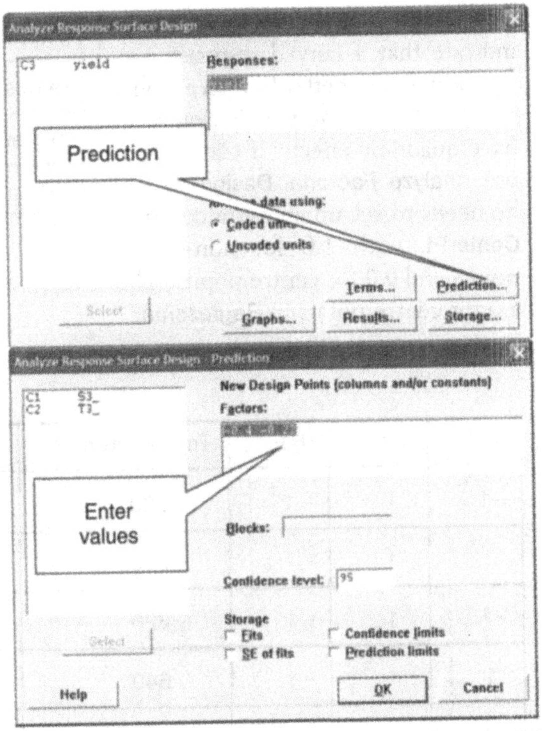

The predicted values, calculated using the fitted equation, appear in the session window. If she also checks the storage items, the values also appear in the worksheet.

Figure 10.5

```
Predicted Response for New Design
Points Using Model for yield

              SE
Point    Fit   Fit       95% CI              95% PI
  1   48.2948    0  (48.2948, 48.2948)  (48.2948, 48.2948)
```

The mean yield over the past six months has been 44.1 kg so Petra has identified a potential increase of nearly 10% in the process yield.

☐ Case 10.2 (SeaDragon)

Cicero Copper was SeaDragon's chief metallurgist, and co-founder of the original company. He thought that degas had other benefits, apart from reducing porosity, and advised Olivia that the process should be run with degas. He had read Wyanet's report and was sceptical about the lack of a pressure effect when degas was used.

Cicero was familiar with evolutionary operation and decided to conduct his own experiment, with the help of the process operator. He would keep **Press A, paint mould A** and **degas** and make small changes in **temperature, pressure** and **recycle**. The ranges for these variables and their small change increments are shown in Table 10.4.

The most economical process settings are: low temperature; low pressure; and, if recycled aluminium is available, high recycling. He decided to start from a centre point of 630, 820, 18, which he would replicate and around which he would apply a 2^3 design, using a coded unit for each variable equal to its small change. The coded data are given in Table 10.5 with the corresponding physical values.

The addition of a centre point may indicate that a curved response surface would be a better fit than a plane, but it will not identify which variables have quadratic effects. If Cicero is to use **Analyze Factorial Design** in Pro he needs to set up an extra column, **CenterPt**, with 1.0 for non-centre points and 0.0 for centre points. This is equivalent to using a coding 0.0 for non-centre points and 1.0 for centre points in **Regression**.

Variable	Lower limit	Upper limit	Small change
temperature	620	700	10
pressure	800	1000	20
recycle	0	20	2

Table 10.4

T1	P1	R1	Temperature	Pressure	Recycle	poros1	CenterPt
−1	−1	−1	620	800	16	4.68	1
1	−1	−1	640	800	16	3.17	1
−1	1	−1	620	840	16	4.69	1
1	1	−1	640	840	16	3.14	1
−1	−1	1	620	800	20	4.8	1
1	−1	1	640	800	20	3.29	1
−1	1	1	620	840	20	4.99	1
1	1	1	640	840	20	3.35	1
0	0	0	630	820	18	4.05	0
0	0	0	630	820	18	3.8	0

Table 10.5

Case 10.2 at www.greenfieldresearch.co.uk/doe/data.htm has four spreadsheets with Tables 10.5–10.8.

So, he uses **Stat > DOE > Factorial > Analyze Factorial Design**, with factors **T1, P1, R1** and worksheet data **Coded**. He selects **Designs and** clicks on the **Specify by column** radio button under **Center Points**. The response is **poros1**. He rejects all **Terms** other than main effects.

He also draws a cube plot of the results, using **Stat > DOE > Factorial > Factorial Plots**, ticking **Cube Plot** and specifying **Data Means** as the **Type of Means to Use in Plots**. He specifies poros1 as the response. This must be done after the analysis, otherwise **Factorial Plots** is greyed out and is not available. The cube plot is in Figure 10.6; he has added a bold arrow to indicate the direction of improvement: greatest reduction in porosity.

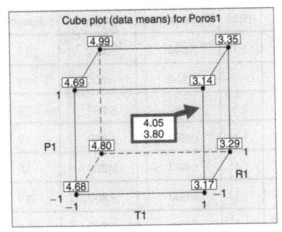

Figure 10.6

The analysis results are:

Factorial Fit: poros1 versus T1, P1, R1

Estimated Effects and Coefficients for poros1 (coded units)

Term	Effect	Coef	SE Coef	T	P
Constant		4.0138	0.03391	118.36	0.000
T1	−1.5525	−0.7762	0.03391	−22.89	0.000
P1	0.0575	0.0288	0.03391	0.85	0.435
R1	0.1875	0.0937	0.03391	2.76	0.040
Ct Pt		−0.0888	0.07583	−1.17	0.295

S = 0.0959166   R-Sq = 99.07%   R-Sq(adj) = 98.33%

You can remove **Ct Pt** from the model if you uncheck **Include center points** in the model. Cicero can use the **Stat > Regression** approach and obtain the same results if he uses **C8 = 1-'CenterPt'** instead of **'centerPt'**.

The temperature effect is so significant that he decides to increase it by two increments of the small change. The effect of each variable is the change in the response (porosity) corresponding to a change of two coded units, as we described in Case 8.2. So, the reduction in porosity corresponding to an increase in temperature of two units is predicted to be 1.5525. He increases the central value of temperature to 650. There is no benefit in changing the values of pressure. The recycle effect is statistically significant and he decides to reduce the central value of recycle by two increments to 14, although this is more than is indicated by the direction of steepest descent. His new centre point is: temperature = 630 + 2 × 10 = 650; pressure = 820; recycle = 18 − 2 × 2 = 14.

He performs a similar experiment with the results given in Table 10.6, and follows the same procedure.

T2	P2	R2	Temperature	Pressure	Recycle	Poros2	CenterPt
−1	−1	−1	640	800	12	3.14	1
1	−1	−1	660	800	12	2.78	1
−1	1	−1	640	840	12	3.13	1
1	1	−1	660	840	12	3.23	1
−1	−1	1	640	800	16	3.21	1
1	−1	1	660	800	16	2.6	1
−1	1	1	640	840	16	3.53	1
1	1	1	660	840	16	2.54	1
0	0	0	650	820	14	2.81	0
0	0	0	650	820	14	3.09	0

Table 10.6

The cube plot (Figure 10.7) indicates that a further increase in temperature would be useful, but does not indicate changes in pressure or recycling. The regression analysis supports this: only temperature is statistically significant. Other factors are not significant, even at the 20% level.

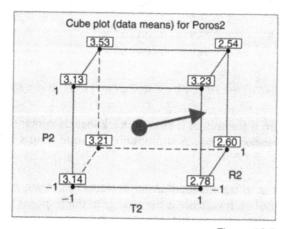

Figure 10.7

Factorial Fit: Poros2 versus T2, P2, R2

Estimated Effects and Coefficients for poros2 (coded units)

```
Term          Effect      Coef    SE Coef         T      P
Constant               3.0200   0.09443     31.98  0.000
T2           -0.4650  -0.2325   0.09443     -2.46  0.057
P2            0.1750   0.0875   0.09443      0.93  0.397
```

```
R2            -0.1000   -0.0500   0.09443   -0.53   0.619
Ct Pt                   -0.0700   0.21116   -0.33   0.754

S = 0.267095   R-Sq = 59.38%   R-Sq(adj) = 26.89%
```

He decides to increase temperature by one coded unit (10°C). His new centre point is

temperature 650 + 10 = 660; pressure = 820; recycle = 14.

The results of his experiment are in the first ten rows of Table 10.7.

T3	P3	R3	Temperature	Pressure	Recycle	Poros3
-1	-1	-1	650	800	12	3.23
1	-1	-1	670	800	12	2.5
-1	1	-1	650	840	12	3.38
1	1	-1	670	840	12	2.42
-1	-1	1	650	800	16	3.09
1	-1	1	670	800	16	2.25
-1	1	1	650	840	16	3.05
1	1	1	670	840	16	2.5
0	0	0	660	820	14	2.84
0	0	0	660	820	14	2.5
0	0	0	660	820	14	2.54
0	0	0	660	820	14	3.15
-1.5	0	0	645	820	14	2.98
1.5	0	0	675	820	14	2.99
0	0	-1.5	660	820	11	2.37
0	0	1.5	660	820	17	3.19

Table 10.7

He repeats the analysis, with only the three main effects, and again finds that the coefficient of temperature is not statistically significant, so he adds the quadratic term: coded units of temperature squared.

Although the third experiment indicated a further increase in temperature, Cicero decides to obtain some confirmatory results. He enlarges the last experiment by adding star points for temperature and recycle at ±1.5 units. He will leave pressure at its mid-value of 820 and have two more runs at the centre. The results have been added in last six rows of Table 10.7. The coefficients in the regression can now be estimated more precisely. The new model again indicates that it would be beneficial to raise the temperature further. It may seem that he is performing a large number of experiments, but they are all producing high-quality wheels.

Response Surface Regression: Poros3 versus T3, P3, R3, T3*T3

The analysis was done using coded units.

Estimated Regression Coefficients for poros3

Term	Coef	SE Coef	T	P
Constant	2.74472	0.11800	23.261	0.000
T3	−0.24520	0.09066	−2.705	0.020
P3	0.03500	0.11333	0.309	0.763
R3	0.04720	0.09066	0.521	0.613
T3*T3	0.08516	0.11086	0.768	0.459

$S = 0.3205$ R-Sq = 42.9%   R-Sq(adj) = 22.2%

At this stage Cicero is convinced that temperature has dominant effect on the process and decides to perform a replicated sequence of runs with temperatures from 645 to 700 in steps of 5.0. He will randomise the order. He will leave the pressure at 820 and the recycle at 14. His results are in Table 10.8.

Temperature	Porosity	Temperature	Porosity
645	2.92	670	2.34
645	3.55	675	2.53
650	3.13	675	2.76
650	3.02	680	2.76
655	2.88	680	2.77
655	2.76	685	2.76
660	2.81	685	2.73
660	2.61	690	3.02
665	2.54	690	2.69
665	2.5	695	3.24
670	2.95	700	3.76

Table 10.8

For analysis, he uses **Stat > Regression > Fitted Line Plot** and asks for a quadratic regression model as in Figure 10.8. This produces the graph in Figure 10.9 which includes the fitted regression equation. He can infer from this that the quadratic effect of the temperature is strong with a

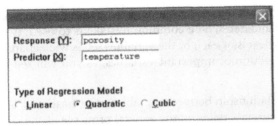

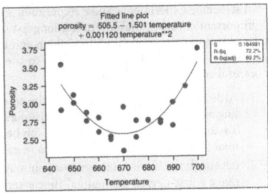

Figure 10.8

Figure 10.9

minimum porosity at about 670°C. He reports to Olivia that he expects to reduce the defects caused by porosity in the aluminium wheels by using **degas** and set values of:

temperature = 670°C; pressure = 820; recycle = 14.

However, he adds that the amount of recycle does not appear to be critical and advises that the process be run with whatever recycle is available. The amount of recycle should be reduced if the process is run at these set values with degassing.

10.1   Linear or quadratic?

In our discussion of cases in this chapter we have used linear and quadratic effects as well as interactions. Here, we offer you some guidance about how you should decide when to include quadratic terms in the model upon which you want to design an experiment. A common mistake arises

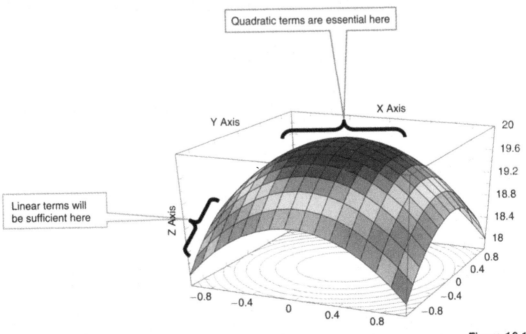

Figure 10.10

in screening experiments: those experiments in which the researcher aims to choose the few most important control variables from a long list of candidates. The common mistake is to use a two-level factorial experiment with the high and low levels of each of the variables set as wide apart as possible. This will almost always lead to the rejection of important variables, as you will see by careful consideration of Figure 10.10.

1. Although we rarely, if ever, know the true relationship between variables, statistical modelling as applied to design and analysis of experiments relies on the general principle of the Taylor series: any smooth function can be approximated about any chosen point as a polynomial (linear and higher power terms in each variable and their products).

2. Although the Taylor series expansion may be a high-degree polynomial, over a very short range a first-degree polynomial (linear terms only) will be a satisfactory representation of the true relationship and over a wider range a second degree polynomial (linear terms, their products, and squared terms) will be a satisfactory representation of the true relationship.

3. Random error may obscure the true relationship in either case. Replication and least squares regression analysis will generally deal with this difficulty.

4. A function that over wide ranges of the predictor variables can be approximated by a quadratic model (second-degree polynomial) may be studied locally, over short ranges, using a linear model (first-degree polynomial) provided that range does not embrace high curvature of the response variable: at or close to the maximum or minimum of the response.

11

Robust design

The aim of *robust design* is to devise products that will perform well over a wide range of environmental conditions. For example, you expect your notebook PC to work over a temperature range from 0°C to 40°C and in relative humidity up to 80%. Few people would be interested in a PC that they could use only in air-conditioned surroundings.

The idea also applies to manufacturing processes. We expect some variation in raw materials, ambient temperature, process operators, and so on. A robust process is one that will provide a product that meets the specification despite such random variation. A domestic example is a recipe for a cake. An easy recipe is one that will lead to a nice cake despite the oven temperature being too low and the proportions of ingredients being somewhat haphazard. For cooks, an easy recipe is a robust recipe.

Noise factors are those factors whose values are hard to control during normal process or use conditions.

In general, we want:

- either to keep the response at a target value and have minimum variability about that target value;
- or to keep the response as low, or as high, as possible.

In either case we aim to minimise the standard deviation of the response. Then, in the first case, we need to bring the mean response to the target value and, in the second case, we aim to minimise, or maximise, the mean response.

Genichi Taguchi is famous for his work on robust design and has had great influence. But some of the statistical detail on which Taguchi methods rely is equivocal, particularly the use of highly fractional factorial designs. These often confound main effects with two-factor interactions and, when the latter are not negligible, can lead to seriously misleading conclusions. We prefer to use the designs we describe in Chapters 7–9. Taguchi also discusses *signal-to-noise ratios*: these combine the mean and standard deviation to give a single response variable to minimise or maximise. We have found that it can be more helpful to treat the mean and standard deviation as separate responses. This is quite in keeping with Taguchi's ideas, as he advocates finding some control variables that primarily affect the mean level and others that also affect variability. The ideal, which may be realisable, is to choose the latter set of controls to minimise the variance, and to choose the former set to bring the mean to some specified value. MINITAB Pro has a **Stat > DOE > Taguchi** suite but we shall use the factorial and central composite designs only.

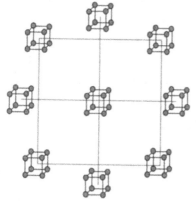

We set up an experiment in which we have some factors to represent environmental variation, *the noise*, and other factors that can be used as control variables. In the experimental situation we can set values for the noise variables but, in usual operation, we shall have no influence over them. We shall vary the factors that correspond to control variables according to a factorial or central composite design following the methods of Chapters 7–9. This design is sometimes known as the *inner array*.

At each factor combination of this design we make runs corresponding to the points of a factorial design for the noise factors. The factorial design for the noise variables is known as the *outer array*.

Figure 11.1

Figure 11.1 shows an inner array for two control variables and an outer array for three noise variables. The inner array is a full composite design: a 2^2 factorial augmented with four star or axial points and one central point. The outer array is a 2^3 factorial that is repeated at every point of the inner array.

We then calculate the mean and standard deviation of the responses from each outer array. Finally, we use regression analysis to look for control variables that reduce the standard deviation and adjust the mean.

▽ Case 11.1 (SeaDragon)

Staff at the UoE have complained that their university smart cards often fail to open security doors around the campus. The smart cards and readers are supplied by SeaDragon. Sam Silver, a designer of electronic equipment, has been asked to investigate. He visits the UoE and finds that the voltage supply to the readers is variable and that some of the cards are badly worn.

The crucial response from the card reader is the amplified signal level from the magnetic head (**y**, measured in decibels), and its target value is 3.0. He has identified three design factors that he can adjust to improve robustness: amplifier gain (**AG**), current limit (**CL**), and spring torque (**ST**).

For a first experiment he will set these at high and low levels and use a 2^3 design for the inner array. In this experiment, he will simulate the environmental noise by setting the voltage supply (**V**) and card wear (**W**) at high and low levels.

He can set precise voltages in the laboratory, and will use a new card and a badly worn card, that had belonged to the Dean of Engineering before he got a replacement, for the two levels of card wear. He will vary these noise factors according to a 2^2 design. Thus, there will be a total of 32 runs: one at each of the four different combinations of **V** and **W**, for each of the eight points of the inner array.

He creates the 2^3 design in a MINITAB worksheet using **Stat > DOE > Factorial > Create Factorial Design**, for the three variables **AG CL ST**, asking for four replicates with no centre points. He orders

Inner array (control variables)			Outer array (the noise)		
AG	CL	ST	V	W	
−1	−1	−1	−1	−1	
1	−1	−1	−1	−1	
−1	1	−1	−1	−1	
1	1	−1	−1	−1	
−1	−1	1	−1	−1	
1	−1	1	−1	−1	
−1	1	1	−1	−1	
1	1	1	−1	−1	
−1	−1	−1	1	−1	
1	−1	−1	1	−1	
−1	1	−1	1	−1	
1	1	−1	1	−1	
−1	−1	1	1	−1	
1	−1	1	1	−1	
−1	1	1	1	−1	
1	1	1	1	−1	
−1	−1	−1	−1	1	
1	−1	−1	−1	1	
−1	1	−1	−1	1	
1	1	−1	−1	1	
−1	−1	1	−1	1	
1	−1	1	−1	1	
−1	1	1	−1	1	
1	1	1	−1	1	
−1	−1	−1	1	1	
1	−1	−1	1	1	
−1	1	−1	1	1	
1	1	−1	1	1	
−1	−1	1	1	1	
1	−1	1	1	1	
−1	1	1	1	1	
1	1	1	1	1	

Table 11.1

the columns by standard order **and** obtains the first three columns in Table 11.1. We have added two extra columns to represent the four combinations of the two noise variables, **V**, and **W**, using **Make Patterned Data**. These variables do not enter the analysis; they are used in the designed experiment to add operational noise in a controlled way. The experimental measurements of the amplified signal level from the magnetic head (**y**), are added in the sixth column. The data are on www.greenfield-research.co.uk/doe/data.htm. Open Case 11.1, where you will find Tables 11.1, 11.2 and 11.3 on three spreadsheets. Copy Table 11.1 starting with the row containing the variable names. Paste into a MINITAB worksheet.

Sam calculates the means and standard deviations with: **Stat > Basic Statistics > Store Descriptive Statistics**. In **Variables** he enters y(dB) and in **By Variables** he enters **AG-ST**. In **Options** he ticks only **Store distinct values of By variables**. In **Statistics** he ticks only **Mean and Standard deviation**. Values appear on the worksheet next to the original values but the variables are named **ByVar1**, **ByVar2** and **ByVar3** instead of **AG**, **CL** and **ST**. He selects and copies the new array of values and pastes it into a new worksheet (**File > New**, then click **Minitab Worksheet**) where he changes the variable names. He labels a blank column as **ln(SD)**. He uses **Calc > Calculator** and enters **ln(SD)** in the **Store result in variable** box. He selects **Natural log** from the **Functions** list. This puts **LOGE(number)** into the **Expression** pane. He replaces 'number' by 'SD' and clicks the **OK** tab. He now has Table 11.2.

Sam creates another column with the logarithm of the standard deviation of response. He uses this as a dependent variable because the errors are then more likely to be

AG	CL	ST	Mean	StDev	ln(SD)
−1	−1	−1	2.50	0.36	−1.03
−1	−1	1	2.83	0.39	−0.95
−1	1	−1	2.78	0.22	−1.51
−1	1	1	1.88	0.54	−0.61
1	−1	−1	2.35	0.98	−0.02
1	−1	1	2.28	0.80	−0.22
1	1	−1	2.63	0.85	−0.16
1	1	1	1.83	1.37	0.31

Table 11.2

approximately normal, so we should get more precise results, and because it ensures that predictions for median standard deviation must be positive.

He analyses these results with **Stat > DOE > Factorial > Analyze Factorial Design**. With this, he is able to ask for analyses for several response variables at a time. He chooses **Mean** and **ln(SD)** and retains the main effects and first-order interactions, excluding the three-factor interaction **ABC**. He looks first at the results for **ln(SD)**.

Response Surface Regression: ln(SD) versus AG, CL, ST

```
The analysis was done using coded units.

Estimated Regression Coefficients for ln(SD)

Term            Coef      SE Coef         T        P
Constant     -0.52309    0.01779     -29.398    0.022
AG            0.50183    0.01779      28.203    0.023
CL            0.03299    0.01779       1.854    0.315
ST            0.15590    0.01779       8.762    0.072
AG*CL         0.06569    0.01779       3.692    0.168
AG*ST        -0.08886    0.01779      -4.994    0.126
CL*ST         0.18609    0.01779      10.458    0.061

S = 0.05033    R-Sq = 99.9%    R-Sq(adj) = 99.3%
```

The fit is almost perfect because the design is almost *saturated*, although not quite because **ABC** was excluded.

Saturated? If you want to study a relationship between two variables X and Y, and you have only two points, you can draw a straight line between those two points and write the relationship as $y = a + bx$. Here you have no information at all about variation. You have two points and you have two coefficients that you can fit from those two points. This design is saturated. If you have three points you may see that there is some variation and there would be one degree of freedom. A *saturated experiment* is one in which the number of observations equals the number of coefficients to estimate.

There are eight observations in Sam's experiment and seven coefficients to estimate: one constant, three main effects and three first-order interactions. So it is almost saturated and there is one degree of freedom. If he were to include the three-factor interaction (**AG*CL*ST**) in the model, the design would be saturated and there would be no degrees of freedom.

Although there is only one degree of freedom for error, the effect of **AG** is statistically significant, beyond the 5% level, and his best strategy is to set **AG** to the low level (−1). It is also desirable to arrange for the product of **AG** and **CL** to be negative, so **CL** should be positive.

He now examines the regression of the mean on the control variables.

Response Surface Regression: Mean versus AG, CL, ST

```
The analysis was done using coded units.

Estimated Regression Coefficients for Mean

Term            Coef      SE Coef         T        P
Constant      2.38125    0.06250      38.100    0.017
AG           -0.11250    0.06250      -1.800    0.323
```

CL	−0.10625	0.06250	−1.700	0.339
ST	−0.18125	0.06250	−2.900	0.211
AG*CL	0.06250	0.06250	1.000	0.500
AG*ST	−0.03750	0.06250	−0.600	0.656
CL*ST	−0.24375	0.06250	−3.900	0.160

S = 0.1768   R-Sq = 96.9%   R-Sq(adj) = 78.2%

None of the coefficients is statistically significant, although two *t*-ratios are reasonably high, because we have only one degree of freedom. If Sam leaves out the **AG*CL** and **AG*ST** interactions, which are negligible, the P-values for the other variables increase. Notice, however, that the coefficients remain the same because, with coded values, the design is orthogonal so the variables and interactions are not correlated.

Sam has little choice of strategy. He needs to raise the mean towards 3.0. From the analysis, he sees that he should set ST at the low level. Apart from this he should take **AG** low and **CL** high, despite its negative coefficient, to take advantage of the **CL** by **ST** interaction. Again, he is fortunate that this combination is associated with a smaller standard deviation.

He repeats **Stat > DOE > Factorial > Analyze Factorial Design**, clicks on **Prediction** and enters values **−1.0, 1.0** and **−1.0** (see Figure 11.2). The predicted values, calculated using the fitted equation, with all three two-factor interactions, appear in the session window. If he also checks the storage items, the values also appear in the worksheet.

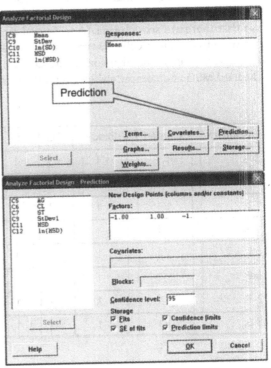

Figure 11.2

Predicted Response for New Design Points Using Model for Mean

Point	Fit	SE Fit	95% CI	95% PI
1	2.7125	0.165359	(0.611409, 4.81359)	(−0.363182, 5.78818)

Predicted Response for New Design Points Using Model for ln(SD)

Point	Fit	SE Fit	95% CI	95% PI
1	−1.48848	0.0470773	(−2.08665, −0.890302)	(−2.36411, −0.612839)

The point predictions are promising, though the mean response is still below target. Assuming a symmetric distribution, the predicted mean logarithm of standard deviation, −1.48848, gives a predicted median standard deviation of 0.23 which is much lower than all but one of the values in Table 11.2.

In this application, the mean squared deviation (MSD), or its logarithm, is also a sensible variable to minimise. Here we describe the deviation as the difference between the observed value of **y** and the target value for **y** of 3.0 dB. For each set of four responses in which the levels of **AG, CL, ST** are the same, it is calculated as

$$MSD = \{(y(1) - 3)^2 + \cdots + (y(4) - 3)^2\}/4$$

You can calculate the **MSD** from the mean and standard deviation:

$$MSD = sd*(n - 1)/n + (target - mean)**2$$

Alternatively, you can make the direct calculation. This is perhaps easier in Excel than in MINITAB. Copy the first six columns of data (forget the top headings) of Table 11.1 into an Excel spreadsheet, so that there are six columns (**A** to **F**) and 32 rows. Into cell **G1** type = (F1-3)^2 and fill down the column to **G32**. Into cell **H33** type = SUM(G1 + G9 + G17 + G25)/4 and fill down the column to **G40**. You will have the eight values of **MSD**. Copy these into the MINITAB worksheet as a column headed **MSD** and compute the natural logarithms into a column headed **ln(MSD)** as in Table 11.3.

RunOrder	CenterPt	Blocks	AG	CL	ST	Mean	StDev	ln(SD)	MSD	ln(MSD)
1	1	1	−1	−1	−1	2.50	0.36	−1.03	0.35	−1.06
2	1	1	1	−1	−1	2.83	0.39	−0.95	0.14	−1.95
3	1	1	−1	1	−1	2.78	0.22	−1.51	0.09	−2.44
4	1	1	1	1	−1	1.88	0.54	−0.61	1.49	0.40
5	1	1	−1	−1	1	2.35	0.98	−0.02	1.15	0.14
6	1	1	1	−1	1	2.28	0.80	−0.22	1.01	0.01
7	1	1	−1	1	1	2.63	0.85	−0.16	0.69	−0.37
8	1	1	1	1	1	1.83	1.37	0.31	2.78	1.02

Table 11.3

Factorial Fit: ln(MSD) versus AG, CL, ST

Estimated Effects and Coefficients for ln(MSD) (coded units)

Term	Effect	Coef	SE Coef	T	P
Constant		−0.5325	0.1690	−3.15	0.051
AG	1.4608	0.7304	0.1690	4.32	0.023
CL	0.3698	0.1849	0.1690	1.09	0.354
ST	0.8048	0.4024	0.1690	2.38	0.097
CL*ST	1.3108	0.6554	0.1690	3.88	0.030

S = 0.477934   R-Sq = 93.12%   R-Sq(adj) = 83.94%

Repeat the analysis with **ln(MSD)** as the response variable. The practical conclusions are unchanged.

Although this first experiment is encouraging, Sam will continue with his experimental program by adding a star design to the inner array. This will increase the precision of estimates, indicate whether higher values of the control variables might increase the mean further, and give an indication of any quadratic effects.

▽ **Case 11.2 (SeaDragon)**

Olivia Orange wants the standard deviation of the wheels process to be investigated. In day-to-day operation, the press, paint mould, and percentage of scrap will vary and Wyanet decides to treat them as noise factors. Since Cicero Copper's intervention, the plant has been running with degassing. So she will use a central composite design for the remaining two control variables, temperature and pressure, as the inner array, and a 2^3 for the noise variables in the outer array. Even so, this will be a large experiment with $9 \times 8 = 72$ runs. Wyanet hopes that it will finally convince everyone about the optimum temperature, at least.

The control factors will take the values -1.4, -1, 0, 1, and 1.4, in coded units, and span the ranges 620 to 700 and 800 to 1000 for temperature and pressure, respectively. This gives, in the central composite design: 620, 631, 660, 689, and 700 for temperature; and 800, 829, 900, 971 and 1000 for pressure (see Figure 11.3). The values of **recycle** used as a noise factor, will be **0%** for low and **20%** for high.

Tables 11.4a, 11.4b and 11.4c are so large that we omit them from this text, but you will find them at www.greenfieldresearch.co.uk/doe/data.htm. Open Case 11.2, where you will find them along with Table 11.5 on four spreadsheets. The first 32 rows (Table 11.4a) represent eight observations at each point of a 2^2 factorial for temperature and pressure. Each of the observations, in each set of eight, is at a point of a 2^3 factorial for **Recycle, Press** and **Paint**, the noise factors. The next 24 rows (Table 11.4b) represent eight observations at each of the three star levels of **temperature** (620, 660, 700) for which **pressure** is at 900. The final 16 rows (Table 11.4c) represent eight observations at the two outer of the three star levels of pressure (800, 900, 1000) for which **temperature** is at 660.

Wyanet calculates the means and standard deviations with **Stat > Basic Statistics > Store Descriptive Statistics**, with **Porosity** in the **Variables** box and **Temp Pressure** in the **By variables** box. The results of these calculations appear in the session window as in Table 11.5.

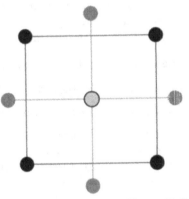

Figure 11.3

StdOrder	T	P	Mean	SD	ln(SD)
1	631	829	4.09	0.198	−1.618
2	689	829	3.21	0.236	−1.443
3	631	971	3.96	0.241	−1.424
4	689	971	3.05	0.203	−1.595
5	620	900	4.88	0.193	−1.646
6	700	900	3.60	0.274	−1.296
7	660	800	2.87	0.213	−1.545
8	660	1000	2.77	0.117	−2.142
9	660	900	2.78	0.199	−1.614

Table 11.5

In a new worksheet, she creates a single replicate of a central composite design for the two variables T and P and copies the mean values and standard deviations from Table 11.5 into results columns. she then used **Analyze Response Surface Design**, taking care to specify the data as uncoded.

Response Surface Regression: mean versus T, P

The analysis was done using uncoded units.

Estimated Regression Coefficients for mean

Term	Coef	SE Coef	T	P
Constant	416.091	14.0973	29.516	0.000
T	−1.227	0.0339	−36.161	0.000
P	−0.006	0.0091	−0.691	0.539
T*T	0.001	0.0000	36.744	0.000
P*P	0.000	0.0000	1.154	0.332
T*P	−0.000	0.0000	−0.504	0.649

This regression of the mean porosity on temperature and pressure confirms Wyanet's earlier findings about temperature. There is still no evidence that raising the pressure will affect the porosity when degassing is applied.

Response Surface Regression: ln(SD) versus T, P

The analysis was done using uncoded units.

Estimated Regression Coefficients for ln(SD)

Term	Coef	SE Coef	T	P
Constant	20.4225	95.0137	0.215	0.844
T	−0.1411	0.2287	−0.617	0.581
P	0.0543	0.0616	0.882	0.443
T*T	0.0001	0.0002	0.812	0.476
P*P	−0.0000	0.0000	−0.575	0.606
T*P	−0.0000	0.0001	−0.738	0.514

This regression of the logarithm of standard deviation on porosity provides no evidence that the standard deviation is affected by the control variables, even after dropping the least significant terms from the analysis as follows:

Response Surface Regression: ln(SD) versus T, P

The analysis was done using uncoded units.

Estimated Regression Coefficients for ln(SD)

Term	Coef	SE Coef	T	P
Constant	84.3541	50.4000	1.674	0.155
T	−0.2592	0.1529	−1.695	0.151
P	−0.0014	0.0010	−1.372	0.228
T*T	0.0002	0.0001	1.710	0.148

The indication from this experiment and its analysis is that, over the range of temperature and pressure used in the experiment, these variables have little effect on variation of the response variable.

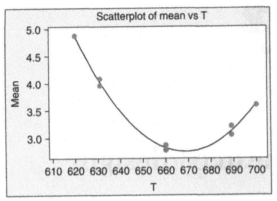

Figure 11.4

Change in the level of pressure has little effect on the mean porosity. However, raising the temperature may be expected to reduce porosity although the size of the quadratic effect suggests we may be close to a minimum. So Wyanet analyses her data again but with only temperature and temperature squared in her model.

The result is striking: both terms are highly significant ($P < 0.001$). A scatterplot (Figure 11.4) confirms the relationship. It also indicates that the best temperature is about 670°C. If you do the analysis in coded units, remember that MINITAB will code ends of the range to -1. and $+1$ respectively.

Response Surface Regression: mean versus T

The analysis was done using coded units.

Estimated Regression Coefficients for mean

Term	Coef	SE Coef	T	P
Constant	2.8094	0.03629	77.416	0.000
T	−0.6281	0.03390	−18.527	0.000
T*T	1.4440	0.06178	23.373	0.000

S = 0.06866   R-Sq = 99.3%   R-Sq(adj) = 99.1%

Estimated Regression Coefficients for mean using data in uncoded units

Term	Coef
Constant	406.306
T	−1.20701
T*T	0.000902510

The fitted equation is

$$\text{Mean} = 406.0 - 1.21 {*} T + 0.0009 {*} T^2$$

Differentiation leads to the estimate that the minimum porosity is when T = 672.

△ Review of Case 11.2

Run the process with degassing at a temperature of 672°C. None of the press, paint mould, percentage recycle, or pressure appears to have a significant effect on porosity. Running the process at the lowest feasible pressure of 800 is recommended, as this will save money. The characteristics of the process may change over time, so the porosity of wheels should be monitored, by taking a random sample from production, on a weekly basis. Further hill climbing experiments about the current estimated optimum, using small increments, could occasionally be performed.

12

Hierarchical (nested) designs

Concrete is one of the most widely used materials in civil engineering construction. The quality of concrete is usually assessed from measurement of crushing strengths of test cubes. These measurements can be made only after the cubes have been thoroughly cured, usually after 28 days. Given the scale of civil engineering works, and the delay until test results are known, rejection of defective concrete is not a practical option. Rather, the quality of the concrete will be assessed continuously with the aim of detecting and rectifying any changes in quality well before any defective batches are produced.

During the construction of the main runway at Heathrow Airport, engineers carefully monitored the quality of the concrete by making test cubes at random times throughout the day. To allow time for curing, they measured the compressive strength of the cubes after 28 days. Strengths of cubes made from the same batch of concrete vary. It is also very likely that different batches of concrete will vary, even when made from the same delivery of cement. Reasons for this include: discrepancies in the proportions of aggregate and sand mixed with the cement; discrepancies in the amount of water added; and differences in ambient temperature. Finally, the quality of cement may vary between deliveries. At the beginning of the construction, the chief engineer performed an experiment to identify the sources of the variation of the strengths of the test cubes. The objectives are: to check if the overall variance is small enough to satisfy the specification; to see if there is scope for reducing any of the components of variance; and to set up a monitoring scheme that would not only detect an increase in variability but also indicate a likely source.

The experimental design was to take a number, which to be general we will refer to as I, of deliveries of cement, randomly selected over a two-week period. From each delivery a number, which we will refer to as J, of batches of concrete were randomly selected from all the batches made. Then, a number, call it K, of test cubes were made from randomly selected concrete from each batch. A diagram for I, J and K equal to 6, 2, and 4 respectively is shown in Figure 12.1.

Cement delivery

Concrete batch

Concrete test cubes

Figure 12.1

The cubes were stored for 28 days before measuring their compressive strength. Our model for this design, writing OM for the overall mean, is

$$\text{strength}(i, j, k) = \text{OM} + \text{delivery}(i) + \text{batch(delivery)}(j(i)) + \text{cube(batch(delivery))}(k(i, j))$$

where delivery(i) is the deviation of the ith delivery from OM, and batch($j(i)$) is the deviation of the mean of the jth batch, made from the ith delivery, from OM + delivery(i). The cube($k(i, j)$) is the deviation of the kth cube, made from this jth batch, from OM + delivery(i) + batch(j). The cube($k(i, j)$) term is equivalent to error(i, j, k) where the errors are independent with mean 0. The i, j and k range from 1 up to I, J and K, respectively.

The design is said to nested because, for example, the deviation for batch 2 made from delivery 1 is independent of the deviation for batch 2 made from delivery 2. This is emphasised by the notation $j(i)$, which indicates that j is specific to its i. The cubes are nested within batches, and the batches are nested within deliveries. An alternative description is that cubes, batches, and deliveries form a hierarchy. The analysis assumes that the components of variance are constant. Thus, the variance of cubes within a batch, which we abbreviate as the within-batch variance, is assumed to be the same for all batches. The within-batch variance is the variance of the errors in our model. Also, the variance of batches within a delivery, known as the within-delivery variance, is assumed to be the same for all deliveries. Furthermore, the variance of deliveries, known as the between-deliveries variance, is assumed to be constant.

In the ANOVA context we refer to batches and deliveries as random effects because the batches are a random sample from all the batches made from a delivery, and the deliveries are considered a random sample of all possible deliveries. The definition of an overall mean implies that the delivery deviations and the batch deviations have means of zero, as do the random errors. Therefore, these effects are defined by their variances or standard deviations. We can use our model to express the variance of strengths as the sum of the variances of the components on the right-hand side, since the variance of the sum of independent components equals the sum of their variances:

$$\text{var[strength]} = \text{var[delivery]} + \text{var[batch(delivery)]} + \text{var[error]}$$

MINITAB Pro will estimate the components of variance for us, but we should understand the principles. So far, we have described a balanced design, with the same numbers of cubes from each batch and batches from each delivery, and this makes the calculations easier. It is straightforward to estimate the variance of the errors. For each sample of K cubes from a batch we estimate the variance with $K - 1$ degrees of freedom. We then average the IJ estimates:

$$\text{est\_var(error)} = \text{average over batches(estimated variance of cubes within batches)}$$

This estimate of the variance of the errors, also known as the within-batches variance, has $IJ(K - 1)$ degrees of freedom.

We need a slightly more subtle argument to estimate the variance of the batches within deliveries. For each delivery, calculate the mean of the K cubes in each batch. Next calculate the variance of these J means using the denominator $(J - 1)$. Now average these variances over I deliveries, and refer to this as the unadjusted estimate of within-delivery variance with $I(J - 1)$ degrees of freedom. Although this is a good start, it is likely to be an overestimate because we do not know the precise means of the sample batches. We have estimated each batch mean from a random sample of K concrete cubes, and these estimates have a standard error equal to the variance of the errors divided by K. Thus

$$\text{unadjusted estimate of within-delivery variance} = \text{est\_var[batch(delivery)]} + \text{est\_var[error]}/K$$

We calculate the term on the left-hand side of the equation and the est_var[error], so we can obtain est_var[batch(delivery)] by subtraction:

est_var[batch(delivery)] = unadjusted estimate of within-delivery variance − est_var(errors)/K

A similar argument for deliveries gives

unadjusted estimate of variance of deliveries = est_var[delivery] + est_varbatch[delivery]/J + est_var[error]/JK

The unadjusted estimate of variance of deliveries has $I − 1$ degrees of freedom.

An estimate of the mean and standard deviation of the population of all cube strengths is required to assess whether the specification is met. The estimate of the mean is the mean of all IJK cube strengths. The overall standard is estimated by

std[strength] = sqrt(est_var[strength]) = sqrt(est_var[delivery] + est_var[batch(delivery)] + est_var[error])

You may ask why this estimate is preferable to the standard deviation of all IJK cube strengths. The answer is that our sample cube strengths are not a simple random sample from the population of all possible cubes; they have been obtained through multi-stage sampling. The standard deviation of all cube strengths will be biased towards low values, and this bias can be appreciable if we have only a few deliveries.

The estimates of the standard deviations of the components of the overall variability are important to the engineers. For example, if the between-batch, within-delivery standard deviation is high we should improve the procedures for mixing the concrete. If the standard deviation between deliveries is large we should investigate our supply. Before demonstrating an analysis for cubes in Case 12.3 we consider two cases with just two hierarchical levels.

▽ Case 12.1 (UoE)

The data for this case are an excerpt from a more extensive study by Hanna *et al.* (2005).

Our UoE psychologist, Ingrid Indigo, is a member of a team investigating methadone replacement for opiate dependence. Euphoria scores are measured on the Methadone-Benzedrine Group (MBG) Scale of the Addiction Research Center. How variable are these scores?

The data in Table 12.1 are the euphoria scores, measured on three occasions, for a random sample of six patients on a maintenance dose of methadone. Open Case 12.1 at http://www.greenfield-research.co.uk/doe/data.htm where there are three spreadsheets, Tables 12.1, 12.2 and 12.3. At this early stage of the study, Ingrid would like to make preliminary estimates of: the standard deviation of euphoria scores, obtained on different occasions, for each individual patient; and the standard deviation of mean euphoria scores for the hypothetical population all possible patients. This information should help clinicians to treat such patients because they will have some idea of how much difference to expect patients to report between consultations. She also hopes it will help psychiatrists to understand withdrawal symptoms if they know to what extent euphoria varies between patients and to what extent it varies between consultations for individual patients. Ingrid's model for this design, writing OM for the overall mean, is:

patient(*i, j*) = OM + patient(*i*) + occasion(*j*(*i*))

where patient(i) is the deviation of the hypothetical population mean for the ith patient from OM, and occasion(j(i)) is the deviation of the jth occasion for patient i from the mean of all the measures of patient i. This last term is equivalent to error(i, j), where the errors are independent with mean zero. The i and j range from 1 to 6, and up to 3, respectively. If you use Pro you can use several of the **ANOVA** routines for the analysis. **General Linear Model** provides the most detail.

In this analysis Ingrid is not concerned with the means for individual patients. Her purpose is to estimate the standard deviation of the distribution of means for all possible patients. So, she wishes to consider her patients as a random sample of all possible patients and declares them to be a random effect. Note, therefore, that in Figure 12.2 **patient** must be declared twice: once as in the model and once as a random factor. She used **Stat > ANOVA > General Linear Model** and, as in Figure 12.2, selected **MBG** as the response variable with **patient** as the model variables. She chose to display results except for coefficients of all terms.

The annotated results start with the following:

General Linear Model: MBG versus patient

```
Factor    Type    Levels  Values
patient   random     6    1,2,3,4,5,6
```

This is a summary of Ingrid's design with six patients representing random effects. The patients are identified by numbers from 1 to 6.

In the following analysis of variance table the **Total Seq SS** is the sum of squared deviations of the 18 **MBG** scores from their mean. The **Total Seq SS** has 17 degrees of freedom because it is based on 18 deviations which have one constraint: they add to zero. This constraint is a consequence of calculating deviations from the sample mean.

Patient	MBG	Patient	MBG
1	54.0	4	61.5
1	75.0	4	25.5
1	81.0	4	0.0
2	72.0	5	88.0
2	70.5	5	102.5
2	83.5	5	71.0
3	18.5	6	53.0
3	0.0	6	90.0
3	27.0	6	87.0

Table 12.1

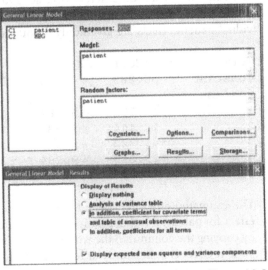

Figure 12.2

Analysis of Variance for MBG, using Adjusted SS for Tests

```
Source    DF   Seq SS    Adj SS    Adj MS     F      P
patient    5   12943.6   12943.6   2588.7   7.51   0.002
Error     12    4135.7    4135.7    344.6
Total     17   17079.3
```

This **Total Seq SS** is the sum of **Error Seq SS** and **patient Seq SS**. The **Error Seq SS** is the sum of the squared residuals calculated after fitting the model which allows for differences between the six patients.

When we calculate residuals we first estimate a further five parameters from the data: the deviations of the six patients from the overall mean, with a constraint that these deviations add to zero. The patient **Seq SS** is three times the sum of the six patient means from the overall sample mean. There are five degrees of freedom because the sum is based on six deviations with a constraint that they add to zero. The multiplier of 3 arises because all the **Seq SS** are sums over six patients and three occasions, and patient means are averaged over the three occasions. The next column in the ANOVA table is the adjusted sum of squares (**Adj SS**) which, in this case, is identical to the sequential sum of squares (**Seq SS**). In general **Adj SS** is the sums of squares with only previous terms in the model, and **Seq SS** is sums of squares for terms with all the other terms in the model. One advantage of an orthogonal design is that the **Seq SS** and **Adj SS** are identical. The adjusted mean square (**Adj MS**) is the **Adj SS** divided by the degrees of freedom.

```
S = 18.5645 R-Sq = 75.79% R-Sq(adj) = 65.70%
```

The **S** is the square root of the error (**Adj**) mean square. It is an estimate of the standard deviation of withdrawal scores on different occasions for the same patient. The **R-Sq** is the proportion of the total sum of squares accounted for by including patients in the model. The estimates of the individual patient effects are given only if you select **Coefficients for all terms**.

```
Term             Coef    SE Coef       T        P
Constant       58.889     4.376    13.46    0.000
patient
1              11.111     9.784     1.14    0.278
2              16.444     9.784     1.68    0.119
3             -43.722     9.784    -4.47    0.001
4             -29.889     9.784    -3.05    0.010
5              28.278     9.784     2.89    0.014
```

The estimated effect for patient 6 is not listed. Since the sum of the six effects equals zero, the effect for patient 6 is the negative of the sum of the effects for patients 1–5. This can be calculated by copying the columns to the worksheet and is 17.78. Small P-values indicate that the corresponding patient effect is significantly different from the average and these will lead to a substantial variance between patients.

```
Unusual Observations for MBG

Obs      MBG       Fit    SE Fit   Residual   St Resid
 10    61.500    29.000    10.718     32.500      2.14 R

R denotes an observation with a large standardized residual.
```

A studentised residual as large as 2.14 in a sample of 18 is not surprising. It is hardly enough to reject the value, but perhaps large enough to encourage you to see if there was something unusual about that patient on that occasion.

The following expected values of the mean squares are the basis of the *F*-test in the original ANOVA table.

```
Expected Mean Squares, using Adjusted SS

                Expected Mean
                Square for Each
    Source      Term
1   patient     (2) + 3.0 (1)
2   Error       (2)
```

If we imagine repeating the experiment millions of times the expected (population average) value of the error mean square is the variance of the errors, that is the within-patient variance (denoted by **(2)** as it is the second component of variance). Note that **(2)** represents an unknown constant rather than its estimate, which is 344.6.

The expected value of the patient mean square is 3 × (between-patient variance + within-patient variance/3) which is denoted by **3*(1) + (2)**, where the between-patient variance is considered as the first component of variance **(1)**. The factor of 3 arises because the SS are sums over patients and occasions; the sum of an average, itself over occasions, over occasions is the product of that average with the number of occasions. A test of the hypothesis that **(1)** is zero is provided by the ratio of the estimated patient mean square to the estimated error mean square. If this ratio is significantly greater than 1.0 we have sufficient evidence to reject the hypothesis. In this case the ratio is 7.51 and the associated P-value is 0.002 so we have strong evidence that **(1)** is positive rather than zero, representing variation between patients. As is often the case, the null hypothesis is hardly credible, and a more useful interpretation of our ability to reject it is that the experiment has been powerful enough to provide plausible estimates of the components of variance. You will see this pattern again, with more terms, in later examples.

The components of variance are estimated by equating the expected values of the mean squares to the sample values:

$$(2) + 3.0 * (1) = 2588.7$$
$$(2) =\ 344.6$$

MINITAB gives the solution to these equations if, as in Figure 12.2, you tick **Display expected mean squares and variance components**.

```
Variance Components, using Adjusted SS

Source              Estimated Value
patient             748.0
Error               344.6
```

Paste these values into a spare column of the worksheet, such as **C10**, and use **Calc** to provide the standard deviations of scores

$$C11 = sqrt(C10)$$

and the overall standard deviation

$$C12 = sqrt(sum(C10))$$

	Variance components	Standard deviations of scores	Overall standard deviation
Between patients	748.0	27.35	33.055
Within patients	344.6	18.56	

The estimated standard deviation of within-patient scores is 18.6. The standard deviation of the difference in scores for the same patient on different occasions is the square root of twice the within-patient variance and equals 26.3. A practical interpretation is that a psychiatrist can estimate a one in three chance that the difference in a patient's scores obtained on separate occasions will exceed 26. Similarly, we can estimate a one in three chance that scores for different patients will differ by more than 47 ($33\sqrt{2}$). Although the within-patient standard deviation is less than that between patients it is far from negligible and psychiatrists should expect considerable variation in this measure.

If you are relying on Student, you can use **Stat > ANOVA > One-Way**. You can perform the components of variance analysis either by using the one-way analysis of variance with some reinterpretation and additional arithmetic or from first principles. The MINITAB one-way ANOVA tests the hypothesis that the means of several populations are equal. The method is an extension of the two-sample t-test, specifically for the case where the population variances are assumed to be equal. A one-way ANOVA requires the following:

- a response or measurement taken from the units sampled (**MBG** in Ingrid's case);
- a factor or discrete variable that is altered systematically (**patient number**). The different values of the factor are its levels.

A one-way ANOVA can be used to tell you if there are statistically significant differences among the level means. The null hypothesis for the test is that all population means (level means) are the same. The alternative hypothesis is that one or more population means differ from the others.

Ingrid's model reduces to:

$$\text{patient}(i, j) = \text{OM} + \text{patient}(i) + \text{error}(i, j)$$

Ingrid will explore this model with **Stat > ANOVA > One-Way**, but she would need to interpret the output in the context of patients being random effects. The MINITAB output assumes that the measurements are of specific patients, so patients are treated as fixed effects. She follows **Stat >ANOVA > One-Way (stacked)**. The word 'stacked' indicates that the patient labels (Table 12.1) are stacked in one column with the MBG values in one corresponding rather than having a separate column for each patient.

The MINITAB output is for patients as a fixed effect.

```
One-way ANOVA: MBG versus patient

Source      DF      SS     MS      F       P
patient      5   12944   2589   7.51   0.002
Error       12    4136    345
Total       17   17079

S = 18.56    R-Sq = 75.79%    R-Sq(adj) = 65.70%
```

				Individual 95% CIs For Mean Based on Pooled StDev
Level	N	Mean	StDev	
1	3	70.00	14.18	(------*-------)
2	3	75.33	7.11	(------*-------)
3	3	15.17	13.81	(------*----------)
4	3	29.00	30.90	(------*----------)
5	3	87.17	15.77	(------*-------)
6	3	76.67	20.55	(------*-------)

```
                            0        30        60        90

Pooled StDev = 18.56
```

You will obtain the same results if you use the unstacked data of Table 12.2 but with different labels. If you start with the stacked data of Table 12.1, follow **Data > Unstack columns**; enter **MBG** in the **Unstack the data in** box and **patient** in the **Using subscripts in** box. Then follow **Stat > ANOVA > One-Way (unstacked)**.

However, when there is a hierarchy of only two levels, here **Occasions Within Patients**, the test of the hypothesis that the between patients variance is zero is identical to the hypothesis that the means for the six patients, considered as fixed effects, are identical. But, you do need to know the expected

MBG_1	MBG_2	MBG_3	MBG_4	MBG_5	MBG_6
54	72	18.5	61.5	88	53
75	70.5	0	25.5	102.5	90
81	83.5	27	0	71	87

Table 12.2

values of the mean squares to estimate the components of variance. In the notation of our general linear model analysis, you now have to solve the equations equating the expected values of mean squares to their estimates for yourself.

$$3(1) + (2) = 2589$$
$$(2) = 345$$

Whichever way you choose to do the calculations, a components of variance analysis, as for all ANOVA, assumes components of variance are constant. MINITAB will also provide tests for equality of variances within patients with **Stat > ANOVA > Test for Equal Variances**. In this case both Bartlett's test, which assumes a normal distribution, and Levene's test, which assumes any continuous distribution, tell us that there are no significant differences (P-values of 0.61 and 0.76, respectively) between the variances within patients. We take this as support for the assumption that the within-patient variance is the same for all patients.

An analysis from first principles is no more complicated. Follow **Stat > Basic Statistics > Store Descriptive Statistics** and enter **MBG** as a variable By **patient** (Figure 12.3). Click on **Statistics** and choose **Mean, Standard deviation, Variance** and **N total**. Click on **Options** and **Store distinct values of By variables** (Figure 12.4). The results appear on the worksheet as extra columns (C3 to C7); see Table 12.3.

Figure 12.3

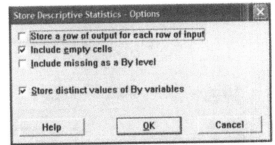

Figure 12.4

In MINITAB's calculator, create four constants, **K1** to **K4**, with values

    K1 = SUM('Variance1')/6
    K2 = STDEV('Mean1')
    K3 = SQRT(K1)
    K4 = K2 * K2 − K1 / 3

See Figure 12.5. Follow **Data > Display Data** and enter **K1 K4** (Figure 12.6). These values are displayed in the session window as

Data Display

K1   344.639
K4   748.028

C3	C4	C5	C6	C7
ByVar1	Mean1	StDev1	Variance1	Count1
1	70.0	14.2	201.0	3
2	75.3	7.1	50.6	3
3	15.2	13.8	190.6	3
4	29.0	30.9	954.8	3
5	87.2	15.8	248.6	3
6	76.7	20.6	422.3	3

Table 12.3

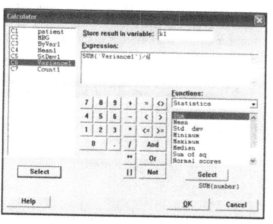

Figure 12.5

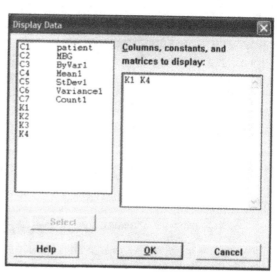

Figure 12.6

K1 is the within-patient variance and **K4** is the between-patient variance. These are identical to those given as the variance components in the general linear model.

△ Review of Case 12.1

The MBG score, an assessment of euphoria, was measured on three occasions for six patients on a maintenance dose of methadone replacement for opiate dependence. The mean was 59, the lowest score was 0 and the highest was 102. The estimated standard deviation of MBG measured at different times on the same patient is 19, and the estimated standard deviation of underlying mean scores of different patients is 27. Doctors should expect considerable variation in MBG between visits; there is an estimated one in three chance that the difference will exceed 26 ($=19\sqrt{2}$). They should also expect about two-thirds of patients to have a mean MBG between 33 and 86. The overall standard deviation of a single MBG score is estimated to be 33. If MBG has an approximate normal distribution, an estimated one in ten MBG scores will exceed 101 ($59 + 1.28 \times 33$), and very low scores will not be particularly unusual. More data would be needed before suggesting a more realistic probability distribution for MBG.

▽ Case 12.2 (AgroPharm)

AgroPharm manufactures a gold solution that is advertised as containing 20 g of potassium aurocyanide per litre. Jasmine Jade is the manager of the division that produces this, and other, metal solutions. AgroPharm has many regular customers who occasionally send aliquots of the solution to test laboratories for assay. The production process is carefully controlled but Jasmine recently received a complaint, that the gold concentration was too low, from a large manufacturer of printed circuit boards called SeaDragon. The complaint was based on an assay by an accredited laboratory working to the agreed international standard. Despite this assurance Jasmine decides to carry out an inter-laboratory trial. The ideal for an international standard is a measurement procedure that will give the same result, for identical material, in any accredited laboratory with any competent technician. In practice the committees responsible for international standards realise that there will be slight differences between assays made by the same technician using the same apparatus on the same day, and somewhat larger, but still small, differences between assays made by different technicians in different laboratories. There is even a standard for running inter-laboratory trials to investigate these discrepancies (ISO 5725). The standard is called *Precision of test methods* and it describes procedures for determination of *repeatability* and *reproducibility* for a standard test method. We give the formal definitions of these at the end of this case, but the standard deviation of the assays by the same technician is a measure of repeatability, and the standard deviation of assays between different laboratories is a measure of reproducibility.

Jasmine is particularly concerned about the reproducibility. She writes back to SeaDragon, explaining that she is surprised about the laboratory result, since AgroPharm aimed for a target of 21 g per litre and had measured the standard deviation of gold content in litre containers to be 0.18 g. She points out that even if the process mean drifted down by two standard deviations, fewer than one in 1000 containers would be below 20 g per litre, and says she is planning an inter-laboratory trial to determine the reproducibility of assays. However, Jasmine upholds the complaint and sends two complimentary litres of gold solution.

She now has to decide how many laboratories to include in the trial. She has a table (Table 12.4) that provides 90% and 95% confidence intervals for the population standard deviation as a multiple of the sample standard deviation for sample sizes of 5, 10, 15 and 20. The table is based on an assumption of a random sample from a normal distribution. If she takes a random sample of 15 laboratories, an approximate 90% confidence interval for the reproducibility standard deviation will be between 0.77 and 1.46 times that point estimate. It is only a rough guide, which overstates the confidence, because she is ignoring the uncertainty about the within laboratory standard deviation, but Jasmine is prepared to accept this level of precision.

Sample size	Lower 95%	Lower 90%	Upper 90%	Upper 95%
5	0.60	0.65	2.37	2.87
10	0.69	0.73	1.65	1.83
15	0.73	0.77	1.46	1.58
20	0.76	0.79	1.37	1.46

Table 12.4

The second sample size decision is the number of replicates from each laboratory. Two is a common choice as this provides an estimate of the within-laboratory standard deviation with 15 degrees of freedom and it does not seem helpful to estimate this much more precisely than the between-laboratory standard deviation. However, it doesn't provide much data to check the assumption that the within-laboratory standard deviation is constant for all laboratories.

The laboratories work with 50 ml samples. She draws a random sample of 15 laboratories from a list of accredited laboratories in the country. She prepares 1.5 litres of the gold solution, mixes it thoroughly, and dispenses it into 50 ml bottles, 10 ml at a time. She numbers the bottles and randomly assigns two bottles to each of the 15 laboratories.

Lab	Replicate	Assay	Replicate	Assay
1	1	21207	2	21276
2	1	21019	2	20996
3	1	21050	2	21045
4	1	21255	2	21255
5	1	21299	2	21281
6	1	21033	2	21093
7	1	21119	2	21174
8	1	21045	2	21077
9	1	21158	2	21142
10	1	21077	2	21017
11	1	21045	2	21033
12	1	20996	2	21038
13	1	21244	2	21341
14	1	21229	2	21012
15	1	21239	2	21341

Table 12.5

She has all the results within two weeks and they are given in units of milligrams per litre in Table 12.5. You can copy this table from Case 12.2 on http://www.greenfieldresearch.co.uk/doe/data.htm. Where columns have the same name, MINITAB automatically suffixes a number to the second instance (**replicate_1**). She needs to stack the columns before she can plot the data. She follows **Data > Stack > Blocks of Columns**, chooses to stack **lab replicate assay** on **lab 'replicate_1' 'assay_1'**. She makes sure that **Use variable names in subscript column** is not ticked. In the new worksheet, there are no labels on the columns except for column 1 which MINITAB automatically heads **Replicate**. The replicate number refers to the block of columns. So Jasmine labels columns 2 to 4 as **lab replicate assay**. Jasmine plots the data (Figure 12.7) using **Graph > Scatterplots** and selecting **With groups**. She writes the title of the plot with **Labels** and then clicks on **Data View** and asks for **Symbols**

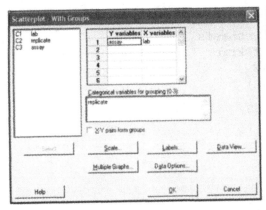

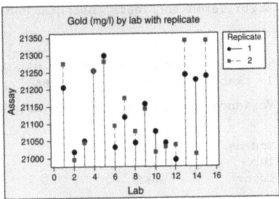

Figure 12.7

and **Project lines**. The project lines link the pairs of replicates from each laboratory. The plot suggests that differences between laboratories are more substantial than differences between the replicates within laboratories. There are no outlying results that warrant following up with any particular laboratory.

The analysis is straightforward in MINITAB. If you have Pro, you have a choice. We prefer the versatile **Stat > ANOVA > General Linear Model**. Jasmine uses this and asks, with **Results**, for **Display expected mean squares and variance components**.

General Linear Model: assay versus lab

```
Factor   Type     Levels  Values
lab      random       15  1, 2, 3, 4, 5, 6, 7, 8, 9, 10, 11, 12, 13, 14, 15
```

Analysis of Variance for assay, using Adjusted SS for Tests

Source	DF	Seq SS	Adj SS	Adj MS	F	P
lab	14	328998	328998	23500	8.20	0.000
Error	15	42977	42977	2865		
Total	29	371975				

S = 53.5269   R-Sq = 88.45%   R-Sq(adj) = 77.66%

Unusual Observations for assay

Obs	assay	Fit	SE Fit	Residual	St Resid
14	21229.0	21120.5	37.8	108.5	2.87 R
29	21012.0	21120.5	37.8	−108.5	−2.87 R

R denotes an observation with a large standardized residual.

Expected Mean Squares, using Adjusted SS

	Source	Expected Mean Square for Each Term
1	lab	(2) + 2.0000 (1)
2	Error	(2)

```
Error Terms for Tests, using Adjusted SS

                                    Synthesis
                                    of Error
        Source     Error DF    Error MS   MS
  1     lab           15.00       2865    (2)

Variance Components, using Adjusted SS

             Estimated
  Source       Value
  lab          10317
  Error         2865
```

The estimate of the within-laboratory standard deviation is $\sqrt{2865} = 54$. The estimate of the between–laboratory standard deviation is the $\sqrt{10317} = 102$. The estimated standard deviation of assays from all possible laboratories is $\sqrt{2865 + 10317} = 115$. Thus the reproducibility standard deviation, 115, is substantially higher than the repeatability standard deviation, 54. But, it is not high enough to account for the complaint if the AgroPharm process was on its target of 21,000 mg/l.

We can estimate the probability of the assay being below 20,000 if the process mean is 21,000. The standard deviation of gold content in litre containers is 180 mg/l. The standard deviation of an assay is 115, so the assay on a randomly selected jar is distributed with a mean of 20,000 and a standard deviation of $\sqrt{180^2 + 2865 + 10317} = 213$. A normal distribution is likely to be realistic for assays so the probability that the assay on a random jar is less than 20,000 if the process is on target can be found from **Calc > Probability Distribution > Normal**, with **Cumulative probability, mean = 21000, standard deviation = 213, input constant = 20000**, and is less than 2 in a million.

She concludes that the process target is high enough for complaints about gold concentration to be very rare events. Furthermore, AgroPharm monitors the process and there is no evidence that it has been off-target. Nevertheless, for the moment, she is prepared to attribute the complaint to some unidentified special cause variation in the AgroPharm process, but if there are any more complaints she will investigate the test laboratories involved. The concept of reproducibility R, discussed below, will be relevant.

ISO 5725 defines *repeatability*, *r*, as the value below which the absolute difference between two single test results obtained with the same method on identical test material under the same conditions (same operator, same apparatus, same laboratory and a short interval of time) may be expected to lie with a specified probability. In the absence of other indications, the probability is usually chosen to be 95% and the difference is **2*sqrt(Sw^2 + Sw^2)**, where **Sw** is the standard deviation within laboratories. The repeatability for the potassium aurocyanide assays is 0.151 g/l.

ISO 5725 further defines *reproducibility*, *R*, as the value below which the absolute difference between two single test results obtained with the same method on identical test material under different conditions (different operators, different apparatus, different laboratories and/or different time) may be expected to lie with a specified probability. In the absence of other indications the probability is usually chosen to be 95% and the difference is **2*sqrt(Sb^2 + Sw^2 + Sb^2 + Sw^2)**, where **Sb** is the standard deviation between laboratories. The reproducibility for the potassium aurocyanide assays is 0.325 g/l.

In both cases the results are a direct consequence of independent components of variance being additive.

Review of Case 12.2

An inter-laboratory trial of assay of a 20 g/l potassium aurocyanide solution was performed. Two 50 ml samples were sent to 15 laboratories. The mean assay was 21.138 g/l. The estimated within-laboratory standard deviation of assay is 0.054 g/l. and the estimated standard deviation of underlying mean results for different laboratories is 0.102 g/l. The overall standard deviation of an assay is estimated to be 0.12. The target content for the AgroPharm potassium aurocyanide solution is 21 g/l and the standard deviation of content of litre containers is 0.18 g. If the process is on target, the estimated probability that a customer's assay will be below 20 g/l is about 2 in a million.

▽ Case 12.3 (SeaDragon)

Tarquin Turquoise is the site engineer in charge of a new runway construction at Arundel Airport. SeaDragon are supplying the cement, from the plant at which Mercedes Maroon advised about the rotary kiln. At the start of the contract, Tarquin decides to perform a similar experiment to that undertaken at Heathrow Airport. Mercedes will be interested to see how well the expert system is performing.

Cement is delivered each day. The contractor mixes 30 batches of concrete during the day, and Tarquin will randomly select two batches from the 30. He will ask for four cubes to be made from each batch at haphazard times during the fifteen minutes it is being poured. The compressive strengths of the test cubes were measured after 28 days. The results are given in units of mega-pascals (MPa) in Table 12.6 You can copy this table from Case 12.3 at www.greenfield-research.co.uk/doe/data.htm. Where columns have the same name, MINITAB automatically suffixes a number to the second instance (**delivery_1**). Tarquin needs to stack the columns before he can plot the data. He follows **Data > Stack > Blocks of Columns**, chooses to stack **deliverybatch cube strength** on '**delivery_1**' '**batch_1**' '**cube_1**' '**strength_1**' and on '**delivery_2**' '**batch_2**' '**cube_2**' '**strength_2**'. He makes sure that **Use variable names in subscript column** is not ticked. Tarquin labels columns 2 to 5 as delivery batch cube strength.

Delivery	Batch	Cube	Strength	Delivery	Batch	Cube	Strength	Delivery	Batch	Cube	Strength
1	1	1	35.6	2	1	1	30.7	3	1	1	30
1	1	2	33.6	2	1	2	30.5	3	1	2	35
1	1	3	34.1	2	1	3	27.2	3	1	3	35
1	1	4	34.5	2	1	4	26.8	3	1	4	32.6
1	2	1	38.6	2	2	1	31.7	3	2	1	27.9
1	2	2	41.6	2	2	2	30	3	2	2	27.7
1	2	3	40.7	2	2	3	33.8	3	2	3	29
1	2	4	39.9	2	2	4	29.6	3	2	4	32.8

Table 12.6

4	1	1	34.3	5	1	1	33.2	6	1	1	39.5
4	1	2	36.4	5	1	2	35.2	6	1	2	42.1
4	1	3	33.4	5	1	3	37.8	6	1	3	38.5
4	1	4	33.4	5	1	4	35.4	6	1	4	40.2
4	2	1	38.7	5	2	1	35.8	6	2	1	38.7
4	2	2	38.5	5	2	2	37.1	6	2	2	36.1
4	2	3	43.3	5	2	3	37.1	6	2	3	35.9
4	2	4	36.7	5	2	4	39.5	6	2	4	42.8

Table 12.6 (Continued)

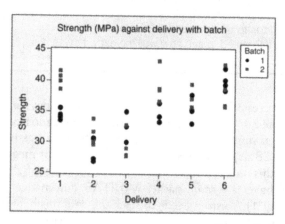

Figure 12.8

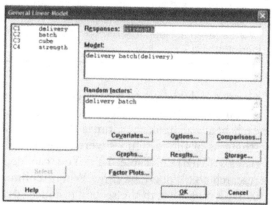

Figure 12.9

As usual, Tarquin starts with a plot (Figure 12.8). There appears to be variation at all levels, but there are no obvious outlying points. He uses **Stat > ANOVA > General Linear Model** for the analysis described in the introduction to this chapter.

Notice in Figure 12.9 how Tarquin had to specify the nesting of batches within deliveries in MINITAB. It matches the mathematical model nicely. He clicks **Results** and selects **Display expected mean square and variance components**.

```
General Linear Model: strength versus delivery, batch

Factor             Type      Levels   Values
delivery           random         6   1, 2, 3, 4, 5, 6
batch(delivery)    random        12   1, 2, 1, 2, 1, 2, 1, 2, 1, 2, 1, 2

Analysis of Variance for strength, using Adjusted SS for Tests
Source             DF      Seq SS      Adj SS      Adj MS      F       P
delivery            5     536.469     536.469     107.294    3.80    0.067
batch(delivery)     6     169.349     169.349      28.225    6.75    0.000
Error              36     150.468     150.468       4.180
Total              47     856.285

S = 2.04442    R-Sq = 82.43%    R-Sq(adj) = 77.06%
```

```
Unusual Observations for strength
Obs    strength       Fit    SE Fit    Residual    St Resid
 31    43.3000    39.3000    1.0222     4.0000        2.26 R
 48    42.8000    38.3750    1.0222     4.4250        2.50 R
R denotes an observation with a large standardized residual.

Expected Mean Squares, using Adjusted SS
       Source            Expected Mean Square for Each Term
1      delivery          (3) + 4.0000 (2) + 8.0000 (1)
2      batch(delivery)   (3) + 4.0000 (2)
3      Error             (3)

Error Terms for Tests, using Adjusted SS
                                                   Synthesis
                                                   of Error
       Source            Error DF    Error MS    MS
1      delivery              6.00      28.225    (2)
2      batch(delivery)      36.00       4.180    (3)

Variance Components, using Adjusted SS
                          Estimated
Source                      Value
delivery                    9.884
batch(delivery)             6.011
Error                       4.180
```

Finally, Tarquin calculates the corresponding standard deviations, by taking square roots of the estimated variances, and he estimates the overall standard deviation of cube strengths (Table 12.7). The variance attributed to cubes within batches is the same as the residual error variance (4.18). The overall variance is the sum of the variances of independent components of variability. The overall standard deviation is the square root of the sum of variances.	**Estimate of standard deviation**	
	Delivery	3.14
	Batch within delivery	2.45
	Cube within batch	2.04
	Overall	4.48

Table 12.7

Tarquin uses **Stat > Descriptive Statistics** to find the mean and standard deviation of all 48 cubes: mean = 35.18; standard deviation = 4.27. The latter is slightly less than the estimate in the table, 4.48, although it is calculated from the same data. The explanation for this is that the corresponding estimator is slightly biased downwards, because we have a multi-stage sample rather than a simple random sample, so the estimate of overall standard deviation given in the table is preferable.

From the estimates in the table the most effective way to reduce the overall standard deviation would be to reduce variability between deliveries. If this estimate were zero, the estimate of overall standard deviation would be reduced to 3.19. However, Tarquin has a sample of 28-day strengths from only six deliveries of cement. The evidence against a hypothesis that the between delivery standard deviation is zero does not reach the 5% level (P = 0.067) and this indicates the lack of precision of the estimate. Also, the estimate of overall standard deviation is within the specified limits. He decides he will wait another week and analyse strengths for cubes made in the first two weeks, before raising the matter with SeaDragon.

Although it has not happened here, it is possible to obtain negative estimates for components of variance, when F-ratios are less than one. In these cases estimates should be replaced by zero.

A feature of the experimental design is that it is nicely balanced. Tarquin took four cubes from each batch, and two batches from every delivery. If, for example, one cube was lost the **Seq SS** and **Adj SS** for deliveries would differ slightly.

If you have Student it is straightforward to obtain the components of variance for a balanced design using **Stat > Basic Statistics > Store Descriptive Statistics** and enter **Strength** by delivery batch similarly to Figure 12.3. Then in **Statistics** choose **Mean, Standard deviation** and **Variance**. This gives the means and variances within batches within deliveries. (Table 12.8). The mean of the variances is the estimate of the variance of the errors.

C5	C6	C7	C8	C9	C10
ByVar1	ByVar2	Mean1	StDev1	Variance1	Count1
1	1	34.45	0.85	0.72	4
1	2	40.20	1.27	1.62	4
2	1	28.80	2.09	4.35	4
2	2	31.28	1.91	3.66	4
3	1	33.15	2.39	5.69	4
3	2	29.35	2.37	5.62	4
4	1	34.38	1.42	2.00	4
4	2	39.30	2.81	7.92	4
5	1	35.40	1.88	3.55	4
5	2	37.38	1.54	2.38	4
6	1	40.08	1.52	2.31	4
6	2	38.38	3.21	10.33	4

Table 12.8

Now repeat with **Stat > Basic Statistics > Store Descriptive Statistics** and enter **Mean1** by **ByVar1**, and in **Statistics** choose **Mean, Standard deviation** and **Variance**. The results are given in Table 12.9.

We now have the means for the six deliveries and the variances of the batch means within deliveries. Averaging these variances gives the variance of batch means, and calculating the variance of the means gives the variance of delivery means.

C11	C12	C13	C14	C15
ByVar3	Mean3	StDev3	Variance3	Count3
1	37.33	4.07	16.53	2
2	30.04	1.75	3.06	2
3	31.25	2.69	7.22	2
4	36.84	3.48	12.13	2
5	36.39	1.40	1.95	2
6	39.23	1.20	1.45	2

Table 12.9

In **Calc**, create three constants, **K1** to **K3**, with values

K1 = MEAN('Variance1')
K2 = MEAN('Variance3')
K3 = STDEV('Mean3')**2

Follow **Data > Display Data** and enter **K1-K3.** The results are in the session window:

Data Display

K1	4.17965
K2	7.05620
K3	13.4117

Now, using the notation of MINITAB for the between-deliveries variance **(1)**, between-batches within-deliveries variance **(2)**, and error (between cubes within batch within delivery) **(3)**, we solve:

$$13.4117 = (1) + (2)/2 + (3)/8$$
$$7.0562 = (2) + (3)/4$$
$$4.1796 = (3)$$

The solutions **(4.18, 6.01, 9.88)** are the same estimates as the variance components given by the general linear model.

△ Review of case 12.3

Two batches of concrete were randomly selected from batches made using each of six deliveries of cement. Four test cubes were made from randomly selected concrete from each batch. The crushing strengths of the test cubes were measured after 28 days. The mean strength of all the cube was 35.2 MPa. The estimated standard deviation of cubes strengths within a batch is 2.0 MPa. The estimated standard deviation of batch strength within a delivery of cement is 2.5 MPa. The estimated standard deviation of the means of all cubes made from different deliveries of cement is 3.1 MPa. The overall standard deviation of test cubes is estimated to be 4.5 MPa. On the basis of these estimates, the most effective way of reducing the overall standard deviation would be to reduce the variability of the cement deliveries. However, the estimates of the standard deviations are not very precise. The process will be monitored throughout the construction and improved estimates of the standard deviations will be made.

13

Two factors at several levels

In Chapter 7 we introduced factorial experiments in which the factors were at one of two levels described as high or low. One limitation of these designs is that categorical factors are restricted to two categories. Another limitation is that the effect of a factor, which can vary on a continuous scale, is assumed to be linear, at least over the range investigated in the experiment. The central composite design, discussed in Chapter 9, is a neat way around the second limitation; it enables us to investigate possible quadratic effects. In this chapter, we consider designs with two factors, which can be categorical, that are not restricted to only two categories.

13.1 Two factors at several levels

▽ Case 13.1 (SeaDragon)

This case is based on Miller and Freund (1977), and JFE 21st Century Foundation (2003) gives some useful background on coke making. The mining division of Sea Dragon manufactures coke for industrial use in iron making and in power stations, and for domestic use in appliances that burn solid fuel. SeaDragon now uses modern coke ovens of three widths: 200 mm, 400 mm and 700 mm. The company intends to install some more ovens of the same designs and wants to choose the width that provides the most efficient process.

Apart from the initial cost of the ovens, the company needs to consider the fuel costs per tonne of coke. To begin with, Mercedes Magenta has been asked to investigate the effects of oven width and flue temperature on the coking time. She decides to investigate two flue temperatures, the usual temperature of 1000°C and a lower temperature of 700°C. The coking process can be run at the lower temperature if the coal particles are preheated to around 350°C. So she plans an experiment in which one variable, temperature (**temp**), has two levels and the other variable, oven width (**width**), has three levels: a 2 × 3 experiment. She can install a preheater for the feed to the ovens. A single replicate of the full factorial experiment involves six runs. The results from her first replicate of the experiment are in the first six rows of Table 13.1.

A model for these data is

$$time(i, j) = OM + width(i) + temp(j) + error(i, j)$$

where OM is the overall mean. Mercedes fitted the model using **Stat > ANOVA > Two-Way** using only the first six rows of Table 13.1. She asked for the means to be displayed and for an additive model to be fitted, as in Figure 13.1. This automatically fits a model without an interaction. If she did not ask for an additive model, the interaction would be included. The data are also in www.greenfieldresearch.co.uk/doe/data.htm.

The results are as follows:

Two-way ANOVA: time versus width, temp

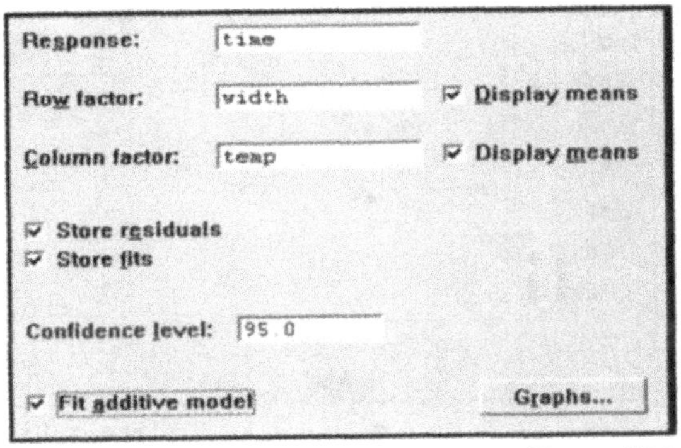

Figure 13.1

width	temp	time	replicate
200	700	7.1	1
200	1000	4.5	1
400	700	13.9	1
400	1000	10.8	1
700	700	21.6	1
700	1000	15.3	1
200	700	6.1	2
200	1000	4.7	2
400	700	13.9	2
400	1000	9.3	2
700	700	21.2	2
700	1000	14.3	2
200	700	5.5	3
200	1000	4.7	3
400	700	15.1	3
400	1000	13.7	3
700	700	22	3
700	1000	14.5	3

Table 13.1

```
Source   DF      SS       MS       F        P
width     2   160.09   80.045   39.72    0.025
temp      1    24.00   24.000   11.91    0.075
Error     2     4.03    2.015
Total     5   188.12

S = 1.420    R-Sq = 97.86%    R-Sq(adj) = 94.64%

                 Individual 95% CIs For Mean Based on
                 Pooled StDev
width    Mean    ------+---------+---------+---------+-
200      5.80    (-----*-----)
400     12.35            (------*-----)
700     18.45                     (-----*-----)
                 ------+---------+---------+---------+-
                     6.0       12.0      18.0      24.0
```

```
            Individual 95% CIs For Mean Based on
            Pooled StDev
temp    Mean    ------+---------+---------+---------+-
700     14.2                    (-----------*---------)
1000    10.2    (----------*----------)
                ------+---------+---------+---------+-
                   9.0       12.0      15.0      18.0
```

These results are much as she expected: the coking time appears to increase with oven width and decrease with temperature. However, she requires more precise estimates of the effects and will replicate the experiment. Also, the above analysis relies on a substantial assumption that there is no interaction between temperature and oven width. The so-called error is confounded with this interaction. If she replicates the design she will be able to fit the model

$$time(i, j, k) = OM + width(i) + temp(j) + width*temp\_interact(i, j) + error(i, j, k)$$

which allows for an interaction. She decides to complete three replicates of the experiment (Table 13.1 in full). Mercedes begins by plotting the data: **Graph > Scatterplot**, selecting **With Groups**. She designates **time** as the **Y variable**, and **width** as the **X variable** and **temp** as the **group variable**.

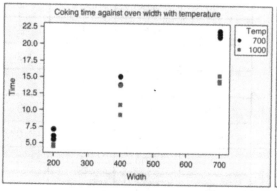

Figure 13.2a

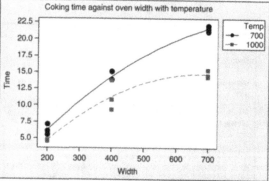

Figure 13.2b

The graph first appears as in Figure 13.2a. There is a clear interaction. The lower flue temperature is as effective as the higher temperature with the narrowest oven, but there is a substantial increase in coking time if the widest oven is used. It also seems as if the coking time starts to level off with higher width: a quadratic effect. So she repeats the graph but also clicks on **Data View > Regression > Quadratic**. The graph reappears as Figure 13.2b.

Here the continuous variable **width** has been restricted to three levels in the experiment. A model that allows for three categories, conveniently referred to as low, medium and high, is equivalent to a model that includes width and width squared. Mercedes chooses to use categories because the manufacturer sells ovens of only these three widths. She includes the interaction in the model for analysis of variance by unchecking the box to **Fit additive model**. The results are as follows:

Two-way ANOVA: time versus width, temp

Source	DF	SS	MS	F	P
width	2	489.074	244.537	222.64	0.000
temp	1	66.509	66.509	60.55	0.000

Interaction	2	22.548	11.274	10.26	0.003
Error	12	13.180	1.098		
Total	17	591.311			

Because there is an important interaction it would be misleading to give estimates for the main effects only. The ANOVA general linear model does provide estimated coefficients for all the terms, as would regression analysis. Mercedes follows **Stat > ANOVA > General Linear Model**. See Figure 13.3. Note that in the **Model** frame she entered **width temp width*temp**. The inclusion of **width*temp** tells MINITAB to create the interaction term for the analysis. If she had used regression analysis instead, she would first have had to calculate two extra columns of values in the worksheet for the interactions. Note also that she asked for main effects and interaction plots which are shown in Figure 13.4.

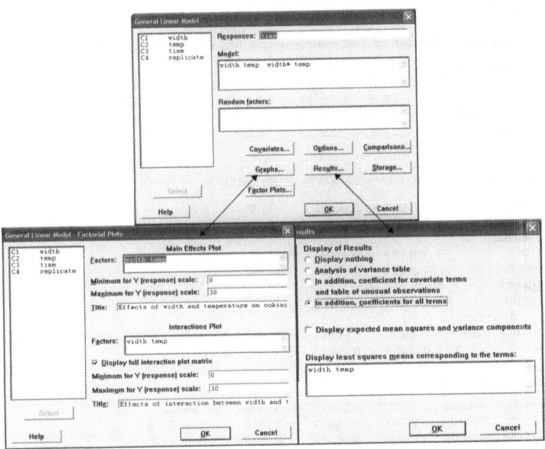

Figure 13.3

General Linear Model: time versus width, temp

Factor	Type	Levels	Values
width	fixed	3	200, 400, 700
temp	fixed	2	700, 1000

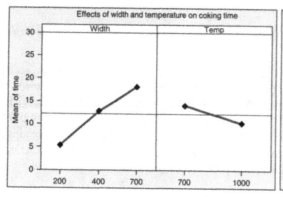

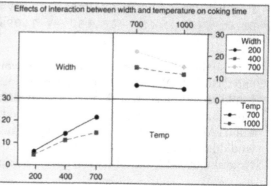

Figure 13.4a

Figure 13.4b

Analysis of Variance for time, using Adjusted SS for Tests

Source	DF	Seq SS	Adj SS	Adj MS	F	P
width	2	489.07	489.07	244.54	222.64	0.000
temp	1	66.51	66.51	66.51	60.55	0.000
width*temp	2	22.55	22.55	11.27	10.26	0.003
Error	12	13.18	13.18	1.10		
Total	17	591.31				

S = 1.04801   R-Sq = 97.77%   R-Sq(adj) = 96.84%

Term		Coef	SE Coef	T	P
Constant		12.1222	0.2470	49.07	0.000
width					
200		−6.6889	0.3493	−19.15	0.000
400		0.6611	0.3493	1.89	0.083
temp					
700		1.9222	0.2470	7.78	0.000
width*temp					
200	700	−1.1222	0.3493	−3.21	0.007
400	700	−0.4056	0.3493	−1.16	0.268

Unusual Observations for time

Obs	time	Fit	SE Fit	Residual	St Resid
10	9.3000	11.2667	0.6051	−1.9667	−2.30 R
16	13.7000	11.2667	0.6051	2.4333	2.84 R

R denotes an observation with a large standardized residual.

Least Squares Means for time

width	Mean	SE Mean
200	5.433	0.4278
400	12.783	0.4278
700	18.150	0.4278

```
temp
  700  14.044  0.3493
 1000  10.200  0.3493
```

She also tried including a replicate term – that is, she used the replicate number as a variable value to see if there was any change in process performance over time. This was not statistically significant.

13.1.1 Analysis using regression

There are two approaches to regression analysis of these data. One is to use the raw values: width in millimetres and temperature in degrees Celsius. The other is to create coded variables to represent the levels of factors.

Using raw values, the first step is to create a new variable, representing the interaction **width*temp**, using **Calc**. Two possible problems can arise with this. One is that with large numbers any derived variables, especially interactions and squares, will lead to excessively large numbers which in turn can lead to numerical errors. For example, the minimum value of width*temp is 140000 and the maximum value 700000. Perhaps these values are not too large, but sometimes derived values can be much larger. This problem can be solved by scaling, for example by changing millimetres to metres and degrees to hundreds of degrees so that we would have a minimum of 1.4 and a maximum of 7.0. The second problem is that derived values are commonly highly correlated with main variables and this can lead to analytic problems, especially in ranking of regressor variables according to their significances and hence to selection of subsets of regressor variables, and more especially if the usual stepwise procedure with partial *F*-tests is used. This correlation problem can be much reduced by centring the data values, or approximately centring them by subtracting a value close to mid-point such as 400 for coke oven width.

Mercedes first scaled the data but, before centring, she looked at the correlations between the main variables and their interaction, using **Stat > Basic Statistics > Correlation**. The results were as follows:

```
Correlations: width, temp, width*temp

            width   temp
temp        0.000
width*temp  0.925   0.344
```

There is no correlation between temperature and width. This follows from the balance of the experimental design. But there is a very high correlation (0.925) between width and the interaction. The mid-range of scaled width is (0.2 + 0.7)/2 = 0.45, and of temperature it is (10 + 7)/2 = 8.5, so Mercedes computed three new variables: **widC = width – 0.45**, **temC = temp – 8.5** and **widC*temC**. The high correlation between width and the interaction was reduced to zero:

```
Correlations: widC, temC, widC*temC

            widC    temC
temC        0.000
widC*temC   0.000  −0.081
```

Now, regression analysis gave:

Regression Analysis: time versus widC, temC, widC*temC

The regression equation is
time = 12.5 + 24.8 widC − 1.34 temC − 3.59 widC*temC

```
Predictor        Coef  SE Coef       T      P
Constant      12.5362   0.3678   34.08  0.000
widC           24.838    1.784   13.92  0.000
temC          -1.3414   0.2452   -5.47  0.000
widC*temC      -3.594    1.189   -3.02  0.009
```

S = 1.55532   R-Sq = 94.3%   R-Sq(adj) = 93.0%

Analysis of Variance

```
Source          DF       SS      MS       F      P
Regression       3   557.44  185.81   76.81  0.000
Residual Error  14    33.87    2.42
Total           17   591.31
```

```
Source      DF  Seq SS
widC         1  468.85
temC         1   66.51
widC*temC    1   22.08
```

whereas with scaled but uncentred data, the results were:

Regression Analysis: time versus width, temp, width*temp

The regression equation is
time = −0.98 + 55.4 width + 0.276 temp − 3.59 width*temp

```
Predictor        Coef  SE Coef       T      P
Constant      -0.985    4.923   -0.20  0.844
width          55.38    10.27    5.39  0.000
temp          0.2757   0.5704    0.48  0.636
width*temp    -3.594    1.189   -3.02  0.009
```

S = 1.55532   R-Sq = 94.3%   R-Sq(adj) = 93.0%

Analysis of Variance

```
Source          DF       SS      MS       F      P
Regression       3   557.44  185.81   76.81  0.000
Residual Error  14    33.87    2.42
Total           17   591.31
```

Clearly, with the regression analysis from the centred data, you must remember to rescale the equation before you can apply it. The analysis with the uncentred data gives just the same overall fit, but the coefficient for temperature has P = 0.636 which might mislead you into excluding temperature.

It was clear from Figure 13.4b that there was curvature in the relationship between time and width. But, further, that curvature was different for different levels of temperature. This suggested to Mercedes that she should add terms to the regression analysis to represent width^2 and temp*width^2. Now, consider values of the latter. With a width of 700 mm and a temperature of 1000°C we should have a value of 490,000,000. The calculations involved in the regression analysis include the sum of squares of such values. With 18 observations that gives a figure of 4,321,800,000,000,000,000. So Mercedes shared our concern about possible numerical problems. She used the scaled and centred values.

Regression Analysis: time versus widthS, tempS, . . .

```
The regression equation is
time = 12.8 + 29.2 widthS - 1.01 tempS - 3.15 widS*temS - 3.81
temS*widS^2 - 37.7 widS^2
```

Predictor	Coef	SE Coef	T	P
Constant	12.7833	0.4278	29.88	0.000
widthS	29.206	1.575	18.54	0.000
tempS	-1.0111	0.2852	-3.54	0.004
widS*temS	-3.152	1.050	-3.00	0.011
temS*widS^2	-3.815	5.861	-0.65	0.527
widS^2	-37.722	8.791	-4.29	0.001

The regression analysis showed that the square of the width was highly significant in predicting coking time but, with the available data, the interaction temp*width^2 was not.

Following the procedure for obtaining a suitable set of indicator variables discussed in Chapter 8, Mercedes creates seven new columns that she heads widthL, widthM, widthH, tempL, tempH, wM*tH and wH*tH. She applies Calc > Make Indicator Variables to create dummy codes for width in widthL, widthM and width, followed by dummy codes for temp in tempL and tempH. She deletes the columns of widthL and tempL and then uses Calc to make interaction values in wM*tH and wH*tH. The results are given in Table 13.2.

Regression analysis of time versus widthM to wH*tH gives:

Regression Analysis: time versus widthM, widthH, tempH, wM*tH, wH*tH

```
The regression equation is
time = 6.23 + 8.07 widthM + 15.4 widthH - 1.60 tempH - 1.43
wM*tH - 5.30 wH*tH
```

Predictor	Coef	SE Coef	T	P
Constant	6.2333	0.6051	10.30	0.000
widthM	8.0667	0.8557	9.43	0.000
widthH	15.3667	0.8557	17.96	0.000

Width	Temp	Time	widthM	widthH	tempH	wM*tH	wH*tH
200	700	7.1	0	0	0	0	0
200	1000	4.5	0	0	1	0	0
400	700	13.9	1	0	0	0	0
400	1000	10.8	1	0	1	1	0
700	700	21.6	0	1	0	0	0
700	1000	15.3	0	1	1	0	1
200	700	6.1	0	0	0	0	0
200	1000	4.7	0	0	1	0	0
400	700	13.9	1	0	0	0	0
400	1000	9.3	1	0	1	1	0
700	700	21.2	0	1	0	0	0
700	1000	14.3	0	1	1	0	1
200	700	5.5	0	0	0	0	0
200	1000	4.7	0	0	1	0	0
400	700	15.1	1	0	0	0	0
400	1000	13.7	1	0	1	1	0
700	700	22	0	1	0	0	0
700	1000	14.5	0	1	1	0	1

Table 13.2

```
tempH        -1.6000    0.8557   -1.87   0.086
wM*tH        -1.433     1.210    -1.18   0.259
wH*tH        -5.300     1.210    -4.38   0.001

S = 1.04801   R-Sq = 97.8%   R-Sq(adj) = 96.8%
Analysis of Variance

Source           DF      SS      MS       F       P
Regression        5   578.13  115.63   105.27   0.000
Residual Error   12    13.18    1.10
Total            17   591.31
```

A clear advantage of this analysis with *indicator variables* is that Mercedes is able to read predicted values from the regression equation using no more than simple addition. If width and temperature are both at their low levels, the predicted coking time is 6.23. If width is at its low level and temp at its high level, predicted coking time is $6.23 - 1.6 = 4.63$. If they are both at their high levels, predicted coking time is $6.23 + 5.4 - 1.6 - 5.3 = 14.73$. However it is expressed, our final model includes all the categories of both variables, and their interactions, and is fitting a separate mean to each of the six combinations.

△ Review of Case 13.1

Ovens are available from the manufacture in three widths, 200 mm, 400 mm and 700 mm, and the coking can be carried out at the usual temperature of 1000°C, or at a lower temperature of 700°C if the coal particles are preheated. Intermediate values of oven width and coking temperature are not practical, and the coking time depends on the specific combination of width and temperature. The estimated mean coking times for the six combinations are in the Table 13.3. The standard errors for these means are approximately 0.6. These mean times will be used in costing various options for the coke making. The precision of the estimates is adequate given the uncertainty about other economic data needed to make the decision about the new plant.

	Temperature	
Width	700°C	1000°C
200 mm	6.2	4.6
400 mm	14.3	11.3
700 mm	21.6	14.7

Table 13.3

13.2 Fixed effects and random effects

▽ Case 13.2 (AgroPharm)

AgroPharm has formulated a fertiliser to increase the yield from home-grown tomato plants. Chantrea Cyan has been asked to compare the effect of no (**N**), low (**L**) and medium (**M**) levels of fertiliser on Oregon spring tomato plants from a random sample of nurseries.

Chantrea obtained 12 tomato plants from each of three nurseries, randomly selected from a list of nurseries in the region. She grew the plants in small tubs near the centre of a large greenhouse, and randomised the positions of the 36 plants over a 6 × 6 square grid. She then randomised the allocation of **N**, **L** and **M** levels of fertiliser subject to four plants from each nursery receiving each level. The yields in grams are in Table 13.4. The data are also in www.greenfieldresearch.co.uk/doe/data.htm. This is a **3 x 3** factorial experiment in which both the control variables are qualitative factors.

Chantrea starts her analysis by plotting the data. This calls for care in dealing with the qualitative factors. How can she plot the quantitative response (**yield**) against these factors? Even if she can plot the yield values against the symbols **M**, **L** and **N**, there is the added problem that MINITAB, by default, will select text values in alphabetic order, **L M N**, which does not reflect the ordinal quantities of fertiliser.

Nursery	Fertiliser	Yield	Nursery	Fertiliser	Yield
nurs1	M	1291	nurs2	L	1059
nurs1	M	1545	nurs2	L	1169
nurs1	M	723	nurs2	N	235
nurs1	M	1774	nurs2	N	683
nurs1	L	338	nurs2	N	577
nurs1	L	414	nurs2	N	429
nurs1	L	814	nurs3	M	1613
nurs1	L	764	nurs3	M	1367
nurs1	N	214	nurs3	M	1098
nurs1	N	690	nurs3	M	1684
nurs1	N	826	nurs3	L	1733
nurs1	N	592	nurs3	L	1174
nurs2	M	1524	nurs3	L	1505
nurs2	M	1862	nurs3	L	1419
nurs2	M	1602	nurs3	N	972
nurs2	M	1255	nurs3	N	1027
nurs2	L	1364	nurs3	N	819
nurs2	L	1231	nurs3	N	607

Table 13.4

MINITAB has the answer. Select the column for fertiliser (C2) and follow **Editor > Column > Value Order**. Check that you have selected the correct column (as in Figure 13.5). Select **User-specified order** and then, in the right-hand box, define the order, one value per line, as **N L M**.

Chantrea uses **Graph > Individual Value Plot**, selecting **With Groups** under **One Y**, and enters **yield** as the graph variable and **fertiliser** and **nursery** (in that order) as categorical variables, as in Figure 13.6. This produces the graph of Figure 13.7.

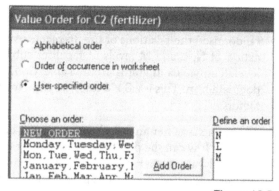

Figure 13.5

A strange feature of Figure 13.7 is that the columns of dots are not in straight lines. This is because MINITAB has, by default, added jitter to the display. Chantrea was able to remove the jitter. At the same time, she was able to change the dots to distinct shapes and colours. She clicked on a dot in the chart (any will do) and then right-clicked to get the drop-down menu as in Figure 13.8. She chose **Edit Individual Symbols** and, as in Figure 13.9, she removed the check from **Add jitter**. She also clicked on the **Groups** tab and chose **nursery** as **Categorical variable for attribute assignment**. Finally, the graph she wants appears as in Figure 13.10. Note that MINITAB automatically added a legend for the styles of dots for nursery.

The increased use of fertiliser appears to increase the yield, though the benefit of going from low to medium looks rather less than obtained by using low level rather than none. It seems that at the low level there is a strong difference between the yields of the three nurseries, although this is not

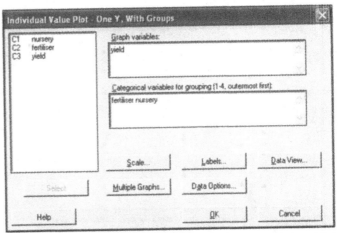

Figure 13.6

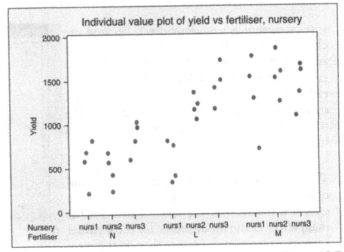

Figure 13.7

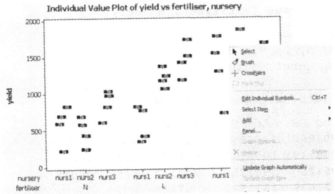

Figure 13.8

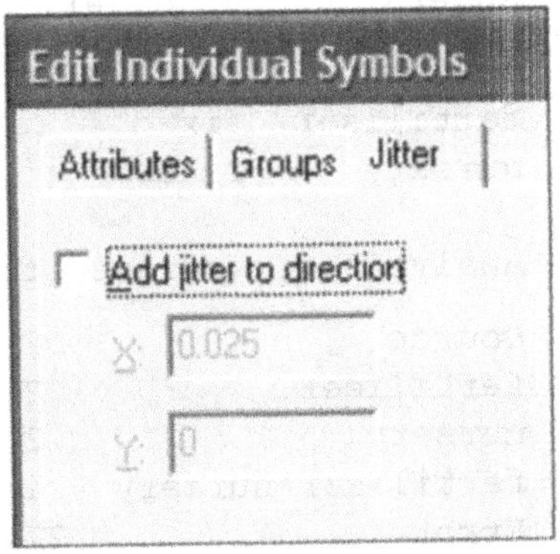

Figure 13.9

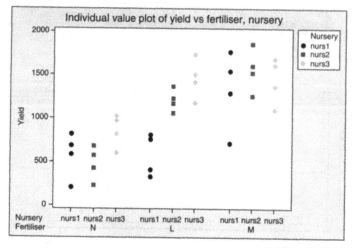

Figure 13.10

so at the other two levels of fertiliser. This suggests that there may be interactions between nurseries and amount of fertiliser.

Chantrea uses **ANOVA > General Linear Model** for her analysis. She wants to treat the three nurseries as a random effect because they are a random sample of nurseries in the region, rather than being of any special interest in themselves. The alternative view, that nurseries are a fixed effect, would be taken by a garden centre which bought tomato plants from these three nurseries. She takes care to

specify **nursery** as a random effect, because the choice between fixed and random affects the analysis. It is easier to follow the output if you give categories of variables suitable text names rather than numbers. Note (Figure 13.11) that **nursery** must also be entered as a term in the model. She also enters **nursery* fertiliser** as an interaction in the model.

The consequence of declaring **nursery** as a random effect is that the *F*-ratio for the main effects is the ratio of the mean square for the main effect to the mean square for the fertiliser by nursery interaction, rather than to the error

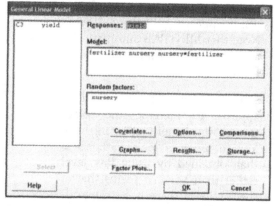

Figure 13.11

mean square. The mathematical justification follows from considering the expected values of the mean squares, which are displayed by MINITAB and derived in more advanced textbooks.

General Linear Model: yield versus fertiliser, nursery

```
Factor       Type     Levels   Values
fertiliser   fixed        3    N, L, M
nursery      random       3    nurs1, nurs2, nurs3
```

Analysis of Variance for yield, using Adjusted SS for Tests

Source	DF	Seq SS	Adj SS	Adj MS	F	P
fertiliser	2	3906560	3906560	1953280	8.13	0.039
nursery	2	1068719	1068719	534360	2.22	0.224
fertiliser*nursery	4	961250	961250	240312	3.56	0.019
Error	27	1821639	1821639	67468		
Total	35	7758168				

Unusual Observations for yield

```
Obs   yield      Fit    SE Fit  Residual  St Resid
  3  723.00  1333.25  129.87   -610.25    -2.71 R
```

R denotes an observation with a large standardized residual.

Expected Mean Squares, using Adjusted SS

```
    Source               Expected Mean Square for Each Term
1   fertiliser           (4) + 4.0000 (3) + Q[1]
2   nursery              (4) + 4.0000 (3) + 12.0000 (2)
3   fertiliser*nursery   (4) + 4.0000 (3)
4   Error                (4)
```

Error Terms for Tests, using Adjusted SS

```
    Source               Error DF   Error MS  Synthesis of Error MS
1   fertiliser             4.00      240312   (3)
2   nursery                4.00      240312   (3)
3   fertiliser*nursery    27.00       67468   (4)
```

Variance Components, using Adjusted SS

```
                         Estimated
Source                     Value
nursery                    24504
fertiliser*nursery         43211
Error                      67468
```

A heuristic explanation will help. Compare the analysis with **nursery** considered as a fixed effect.

General Linear Model: yield versus fertiliser, nursery

```
Factor       Type    Levels  Values
fertiliser   fixed      3    N, L, M
nursery      fixed      3    nurs1, nurs2, nurs3
```

Analysis of Variance for yield, using Adjusted SS for Tests

```
Source               DF    Seq SS    Adj SS    Adj MS      F      P
fertiliser            2   3906560   3906560   1953280  28.95  0.000
nursery               2   1068719   1068719    534360   7.92  0.002
fertiliser*nursery    4    961250    961250    240312   3.56  0.019
Error                27   1821639   1821639     67468
Total                35   7758168
```

S = 259.746   R-Sq = 76.52%   R-Sq(adj) = 69.56%

```
Term                   Coef   SE Coef     T      P
Constant            1055.36     43.29  24.38  0.000
fertiliser
N                   -416.11     61.22  -6.80  0.000
L                     26.64     61.22   0.44  0.667
nursery
```

```
nurs1                   -223.28    61.22   -3.65   0.001
nurs2                     27.14    61.22    0.44   0.661
fertiliser*nursery
N              nurs1      164.53    86.58    1.90   0.068
N              nurs2     -185.39    86.58   -2.14   0.041
L              nurs1     -276.22    86.58   -3.19   0.004
L              nurs2       96.61    86.58    1.12   0.274
```

Unusual Observations for yield

```
Obs    yield      Fit  SE Fit  Residual  St Resid
  3   723.00  1333.25  129.87   -610.25     -2.71 R
```

R denotes an observation with a large standardized residual.

Expected Mean Squares, using Adjusted SS

```
                          Expected Mean
                          Square for
   Source                 Each Term
1  fertiliser             (4) + Q[1, 3]
2  nursery                (4) + Q[2, 3]
3  fertiliser*nursery     (4) + Q[3]
4  Error                  (4)
```

Error Terms for Tests, using Adjusted SS

```
                                              Synthesis
                                              of Error
   Source                Error DF  Error MS   MS
1  fertiliser               27.00     67468   (4)
2  nursery                  27.00     67468   (4)
3  fertiliser*nursery       27.00     67468   (4)
```

Variance Components, using Adjusted SS

```
          Estimated
Source      Value
Error       67468
```

The difference is in the *F*-values and the associated P-values. In the fixed effects case we are interested in the three specific nurseries. There is strong evidence of a difference between the nurseries, and some evidence of an interaction between nurseries and fertilisers which may be due to nurseries applying different fertiliser treatments to the plants before selling them to AgroPharm. There is very strong evidence that the use of fertiliser improves the yield. If we consider **nursery** as a random effect, the interaction between **nursery** and **fertiliser** levels should be considered as part of the error against which we judge the effects of fertiliser. In this case the main effects of fertiliser remain highly statistically significant. The estimated values of effects are the same in both cases, but if nurseries are considered random it is not appropriate to estimate the interaction effects, and **Results > Coefficients** for all terms should not be selected.

All we need is **Stat > Basic Statistics > Display Descriptive Statistics**, with **yield** in the **Variables** box and **fertiliser** in the **By variables** box:

Descriptive Statistics: yield

Variable	fertiliser	N	Mean	SE Mean	TrMean	StDev	Minimum	Q1
yield	N	12	639.3	74.3	643.0	257.3	214.0	466.0
	L	12	1082	123	1091	427	338	777
	M	12	1444.8	91.9	1475.3	318.5	723.0	1264.0

Variable	fertiliser	Median	Q3	Maximum
yield	N	645.0	824.3	1027.0
	L	1172	1405	1733
	M	1534.5	1666.3	1862.0

△ Review of Case 13.2

The yield of Oregon Spring tomato plants can be increased by the addition of fertiliser. The mean yield with no fertiliser can be increased by an estimated factor of 1.69 if a low level of fertiliser is applied and by an estimated factor of 2.26 if a medium level of fertiliser is applied.

▽ Case 13.3 (UoE)

The *Driver's Handbook* issued by the Government of South Australia states: 'It is believed that the presence of friends in the vehicle places unnecessary pressure on young drivers to show off their driving skills which often can exceed their true ability.'

Ingrid Indigo decided to research this issue in Erewhon. She asked for volunteers, from male students who had held a full driving licence for at least one year, to take part in a driving experiment. There were many volunteers and Ingrid sorted them into four age groups: turned 18 within the year before the date the experiment begins, and similarly for 19, 20 and 21. She then randomly chose four of the volunteers within each age group to be drivers in the experiment. She would ask each driver to drive a specified route on four occasions with: no passenger, one student passenger, two student passengers and three student passengers. She had four cars, all the same model, from a car hire company. She asked the drivers to drive a hire car along the test route twice before the experiment began. The aim of this was to make drivers familiar with the test route. Nevertheless, she thought that increasing familiarity with the test route might reduce driving times. So, she recorded the order of driving the route and included it as a variable in the analysis.

She asked each driver to muster a group of three friends, of the same age, to be passengers. She randomly chose one from each passenger group of three to be a single passenger, and two from each group of three to be a pair of passengers. Also the number of passengers was randomised, relative to the order the route was driven, subject to a restriction. The restriction was that, for each age group, each number of passengers would be driven first by one driver, second by another driver, third by a third driver, and fourth by the remaining driver. This can be achieved by using a 4 × 4 *Latin square*, in which each of four Latin letters appears once in each row and once in each column

(Table 13.5). Within each age group, the letters **A**, **B**, **C**, and **D** represent the four drivers, the rows represent the number of passengers, and the columns represent the order for that route. Much has been written about Latin squares, and Su Doku puzzles are based on these ideas. However, it is easy to construct a Latin square of any size, m say. Write down the first m letters of the Latin alphabet in the first row, for the second row, shift the first row by one step to the right and bring the final letter to the front, and so on for m rows. In our case of four drivers we have

ABCD

DABC

CDAB

BCDA

Then we break up the pattern by randomising the order of the rows and having done that randomising the order of the columns. This can be done separately for each age group. For example, for age 18 our random order of rows might be that what are now rows {1234} become rows {4132} and for columns it might be that what are now columns {1234} become columns {3214}. Then we would have the following allocation asia Table 13.5:

Passenger	Occasion			
	1	2	3	4
0	B	A	D	C
1	D	C	B	A
2	A	D	B	C
3	C	B	A	D

Table 13.5

Now randomly allocate driver to letter. Driver **A** drives with two passengers on the first occasion, no passengers on the second occasion, three passengers on the third occasion, and one passenger on the fourth occasion.

Ingrid ran the experiment over four weeks during the summer. There are 16 drivers, call them 1, ... ,16, four in each of four age groups (18, 19, 20, 21). The drivers are nested within age groups because the first mentioned driver in age group 18 has nothing in common with the first mentioned driver in the other age groups. The average effect of drivers 1 to 4 is confounded with the age 18 effect; similarly for drivers 5 up to 8 with the age of 19 effect, 9 to 12 with the age of 20 effect, and 13 up to 16 with the age 21 effect. The analysis takes care of this when we nest the drivers within the ages.

There are also four levels of passenger numbers (0, 1, 2, 3). Although these passengers are different for each driver, we do not nest them because it is the number of passengers that is important, and the predictor variable is the number of passengers. Differences in goading from different groups of passengers and differences in drivers' desire to impress are part of the random variation implicit in declaring drivers to be random effects.

Four drivers drove the test route between 7 pm and 8 pm on Monday, Tuesday, Wednesday and Thursday evenings. The usual time to drive the test route was slightly less than one hour. The drivers' start times were staggered slightly to avoid any possibility of the test developing into a road race. The university employed a driving instructor to drive the route at the same time as the student drivers, in an attempt to allow for variations in traffic and other driving conditions. The response was the ratio of the time taken by a student driver to drive the route to the time taken by the driving instructor. The results are in Table 13.6 and at www.greenfieldresearch. co.uk/doe/data.htm. Ingrid randomised the order of runs as much as was possible, allowing for the drivers' availability. MINITAB makes the analysis easy with **Stat > ANOVA > General Linear Model** (see Figure 13.12).

Age	Driver	Passengers	Order	Ratio	Age	Driver	Passengers	Order	Ratio
18	1	0	1	1.00	18	4	2	1	0.92
18	1	1	2	0.99	18	4	3	2	0.96
18	1	2	3	0.94	19	5	0	1	1.01
18	1	3	4	0.92	19	5	1	2	0.99
18	2	0	2	0.97	19	5	2	3	0.98
18	2	1	1	0.92	19	5	3	4	0.92
18	2	2	4	0.90	19	6	0	2	1.05
18	2	3	3	0.90	19	6	1	3	0.99
18	3	0	3	0.99	19	6	2	4	0.96
18	3	1	4	0.98	19	6	3	1	0.93
18	3	2	2	0.89	19	7	0	3	1.02
18	3	3	1	0.94	19	7	1	4	0.96
18	4	0	4	0.98	19	7	2	1	0.96
18	4	1	3	0.93	19	7	3	2	0.94
19	8	0	4	1.02	20	12	2	2	0.95
19	8	1	1	0.97	20	12	3	1	0.94
19	8	2	2	0.94	21	13	0	1	0.97
19	8	3	3	0.95	21	13	1	2	0.95
20	9	0	1	1.03	21	13	2	3	0.93
20	9	1	2	0.97	21	13	3	4	0.90

Table 13.6

Age	Driver	Passengers	Order	Ratio	Age	Driver	Passengers	Order	Ratio
20	9	2	3	0.92	21	14	0	2	1.01
20	9	3	4	0.97	21	14	1	1	0.96
20	10	0	2	1.02	21	14	2	4	0.95
20	10	1	4	0.96	21	14	3	3	0.91
20	10	2	1	0.98	21	15	0	3	1.00
20	10	3	3	0.97	21	15	1	4	0.99
20	11	0	3	1.05	21	15	2	1	0.96
20	11	1	1	0.97	21	15	3	2	0.91
20	11	2	4	0.96	21	16	0	4	0.97
20	11	3	2	0.94	21	16	1	3	0.96
20	12	0	4	0.99	21	16	2	2	0.93
20	12	1	3	0.95	21	16	3	1	0.90

Table 13.6 (Continued)

Ingrid is investigating the effects, on the ratio of times, of the number of passengers and the driver's age group. She enters **ratio** as the response. The model comprises **passengers** and **age** as main effects, the **order** as a fixed effect, the interaction **passengers*age**, and **driver** nested within **age**. She is allowing for the order of driving with the different number of passengers and the individual driver effects. Ingrid also designates the **driver** effect as a random factor because she is considering the drivers as a random sample of drivers from a wide population. She clicks the **Results** tab and asks for the least squares means of the four factors (**age driver passengers order**) to be displayed.

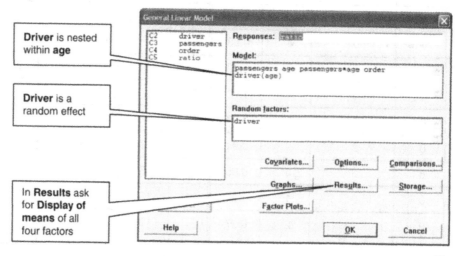

Figure 13.12

General Linear Model: ratio versus passengers, age, order, driver

Factor	Type	Levels	Values
passengers	fixed	4	0, 1, 2, 3
age	fixed	4	18, 19, 20, 21
order	fixed	4	1, 2, 3, 4
driver(age)	random	16	1, 2, 3, 4, 5, 6, 7, 8, 9, 10, 11, 12, 13, 14, 15, 16

Analysis of Variance for ratio, using Adjusted SS for Tests

Source	DF	Seq SS	Adj SS	Adj MS	F	P
passengers	3	0.0512422	0.0512422	0.0170807	52.49	0.000
age	3	0.0109297	0.0109297	0.0036432	5.86	0.011
passengers*age	9	0.0060641	0.0060641	0.0006738	2.07	0.062
order	3	0.0002297	0.0002297	0.0000766	0.24	0.871
driver(age)	12	0.0074563	0.0074563	0.0006214	1.91	0.070
Error	33	0.0107391	0.0107391	0.0003254		
Total	63	0.0866609				

$S = 0.0180396$ R-Sq = 87.61%   R-Sq(adj) = 76.34%

Unusual Observations for ratio

Obs	ratio	Fit	SE Fit	Residual	St Resid	
11	0.89000	0.91922	0.01256	−0.02922	−2.26	R
14	0.93000	0.95797	0.01256	−0.02797	−2.16	R
35	0.92000	0.95297	0.01256	−0.03297	−2.55	R

R denotes an observation with a large standardized residual.

Least Squares Means for ratio

age	Mean
18	0.9456
19	0.9744
20	0.9731
21	0.9500
passengers	
0	1.0050
1	0.9650
2	0.9419
3	0.9313

The interaction effect is not significant at the 5% level so we shall ignore it and discuss the main effects. There is strong evidence that the number of passengers has an effect on the driving time. It seems that even one passenger results in a reduction of the average ratio of time taken to time taken by the driving instructor, and two or three passengers reduce the mean further. The age effect is also significant. The 18- and 21-year-olds seem to drive somewhat faster than the others. This may be due to false confidence for the former, and experience for the latter. It is of less concern than the passenger effect.

General linear model analysis is not available in the student version of MINITAB. You could, however, discover the important relationships by plotting the data. Follow **Graph > Scatterplot > With Connect Line.** Enter **ratio** as the **Y** variable in each of four graphs and each of the factors as the **X** variable. Click on **Multiple Graphs** and choose **Show Pairs of Graph Variables In separate panels of the same graph.** The result is presented in Figure 13.13. This illustrates and supports the inferences from the general linear model analysis in the Pro version.

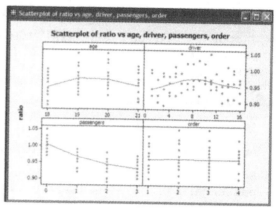

Figure 13.13

What is the corresponding population in this example? It is the population of male volunteers, who have held driving licences for at least a year and are aged between 18 and 21, at UoE. Perhaps volunteers are more extrovert than average students, and UoE students have had more years of education than average young men in the state. If we generalise the findings, we are making a subjective decision that the peer pressure effect does not interact with level of education or, if we generalise to women, the sex of the driver. Typically, the importance of the results is in the generalisation, and it may be prudent to follow potentially important effects with wider-ranging experiments. Also, the results of the experiment are only useful if they lead to some sort of action, such as the warning in the South Australia *Driver's Handbook.*

△ Review of Case 13.2

A study at the UoE estimates that the time taken by young men aged 18–21 to drive a set road route decreases from the time taken by an experienced professional driver with the number of friends in the car. Relative to the experienced driver time of 1.00, the estimated mean times taken with 0, 1, 2, 3 friends in the car are, 1.00, 0.96, 0.94 and 0.93, respectively. Such reductions are a risk to road safety.

13.3 The difference between a completely randomised design with two factors and a randomised block design

Paul Plum is the senior engineer in SeaDragon's bio-engineering division. One of the current projects, in collaboration with the University of Erewhon's medical school, is the design of a prosthetic heart valve. They wish to compare four different designs of heart valves in a test rig that is designed to simulate a human heart. Paul had discussed the experiment with a statistics student, Ishmael Ivory. Ishmael had chosen to help with this research in the medical school for his final year project. They had agreed that it would be sensible to test the valves at six pulse rates between 20 and 220 beats per minute, but this had been followed by a disagreement. Ishmael recommended a completely randomised design that would require six valves of each type. The six valves of each type would be randomly assigned to one of the pulse rates, and the 24 runs would be performed in a random order. Paul said the cost of producing 24 prototype heart valves would be prohibitive. That evening he had consulted an introductory statistics book and read about a randomised block design. The next day Paul rang Ishmael and suggested that they could treat the design of heart valve as a block, reducing the number of valves required from 24 to four. Ishmael had been quick to point out the fallacy. With only one valve of each type it would be impossible to tell whether an apparently superior valve was a better design or simply a particularly good example of its type. Paul had conceded that valves of the same type would vary because of slight differences in machining the minute components. After the phone call, Ishmael continued to think about the experiment. After all, if they restricted the experiment to two pulse rates, two valves of each type would suffice. Why shouldn't they try all six pulse rates with just two valves of each type? A similar example had not yet been covered in his course on the design of experiments, but he referred to the recommended textbook and came across a split-plot design. This is a suitable design for the experiment and we discuss it in the next chapter.

We do sometimes have to compromise, but we should at least be aware of the implications if we do. In an idealised Case 13.1, Mercedes will have used a different randomly selected oven of the given width for every run. But if the company has only one oven of each width, or is prepared to adapt only one for preheating the coal particles, this is not feasible. Strictly, our results would then be applicable only to those three ovens. But we might be prepared to assume that the variability between ovens of the same size and design is negligible compared with the width and temperature effect. However, if we use the result to decide on the width of new ovens we assume that the width effect will be the same as in the present ones. This seems reasonable if the design is unchanged, but could be very misleading if the designs have been improved. Even randomised clinical trials are not free of such limitations, because the patients in the trial may not be a plausible approximation to a random sample from people who may seek the treatments. We may rely on medical advice that the differences between treatments are likely to be similar in different populations. We can also look for any interactions between treatments and the characteristics of patients that we do have in the trial.

14

Crossed and nested factors, and split-plot designs

14.1 Crossed and nested factors

We have already seen one case with nested factors, the distracted young drivers in Case 13.3.

▽ Case 14.1 (UoE)

Nathaniel Navy, of the Department of Social Studies at UoE, is working on a project with the aim of improving access into taxis for people in wheel chairs. Two London style taxis have been modified in different ways but they are located in different cities, Aredon (**A**) and Belhwon (**B**). As well as comparing these modifications, three different designs for securing belts for the chair and passenger are to be assessed (**belts 1**, **2** and **3**). Each design of securing belt can be used in either taxi. Four volunteer passengers have been recruited in **A** and each will try all three designs of securing belt in the **A** modified taxi. A different group of four volunteers will try all three designs in the **B** modified taxi. Each volunteer passenger will give an accessibility score for getting into the taxi for each design of securing belt. The score is calculated on a scale from 0 to 100, with high values indicating good access. Nathaniel attempts to make these scores comparable by constructing a detailed questionnaire. He will help the volunteers complete the questionnaire after they have tried each belt design (a run of the experiment). In each city, the volunteers will try the belt designs in random order, subject to each design being used at least once in each position of the sequence.

If we consider the design of the experiment we see that taxi modification (**A**, **B**) is crossed with belt design (**1**, **2**, **3**). However, the volunteer passengers are not the same for the **A** and **B** modifications. Volunteers are nested within the modification, and this is shown schematically in Table 14.1.

Nathaniel's results are given in Table 14.2 and in www.greenfieldresearch.co.uk/doe/data.htm. The volunteer passengers are **V1** to **V4** in **A**, and **V5** to **V8** in **B**. His first step in the analysis was to plot the

Taxi A			Taxi B		
Belt 1	Belt 2	Belt 3	Belt 1	Belt 2	Belt 3
V1	V1	V1	V5	V5	V5
V2	V2	V2	V6	V6	V6
V3	V3	V3	V7	V7	V7
V4	V4	V4	V8	V8	V8

Table 14.1

Taxi	Belt	Volunteer	Access
A	1	V1	34
A	1	V2	44
A	1	V3	47
A	1	V4	32
A	2	V1	48
A	2	V2	50
A	2	V3	55
A	2	V4	45
A	3	V1	42
A	3	V2	44
A	3	V3	49
A	3	V4	39
B	1	V5	58
B	1	V6	49
B	1	V7	61
B	1	V8	43
B	2	V5	63
B	2	V6	57
B	2	V7	65
B	2	V8	61
B	3	V5	60
B	3	V6	39
B	3	V7	69
B	3	V8	61

Table 14.2

data using the **Individual Value Plot** procedure described in Case 13.1. The plot is in Figure 14.1. The **B** modification appears better than the **A** modification, and **belt 2** appears better than **belt 1**, but he needs to assess if there is any substantial evidence to support his first impression.

He uses the **ANOVA > General Linear Model** procedure. We will specify our model using i (1 for **A** and 2 for **B**) for the taxi modification, j (1, 2, 3) for the

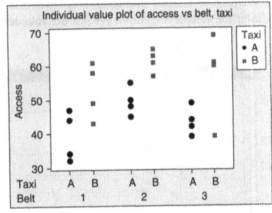

Figure 14.1

belt design, and k (1, 2, 3, 4 within taxi modification) for the volunteer passengers. We assume our volunteers are a random sample from all potential taxi passengers in wheel chairs, and volunteers are therefore a random effect. The errors represent differences in accessibility scores that a volunteer would give the same taxi−belt combination on different occasions, and they are confounded with the volunteer by belt interaction. The model is:

$$\text{access}(i, j, k) = \text{OM} + \text{taxi}(i) + \text{belt}(j) + \text{volunteer(taxi)}(k(i)) + \text{taxi} \times \text{belt}(i, j) + \text{error}(i, j, k)$$

with the following constraints on the fixed effect parameters:

$$\text{taxi}(1) + \text{taxi}(2) = 0$$

$$\text{belt}(1) + \text{belt}(2) + \text{belt}(3) = 0$$

$$\text{taxi} \times \text{belt}(1,1) + \text{taxi} \times \text{belt}(1,2) + \text{taxi} \times \text{belt}(1,3) = 0$$

$$\text{taxi} \times \text{belt}(2,1) + \text{taxi} \times \text{belt}(2,2) + \text{taxi} \times \text{belt}(2,3) = 0$$

$$\text{taxi} \times \text{belt}(1,1) + \text{taxi} \times \text{belt}(2,1) = 0$$

$$\text{taxi} \times \text{belt}(1,2) + \text{taxi} \times \text{belt}(2,2) = 0$$

These six constraints on the interaction parameters imply that

$$\text{taxi} \times \text{belt}(1,3) + \text{taxi} \times \text{belt}(2,3) = 0$$

This constraint depends logically on the other six so six constraints suffice to define the model.

The volunteer effects are a random sample of eight from a distribution with a mean of zero and a variance that we shall call the volunteer variance.

We assume that the errors are randomly and independently distributed with a mean of zero, and we usually assume the errors to have a normal distribution as well.

We need to indicate that volunteers are nested within taxis by specifying the **taxi** column in brackets after the **volunteer** column. We put the **volunteer** column in the **Random factors** box (Figure 14.2).

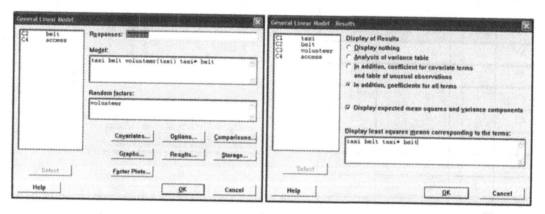

Figure 14.2

If we click on the **Results** button and choose the options shown we obtain the following analysis in the session window. The first few lines are a useful summary of our variables and their levels.

General Linear Model: access versus taxi, belt, volunteer

Factor	Type	Levels	Values
taxi	fixed	2	A, B
belt	fixed	3	1, 2, 3
volunteer(taxi)	random	8	V1, V2, V3, V4, V5, V6, V7, V8

Analysis of Variance for access, using Adjusted SS for Tests

Source	DF	Seq SS	Adj SS	Adj MS	F	P
taxi	1	1027.04	1027.04	1027.04	8.79	0.025
belt	2	361.75	361.75	180.87	7.24	0.009

volunteer(taxi)	6	701.25	701.25	116.87	4.68	0.011
taxi*belt	2	3.58	3.58	1.79	0.07	0.931
Error	12	300.00	300.00	25.00		
Total	23	2393.63				

$S = 5.0$ R-Sq $= 87.47\%$ R-Sq(adj) $= 75.98\%$

The design is orthogonal and the analysis of variance (ANOVA) follows the pattern of dividing the total corrected sum of squares into components. The degrees of freedom for each row correspond to the number of parameters estimated less the number of constraints applied when estimating them. For the fixed effects, the constraints are the same as those on the model parameters. For example, there are six interaction parameters and four constraints, leaving two degrees of freedom. If we estimate the eight volunteer effects, we apply two constraints that do not apply to the model volunteer effects, and thereby lose two degrees of freedom. The constraints are that the sum of the estimated volunteer effects within each taxi modifications is equal to zero:

within taxi(A): est_volunteer(1) + est_volunteer(2) + est_volunteer(3) + est_volunteer(4) = 0

within taxi(B): est_volunteer(5) + est_volunteer(6) + est_volunteer(7) + est_volunteer(8) = 0

Since volunteers are random effects these constraints are not explicitly imposed by the model, but are justified by the model assumption that the expected values of volunteer effects (the average over the population of all volunteers) is zero. From the F-ratios, which we shall justify from expected mean squares, and their associated P-values, we can claim evidence of differences between: the taxi modifications (P = 0.025) and the belt designs (P = 0.009). There is no evidence of any interaction between belt designs and the taxi modification.

MINITAB displays the next section, which summarises the model, only if you click the radio button **In addition, coefficients for all terms**:

Term		Coef	SE Coef	T	P
Constant		50.625	1.021	49.60	0.000
taxi					
A		−6.542	1.021	−6.41	0.000
belt					
1		−4.625	1.443	−3.20	0.008
2		4.875	1.443	3.38	0.005
(taxi)volunteer					
A	V1	−2.750	2.500	−1.10	0.293
A	V2	1.917	2.500	0.77	0.458
A	V3	6.250	2.500	2.50	0.028
B	V5	3.167	2.500	1.27	0.229
B	V6	−8.833	2.500	−3.53	0.004
B	V7	7.833	2.500	3.13	0.009
taxi*belt					
A	1	−0.208	1.443	−0.14	0.888
A	2	0.542	1.443	0.38	0.714

Unfortunately, this summary treats volunteers as a fixed effect and the standard error of the coefficient of taxi is too small. This accounts for the P-values associated with the t-tests for **taxi** being smaller than that given in the original ANOVA table.

Suppose we are asked to predict the access score for an **A** modified taxi with **belt 2**. The constant, 50.625, is the overall mean. The estimated effects are relative to the overall mean and each set is defined with a constraint that they add to zero. This enables us to calculate estimates for the effects that are not listed. For example:

- the effect of modification **B** is the negative of the effect of modification **A**, that is, $-(-6.542) = 6.542$;
- the effect of belt 3 is the negative of the sum of the effects of belts 1 and 2, which is $-(-4.625 + 4.875) = -0.250$;
- the interaction effect of **modification 1** and **belt 3** is the negative of the sum of the interaction effects of **modification 1** with **belt 1** and **2**, which equals $-(-0.208 + 0.542) = -0.334$;
- the prediction for **taxi A, belt 2** is $50.625 + (-6.542) + 4.875 + 0.542 = 49.50$.

We can do this more quickly, and check our arithmetic, by specifying **taxi*belt** in the **Display least squares means corresponding to the terms** box in the **Results** option.

MINITAB also reports unusual observations for **access**:

```
Obs     access      Fit    SE Fit    Residual    St Resid
 16    43.0000   50.5833    3.5355     -7.5833      -2.14 R
 22    39.0000   48.4167    3.5355     -9.4167      -2.66 R
```

R denotes an observation with a large standardized residual.

The largest standardised residual, in absolute magnitude, is 2.66 and this is reasonable for a sample of size 24. Notice that, although we are considering the difference between a single value and its expectation, the **SE Fit** (3.54) is less than the estimated standard deviation of the errors (s), which equals 5.0. The reason for its being less than 5.0 is that the least squares fitting minimises the sum of squared residuals; and as a consequence the residuals are less variable than the unknown errors. This is why we use a denominator that allows for the loss of degrees of freedom when we calculate **S**.

The next part of the MINITAB output is the theoretical basis for the F-tests in the initial analysis of variance. The design is orthogonal and the total corrected sum of squares can be split into five components: **taxi**, **belt**, **volunteer(taxi)**, **taxi*belt**, and **error**. The numerical values obtained in the experiment were at the beginning of MINITAB's output. Their expected values follow:

Analysis of Variance for access, using Adjusted SS for Tests

Source	DF	Seq SS	Adj SS	Adj MS	F	P
taxi	1	1027.04	1027.04	1027.04	8.79	0.025
belt	2	361.75	361.75	180.87	7.24	0.009
volunteer(taxi)	6	701.25	701.25	116.87	4.68	0.011
taxi*belt	2	3.58	3.58	1.79	0.07	0.931
Error	12	300.00	300.00	25.00		
Total	23	2393.63				

Expected Mean Squares, using Adjusted SS

	Source	Expected Mean Square for Each Term
1	taxi	(5) + 3.0000 (3) + Q[1, 4]
2	belt	(5) + Q[2, 4]
3	volunteer(taxi)	(5) + 3.0000 (3)
4	taxi*belt	(5) + Q[4]
5	Error	(5)

The expected mean squares are the population averages from imagined repetitions of the experiment and are in terms of the unknown parameters of the model. MINITAB does not give the precise forms of the expected values, which are derived in standard textbooks such as Montgomery (2004), but the results suffice to justify the F-tests. The numbers refer to the source of the variability given at the starts of the rows.

The expected value of the error mean square is the variance of the independent random errors (5). The expected value of the **taxi*belt** interaction is the sum of the variance of the errors and a non-negative function Q[4] of the interaction parameters. The Q stands for quadratic because the function is based on the squares of the parameters, which accounts for its being non-negative. The function is zero if all the interaction parameters are zero, in which case the **taxi*belt** interaction mean square has the same expected value as the error mean square and their ratio has an expected value of nearly one. Large (meaning substantially greater than one) values of this ratio are evidence against a hypothesis that all the interaction parameters are zero. In this case the F-ratio is 0.07 and, being less than one, certainly provides no evidence of an interaction.

The expected value of the **volunteers** mean square is the sum of the error variance (5) and three times the **volunteer** variance (3). The ratio of **volunteer** mean square to error mean square gives a test of the hypothesis that the **volunteer** variance is zero. Here it is 4.68 with a P-value of 0.011, and we have statistical evidence of a variance of access scores, for the same **taxi** and **belt** combination, in the population. This is unsurprising and a better use of these two expected mean squares is to equate them with the sample values and hence estimate the **volunteer** variance. This appears later in the output under the heading **Variance components**.

The expected value of the **belt** mean square is the error variance plus a function Q[2, 4] which is a non-negative function of the squares of the **belt** main effect parameters. This accounts for the first digit, 2, in the square bracket. The second digit, 4, is a cross-reference to row 4 and reminds us that **belt** design can also have an effect through its interaction with the **taxi** modification. If the main effect parameters of **belt** design are zero then Q[2, 4] is zero. So, the ratio of belt mean square to error mean square provides a test of this hypothesis. This F-ratio is 7.24 with a P-value of 0.009.

The expected value of the taxi mean square is the error variance plus three times the volunteer variance plus a function Q[1, 4] which is a non-negative function of the squares of the taxi main effect parameters. If the main effect parameters of taxi modification are zero then Q[1, 4] is zero. Then the taxi mean square equals the volunteer mean square, and their ratio provides a test of the hypothesis of no difference between the taxi modifications. This F-ratio is 8.79 with a P-value of 0.025. If we had not declared the volunteers to be a random effect, the test would have been based on the ratio of taxi mean square to error mean square. The reason for the difference is that the taxis are tested on different random samples of volunteer passengers, and we wish to draw inferences about the population of all such passengers. In contrast, all eight passengers try all three belts. MINITAB summarises all this with:

```
Error Terms for Tests, using Adjusted SS

                                              Synthesis
                                              of Error
     Source              Error DF   Error MS   MS
1    taxi                    6.00     116.87    (3)
2    belt                   12.00      25.00    (5)
3    volunteer(taxi)        12.00      25.00    (5)
4    taxi*belt              12.00      25.00    (5)
```

Note that the table of error terms for tests is just a summary of what we deduced from the table of expected mean squares. We now have a useful summary of the estimates of the variance components, obtained by equating the expected values to the sample values.

```
Variance Components, using Adjusted SS

                    Estimated
Source                Value
volunteer(taxi)       30.62
Error                 25.00
```

The calculation for the estimate of **volunteer(taxi)** is $(116.87 - 25)/3$.

Finally, we have means for taxi modification and belt design:

```
taxi        Mean
A           44.08
B           57.17
belt
1           46.00
2           55.50
3           50.38
```

Nathaniel has evidence that modification B is better and that belt design 2 is the best. The quickest way for him to predict the mean access score is to use **Results > Least squares means . . . taxi*belt**, which gives 61.50.

```
Least Squares Means for access

taxi        Mean
A           44.08
B           57.17
belt
1           46.00
2           55.50
3           50.38
taxi*belt
A    1      39.25
A    2      49.50
A    3      43.50
B    1      52.75
B    2      61.50
B    3      57.25
```

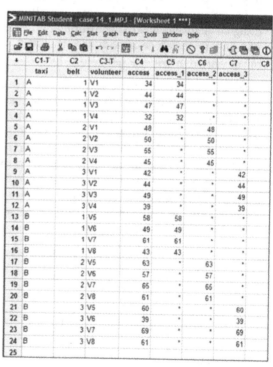

Figure 14.3

If you have only Student, you can compare the taxi modifications by comparing the access scores, using two-sample t-tests, for each belt design. For the first test, select the access scores for **belt 1** and compare the scores between **taxi A** and **taxi B**. Repeat this for **belts 2** and **3**. As an alternative to using **Copy > ... Subset the data**, you can prepare the **data** for this in the MINITAB worksheet by copying the column labelled '**access**' and pasting it into three extra columns. MINITAB automatically extends the labels to '**access_1**', '**access_2**' and '**access_3**'. In each of these columns enter asterisks (*) to denote missing values for those belts not included in the t-test (Figure 14.3). Follow **Stat > Basic Statistics > 2-Sample t**. (Figure 14.4). Choose **Samples in one column** and, for **Samples**, enter '**access_1**'. For **subscripts** enter '**taxi**'. The results are as follows:

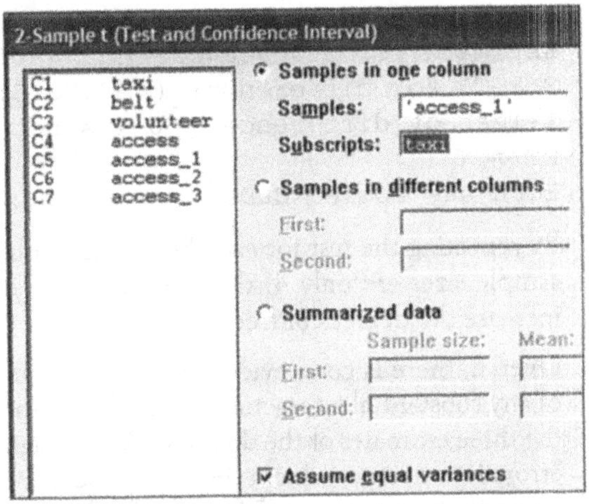

2-Sample t (Test and Confidence Interval)

C1	taxi
C2	belt
C3	volunteer
C4	access
C5	access_1
C6	access_2
C7	access_3

- **Samples in one column**
  - Samples: `'access_1'`
  - Subscripts: `taxi`
- **Samples in different columns**
  - First:
  - Second:
- **Summarized data**
  - Sample size: Mean:
  - First:
  - Second:

☑ **Assume equal variances**

Figure 14.4

Two-Sample T-Test and CI:access_1, taxi

Two-sample T for access_1

taxi	N	Mean	StDev	SE Mean
A	4	39.25	7.37	3.7
B	4	52.75	8.26	4.1

Difference = mu (A) − mu (B)
Estimate for difference: −13.5000
95% CI for difference: (−27.0412, 0.0412)
T-Test of difference = 0 (vs not =): T-Value = −2.44 P-Value = 0.051
DF = 6
Both use Pooled StDev = 7.8262

Repeat this with '**access_2**' and '**access_3**':

Two-Sample T-Test and CI: access_2, taxi

Two-sample T for access_2

taxi	N	Mean	StDev	SE Mean
A	4	49.50	4.20	2.1
B	4	61.50	3.42	1.7

Difference = mu (A) − mu (B)
Estimate for difference: −12.0000
95% CI for difference: (−18.6263, −5.3737)
T-Test of difference = 0 (vs not =): T-Value = −4.43 P-Value = 0.004
DF = 6
Both use Pooled StDev = 3.8297

Two-Sample T-Test and CI: access_3, taxi

Two-sample T for access_3

taxi	N	Mean	StDev	SE Mean
A	4	43.50	4.20	2.1
B	4	57.3	12.8	6.4

```
Difference = mu (A) - mu (B)
Estimate for difference: -13.7500
95% CI for difference: (-30.2515, 2.7515)
T-Test of difference = 0 (vs not =): T-Value = -2.04 P-Value = 0.088
DF = 6
Both use Pooled StDev = 9.5372
```

By repeating the test for each belt design you allow for the possibility of an interaction. Since the sample sizes are only 4, the assumption of equal variances, as in the general linear model, will increase the degrees of freedom.

Overall, there is good evidence that the second taxi modification is the better, and no indication of any substantial interaction. Since there is no evidence of an interaction, it is sensible to combine the three estimates of the difference by averaging. This gives -13.1. If we assume the variance of the errors is the same, whichever belt is used, the mean of the three estimates of the variance of the errors in this analysis is $(7.8262^2 + 3.8297^2 + 9.5372^2)/3.0 = 55.62$. Note that these errors in this analysis include the volunteer effect and the variation for an individual volunteer. You can check this by noticing that 55.62 is the sum of the variance components, **volunteer(taxi)** and error, in the general linear model. Calculation of a standard error for the estimated difference is complicated by the fact that the errors are not independent since each volunteer gives three assessments of ease of access, so the full analysis is needed, unless you use the average of each set of three assessments.

If we wish to compare the belts designs we can analyse the access scores for different belts as a randomised block design with volunteers as blocks. Allow for possible interactions by analysing separately for taxi design A and for taxi design B.

Create a new worksheet. Copy the data for taxi A from worksheet 1 and paste into worksheet 2 Follow **Stat > ANOVA > Two-way** (see Figure 14.5). Enter **access** as the **Response, belt** as the **Row factor** (choose **Display means**) and **volunteer** as the **Column factor**. Results are as follows:

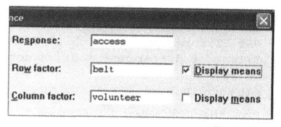

Figure 14.5

Two-way ANOVA: access versus belt, volunteer

```
Source      DF        SS        MS       F       P
belt         2    212.167   106.083   21.34   0.002
volunteer    3    238.917    79.639   16.02   0.003
Error        6     29.833     4.972
Total       11    480.917

S = 2.230   R-Sq = 93.80%     R-Sq(adj) = 88.63%

                Individual 95% CIs For Mean Based on Pooled StDev
belt    Mean    ---------+---------+---------+---------+
1       39.25   (---------*---------)
2       49.50                             (------*------)
3       43.50              (------*------)
                ---------+---------+---------+---------+
                    40.0      44.0      48.0      52.0
```

Repeat this for taxi B:

Two-way ANOVA: access versus belt, volunteer

Source	DF	SS	MS	F	P
belt	2	153.167	76.583	1.70	0.260
volunteer	3	462.333	154.111	3.42	0.093
Error	6	270.167	45.028		
Total	11	885.667			

S = 6.710   R-Sq = 69.50%   R-Sq(adj) = 44.08%

```
                    Individual 95% CIs For Mean Based on Pooled StDev
belt    Mean     ---------+---------+---------+---------+
1       52.75    (----------*----------)
2       61.50              (----------*----------)
3       57.25        (----------*----------)
                 ---------+---------+---------+---------+
                     49.0      56.0      63.0      70.0
```

Overall there is evidence that belt 2 is the best. If we assume the interaction is negligible and combine the two analyses to give eight blocks, the statistical significance increases.

Two-way ANOVA: access versus belt, volunteer

Source	DF	SS	MS	F	P
belt	2	361.75	180.875	8.34	0.004
volunteer	7	1728.29	246.899	11.39	0.000
Error	14	303.58	21.685		
Total	23	2393.63			

S = 4.657   R-Sq = 87.32%   R-Sq(adj) = 79.16%

```
                    Individual 95% CIs For Mean Based on Pooled StDev
belt    Mean      ---------+---------+---------+---------+
1       46.000    (----------*----------)
2       55.500              (----------*----------)
3       50.375        (----------*----------)
                  ---------+---------+---------+---------+
                      45.0      50.0      55.0      60.0
```

It is neater to perform the analysis with the general linear model, but the conclusions will be the same.

> ### △ Review of Case 14.1
>
> The mean access score was 50.6. The volunteer passengers found the best access with taxi design B and belt 2, and there is evidence that this combination would generally be preferred. The estimated mean access score with this combination is 61.5. The estimated difference between the two taxi designs is 13.1, and the estimated differences between belt 2 and belts 1 and 3 are 9.5 and 5.1, respectively.

A small company, Tirang Cycles, manufactures specialist lightweight touring bicycles. The frames are made from a titanium alloy and are assembled and welded by hand. The company can easily sell all it produces in its two factories. SeaDragon supplies the titanium alloy tubing, from the same division that manufactures the aluminium wheels, and is keen to promote its use for bicycle frames. Wyanet's experience with the wheel manufacturing process made her an obvious choice to help the production manager for Tirang Cycles, Patrush Pearl, chose between two workplace layouts and three assembly jigs for the frame production.

Patrush suggests that the most practical way to compare the workplace layouts is to use one in each factory. Wyanet agrees to this suggestion, and decides to ask four craftsmen in each factory to use each of the jigs on two separate occasions. The response variable will be the time a craftsman takes to make a frame. The experimental design is similar to Case 14.1 in as much as the group of four craftsmen is different in the two factories. Formally, craftsmen are nested within workplace layout. The difference between the two cases is that each craftsman is timed using the same jig on two separate occasions. A consequence of this replication is that we can fit an interaction between craftsmen and jigs. This could be significant, because individual craftsmen may be better suited to different jigs. The model is

$$time(i, j, k, l) = OM + layout(i) + jig(j) + craftsman(layout)(k(i)) + layout*jig(i, j)$$
$$+ jig*craftsman(layout)(j, k(i)) + error(i, j, k, l)$$

The craftsmen will be considered a random sample of all possible craftsmen and will be treated as a random effect. The usual constraints on the fixed effect parameters are:

$$layout(1) + layout(2) = 0$$

$$jig(1) + jig(2) + jig(3) = 0$$

The sums of layout(i) × jig(j) over layouts equal zero. Also, the sums of layout(i) × jig(j) over jigs equal zero. The craftsmen effects are assumed to be randomly and independently distributed with a mean of zero and a variance which we will refer to as *craftsmen variance*. The jig by craftsmen interactions are referred to as a *mixed effect* because one factor is fixed and the other is random. We assume that the jig by craftsman interactions are randomly and independently distributed with a mean of zero and a variance which we will refer to as *jig–craftsman variance*. This is known as the *unrestricted model* and is the MINITAB default. MINITAB **Help** tells us that most statistical software assumes the unrestricted form, whereas many textbooks restrict the interaction parameters to sum to zero over the index corresponding to the fixed effect. As usual, the errors are assumed to be randomly and independently distributed with a mean of zero.

The assembly times (in minutes) are in Table 14.3. Note that there are two values in each cell of the table.

	Lay1				Lay2			
	C1	C2	C3	C4	C5	C6	C7	C8
jig1	67, 72	68, 72	84, 87	76, 71	78, 84	82, 76	84, 75	72, 68
jig2	90, 81	87, 84	91, 96	80, 74	87, 85	90, 82	72, 69	84, 90
jig3	75, 63	74, 65	80, 73	77, 70	81, 75	77, 73	72, 81	83, 81

Table 14.3

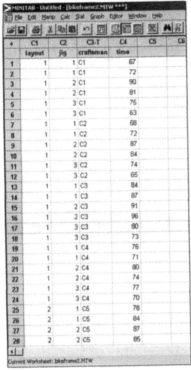

Figure 14.6

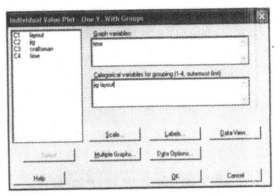

Figure 14.7

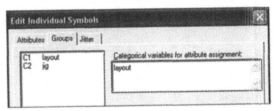

Figure 14.8

Wyanet enters the data into MINITAB as in Figure 14.6. She then plots the data using **Graph > Individual Value Plot**, selecting **With Groups** under **One Y**. She enters **jig** and **layout**, in that order, as categorical variables for grouping (Figure 14.7). To obtain different colours for symbols, and to eliminate jitter, she points at any point on the graph, left-clicks to select all points, right-clicks to show a menu, and selects **Edit Individual Symbols**; she selects the **Groups** tab and chooses **layout** as the **Categorical variable for attribute assignment** (Figure 14.8), and finally selects the **Jitter** tab and unchecks **Add jitter** (Figure 14.9). The individual value plot is in Figure 14.10.

The plot suggests possible differences between jigs. To quantify this, Wyanet again uses the **General Linear Model** procedure, and this time she adds the jig–craftsman interaction in the model (Figure 14.11). She clicks on **Results** and, elects to display least squares means for **layout** and **jig**. The results are as follows:

General Linear Model: time versus layout, jig, craftsman

Factor	Type	Levels	Values
layout	fixed	2	1, 2
jig	fixed	3	1, 2, 3
craftsman(layout)	random	8	C1, C2, C3, C4, C5, C6, C7, C8

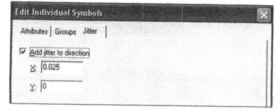

Figure 14.9

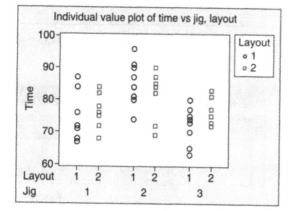

Figure 14.10

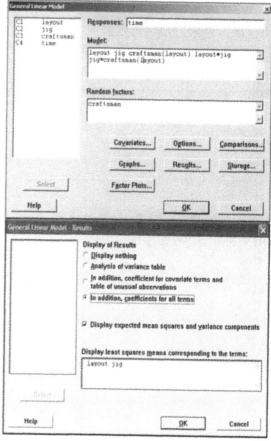

Figure 14.11

```
Analysis of Variance for time, using Adjusted SS for Tests
Source                   DF    Seq SS    Adj SS    Adj MS      F      P
layout                    1     40.33     40.33     40.33   0.40  0.552
jig                       2    756.17    756.17    378.08   6.94  0.010
craftsman(layout)         6    609.92    609.92    101.65   1.87  0.168
layout*jig                2    158.17    158.17     79.08   1.45  0.272
jig*craftsman(layout)    12    653.33    653.33     54.44   2.70  0.019
Error                    24    484.00    484.00     20.17
Total                    47   2701.92

S = 4.49073    R-Sq = 82.09%    R-Sq(adj) = 64.92%

Term                     Coef    SE Coef         T      P
Constant              78.2917     0.6482    120.79  0.000
layout
1                     -0.9167     0.6482     -1.41  0.170
jig
1                     -2.2917     0.9167     -2.50  0.020
2                      5.5833     0.9167      6.09  0.000
```

(layout) craftsman					
1	C1	−2.708	1.588	−1.71	0.101
1	C2	−2.375	1.588	−1.50	0.148
1	C3	7.792	1.588	4.91	0.000
2	C5	2.458	1.588	1.55	0.135
2	C6	0.792	1.588	0.50	0.623
2	C7	−3.708	1.588	−2.34	0.028

layout*jig					
1	1	−0.4583	0.9167	−0.50	0.622
1	2	2.4167	0.9167	2.64	0.014

(layout) jig*craftsman						
1	1	C1	−2.417	2.245	−1.08	0.292
1	1	C2	−2.250	2.245	−1.00	0.326
1	1	C3	3.083	2.245	1.37	0.182
1	2	C1	2.833	2.245	1.26	0.219
1	2	C2	2.500	2.245	1.11	0.277
1	2	C3	0.333	2.245	0.15	0.883
2	1	C5	1.167	2.245	0.52	0.608
2	1	C6	0.833	2.245	0.37	0.714
2	1	C7	5.833	2.245	2.60	0.016
2	2	C5	1.167	2.245	0.52	0.608
2	2	C6	2.833	2.245	1.26	0.219
2	2	C7	−8.167	2.245	−3.64	0.001

Expected Mean Squares, using Adjusted SS

	Source	Expected Mean Square for Each Term
1	layout	(6) + 2.0000 (5) + 6.0000 (3) + Q[1, 4]
2	jig	(6) + 2.0000 (5) + Q[2, 4]
3	craftsman(layout)	(6) + 2.0000 (5) + 6.0000 (3)
4	layout*jig	(6) + 2.0000 (5) + Q[4]
5	jig*craftsman(layout)	(6) + 2.0000 (5)
6	Error	(6)

Error Terms for Tests, using Adjusted SS

	Source	Error DF	Error MS	Synthesis of Error MS
1	layout	6.00	101.65	(3)
2	jig	12.00	54.44	(5)
3	craftsman(layout)	12.00	54.44	(5)
4	layout*jig	12.00	54.44	(5)
5	jig*craftsman(layout)	24.00	20.17	(6)

Variance Components, using Adjusted SS

Source	Estimated Value
craftsman(layout)	7.868
jig*craftsman(layout)	17.139
Error	20.167

```
Least Squares Means for time

layout      Mean
1           77.38
2           79.21
jig
1           76.00
2           83.88
3           75.00
```

The interpretation of the MINITAB output follows the same principles as Case 14.1, so we shall restrict our comments to the practical issues. There is no evidence of a difference between layouts (P = 0.552), but there is strong evidence of a difference between the jigs (P = 0.010). At the end of the output we have estimated mean assembly times of 76, 84 and 75 minutes using jigs 1, 2 and 3, respectively. There is no substantial evidence of an interaction between jigs and layouts (P = 0.272). However, there is evidence of an interaction between craftsman and jig, within a layout (P = 0.019). The practical interpretation of this interaction is that different people find different jigs easier to work with. Given this significant interaction, we should now consider whether to install the jig associated with the least average assembly time, jig 3, or to allow different craftsmen to use different jigs. The components of variance are given before the mean values. The estimated standard deviation of the craftsman–jig interaction within layouts is, $\sqrt{17.139}$ about 4.1, which is considerably larger than the difference between jigs 1 and 3. It is also larger than the standard deviation of craftsmen within layouts, which is $\sqrt{7.868}$, about 2.8. So the interaction is quite substantial, and we have evidence that no one jig is best for all craftsmen.

△ Review of Case 14.2

There seems to be little to choose between workshop layouts, but there is evidence of a difference between jigs. If the company wishes to decide on a single assembly jig it should be 1 or 3. If it is feasible for different craftsmen to use different jigs it would be worthwhile giving them a choice between jigs 1 and 3.

14.2 Split-plot designs

▽ Case 14.3 (AgroPharm)

Chantrea Cyan, whom you may remember from the investigation into the effect of fertiliser on the yield of tomato plants, has been asked to investigate how four irrigation methods and three fertiliser mixtures affect the yield of rice crops.

Two blocks of land were available. Initially, Chantrea thought two replicates of a completely randomised design, which would require 12 plots within each block to be assigned randomly to the 12 combinations of irrigation method and fertiliser mixture, would be suitable for the experiment. But it was not feasible to apply different irrigation methods on such a small scale. Each block was just large enough to be divided into four strips that could be assigned randomly to the four irrigation methods. So, she divided the strips, referred to as main plots, into three sub-plots and allocated these sub-plots randomly to the three fertiliser treatments, as in Figure 14.12. This is an

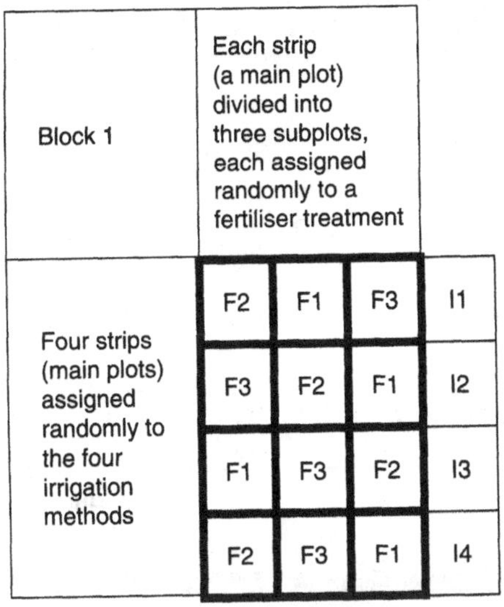

| Block 1 | Each strip (a main plot) divided into three subplots, each assigned randomly to a fertiliser treatment |
| Four strips (main plots) assigned randomly to the four irrigation methods | F2 F1 F3 I1
 F3 F2 F1 I2
 F1 F3 F2 I3
 F2 F3 F1 I4 |

Figure 14.12

Irrigation	Block	Fertiliser	Yield
1	1	1	216
1	1	2	238
1	1	3	277
2	1	1	203
2	1	2	241
2	1	3	268
3	1	1	177
3	1	2	195
3	1	3	201
4	1	1	244
4	1	2	263
4	1	3	312
1	2	1	252
1	2	2	264
1	2	3	323
2	2	1	231
2	2	2	250
2	2	3	248
3	2	1	201
3	2	2	206
3	2	3	209
4	2	1	223
4	2	2	204
4	2	3	233

Table 14.4

example of a split-plot design. The crop yields, in tonnes per hectare (Clarke and Kempson, 1997) are in Table 14.4 and are in www.greenfieldresearch.co.uk/doe/data.htm. She entered the data into a MINITAB spreadsheet.

As usual Chantrea started with plots of the data (Figure 14.13). Irrigation method 1 seems better than method 3 and slightly better than methods 2 and 4. However, irrigation is the main-plot factor and is only replicated twice, once in each block, so we may not have much evidence against a hypothesis that these differences are due to chance. Fertiliser 3 appears better than fertilisers 1 and 2. The different fertiliser types are replicated eight times on the smaller sub-plots, but we still need to determine whether the differences are statistically significant.

The model for this experiment is

$$\text{yield}(i, j, k) = \text{OM} + \text{irrigation}(i) + \text{block}(j)$$
$$+ \text{main\_plot\_error}(i, j) + \text{fertiliser}(k)$$
$$+ \text{irrigation} \times \text{fertiliser}(i, k) + \text{block}$$
$$\times \text{fertiliser}(j, k) + \text{sub\_plot\_error}(i, j, k)$$

Chantrea takes **blocks** to be a random effect. The constraints are:

- sum of irrigation(i) over $i = 1, \ldots, 4$ equals 0;
- sum of fertiliser(k) over $k = 1, 2, 3$ equals 0;
- sum of irrigation $\times$ fertiliser(i, k) over $i = 1, \ldots, 4$ and over $k = 1, 2, 3$ equals 0.

The other assumptions for the unrestricted model are:

- **block(j)** is randomly and independently distributed with mean zero and **block** variance;
- **main_plot_error** is randomly and independently distributed with mean zero and **main_plot_error** variance;
- **block** by **fertiliser** interaction is randomly and independently distributed with mean zero and **blocks*fertiliser** variance;
- **sub_plot_error** is randomly and independently distributed with mean zero and **sub_plot_error** variance.

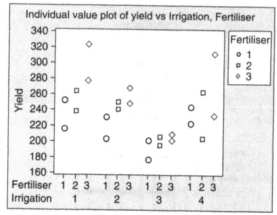

Figure 14.13

There is no term for an interaction between the irrigation and blocks because this is confounded with the main_plot_error, exactly the same as for a single replicate of a randomised block design.

We will use the **General Linear Model** procedure for the analysis (Figure 14.14). The response is the yield. The first two terms in the model are **irrigation** and **block**. Since each irrigation method is tested once in each block the difference between estimated irrigation effects in the two blocks is the only estimate of the variability of the main plots. Thus the mean square associated with the **irrigation** by **block** interaction leads to an estimate of variance of the main-plot error, and any interaction between irrigation method and blocks is confounded with

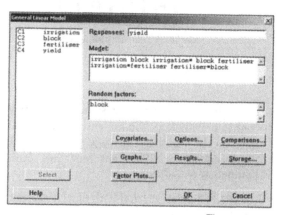

Figure 14.14

this. We are in a better position with the sub-plots and can distinguish between a split-plot design, with its restriction on the randomisation, and replicates of a completely randomised design on two blocks.

General Linear Model: yield versus irrigation, block, fertiliser

Factor	Type	Levels	Values
irrigation	fixed	4	1, 2, 3, 4
block	random	2	1, 2
fertiliser	fixed	3	1, 2, 3

Analysis of Variance for yield, using Adjusted SS for Tests

Source	DF	Seq SS	Adj SS	Adj MS	F	P
irrigation	3	13297.1	13297.1	4432.4	2.04	0.286
block	1	3.4	3.4	3.4	0.00	0.972 x
irrigation*block	3	6510.5	6510.5	2170.2	16.43	0.003
fertiliser	2	6753.0	6753.0	3376.5	8.12	0.110

irrigation*fertiliser	6	2011.0	2011.0	335.2	2.54	0.141
block*fertiliser	2	832.0	832.0	416.0	3.15	0.116
Error	6	792.7	792.7	132.1		
Total	23	30199.6				

x Not an exact F-test.

S = 11.4940   R-Sq = 97.38%   R-Sq(adj) = 89.94%

Expected Mean Squares, using Adjusted SS

	Source	Expected Mean Square for Each Term
1	irrigation	(7) + 3.0000 (3) + Q[1, 5]
2	block	(7) + 4.0000 (6) + 3.0000 (3) + 12.0000 (2)
3	irrigation*block	(7) + 3.0000 (3)
4	fertiliser	(7) + 4.0000 (6) + Q[4, 5]
5	irrigation*fertiliser	(7) + Q[5]
6	block*fertiliser	(7) + 4.0000 (6)
7	Error	(7)

Error Terms for Tests, using Adjusted SS

	Source	Error DF	Error MS	Synthesis of Error MS
1	irrigation	3.00	2170.2	(3)
2	block	3.63	2454.0	(3) + (6) − (7)
3	irrigation*block	6.00	132.1	(7)
4	fertiliser	2.00	416.0	(6)
5	irrigation*fertiliser	6.00	132.1	(7)
6	block*fertiliser	6.00	132.1	(7)

Variance Components, using Adjusted SS

Source	Estimated Value
block	−204.22
irrigation*block	679.35
block*fertiliser	70.97
Error	132.11

Least Squares Means for yield

irrigation	Mean
1	261.7
2	240.2
3	198.2
4	246.5
block	
1	236.3
2	237.0
fertiliser	
1	218.4
2	232.6
3	258.9

The comparison between the main plot treatments has low precision. Despite the apparent difference between yields for the different irrigation methods on the plot, the methods do not give significantly different yield at even a 20% level. The mean yields for the four irrigation methods are: 261.7, 240.2, 198.2 and 246.5. When we calculate a mean for an irrigation method we are averaging 2 main-plot errors and $2 \times 3 = 6$, = sub-plot errors. The variance of a mean is therefore

$$\text{(variance main plot error)}/2 + \text{(variance sub plot error)}/6$$

In terms of the labels in the MINITAB variance components this is

$$\text{(irrigation} \times \text{block)}/2 + \text{error}/6 = 679.35/2 + 132.11/6 = 2170.2/6$$

You can avoid the arithmetic by finding 2170.2 as the irrigation*block Adj MS in the analysis of variance table. Hence, the estimated standard deviation of the difference between two of these means is

$$\sqrt{\frac{2170.2}{6} + \frac{2170.2}{6}} = 26.9$$

If you have followed the argument closely you may have realised that each irrigation mean also involves averaging random block and random block by fertiliser interaction effects. We can ignore this nicety because these averages are identical for all the irrigation methods. The fixed effect averages are zero by definition. An individual 95% confidence interval for the difference between the irrigation method with the highest yield in this experiment, method 1, and that with the lowest, method 3, is

$$261.7 - 198.2 \pm t_{3,\,0.025} \times 26.9 = 63.5 \pm 3.12 \times 26.9 = [-22, 149]$$

A 90% confidence interval is [0, 127].

Fertiliser treatments are compared with greater precision if we are prepared to assume that interaction between fertilisers and blocks is negligible. There is weak evidence of an interaction between fertilisers and blocks and between fertilisers and irrigation methods. The estimated standard deviation of a difference between fertiliser means is

$$\sqrt{\frac{416}{4 \times 2} + \frac{416}{4 \times 2}} = 10.2$$

If we are prepared to disregard the weak evidence of an interaction between fertilisers and blocks and to assume the interaction is negligible, the fertiliser effects become highly significant. Overall, the results from this experiment are equivocal.

△ Review of Case 14.3

There is some weak evidence of a difference between fertilisers, but given the weak evidence of an interaction with the blocks we would hesitate to make any general recommendation. If Chantrea knows that the two blocks have similar soil types and are unlikely to interact with the fertilisers, she would recommend fertiliser 3. The differences between the irrigation methods are substantial but the precision of these estimates is so low that she cannot make firm recommendations on the basis of this experiment alone. Chantrea should recommend a follow-up experiment with more blocks.

▽ Case 14.4 (SeaDragon and UoE)

Following their discussions, described in Section 13.3, Paul Plum and Ishmael Ivory decided to run the experiment with two of each design of prosthetic heart valve. The results from the experiment (Anderson and McLean, 1974) are presented in Table 14.5. The response, the pressure gradient **grad**, is in coded units and the objective is to keep it as low as possible, because the lower the pressure gradient the lower is the resistance to flow of the valve. There are 48 values of **grad** and these are arranged in rows in Table 14.6. You can find the full table in Case 14.4 at www.greenfieldresearch.co.uk/doe/data.htm.

Valve\pulse	20	60	100	140	180	220
A(1)	12	8	4	1	8	14
A(2)	7	5	7	5	13	20
B(1)	20	15	10	8	14	25
B(2)	14	12	7	6	18	21
C(1)	21	13	8	5	15	27
C(2)	13	14	7	9	19	23
D(1)	15	10	8	6	10	21
D(2)	12	14	5	9	14	17

Table 14.5

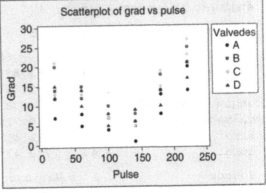

Figure 14.15

valvedes	pulse	engineer	manufacturer	grad
A	20	1	1	12
A	20	2	2	7
A	60	1	1	8
D	140	8	2	9
D	180	7	1	10
D	180	8	2	14
D	220	7	1	21
D	220	8	2	17

Table 14.6

Paul plots the data as in Figure 14.15. Under **Graph > Scatterplot** he clicked on **With Groups** and assigned grad as **Y variable**, pulse as **X variable** and valvedes as **Categorical variable**. The flow gradient varies with pulse rate in a quadratic manner for all four valves, and is lowest within the range of typical human pulse rates. The plot suggests that the **Type A** valve is the best. But we need to quantify this result and assess whether it is statistically, and practically, significant.

Scenario A. Suppose the two valves of each type were manufactured by eight different engineers, using the engineering drawings of the prototypes. Then we have an experimental design with nested and crossed factors, analogous to Case 14.2. There we had jigs crossed with layouts as fixed effects. Craftsmen were imagined to be a random sample from a population of all possible craftsmen and were nested with layout. Here we can consider valve design (A, B, C, D) crossed with pulse rates as fixed effects. The engineers are nested within valve design and can be considered a random sample of all possible engineers. The model is:

$$grad(i, j, k) = valvedes(i) + pulse(j) + engineer(valvedes)(k(i)) + valvedes \times pulse(i, j) + error(i, j, k)$$

We use **General Linear Model**, and under **Results** we select **Display expected mean squares** . . . and **Display means . . . valvedes**. We have edited the output so that the variance components are incorporated into the table of expected mean squares.

ANOVA: grad versus valvedesign, pulse, engineer

Factor	Type	Levels	Values
valvedes	fixed	4	1 2 3 4
pulse	fixed	6	1 2 3 4 5 6
engineer(valvedes)	random	2	1 2

Analysis of Variance for grad

Source	DF	SS	MS	F	P
valvedes	3	261.896	87.299	13.39	0.015
pulse	5	1192.354	238.471	27.50	0.000
engineer(valvedes)	4	26.083	6.521	0.75	0.568
valvedes*pulse	15	55.729	3.715	0.43	0.951
Error	20	173.417	8.671		
Total	47	1709.479			

	Source	Variance component	Error term	Expected Mean Square for Each Term (using unrestricted model)
1	valvedes		3	(5) + 6(3) + Q[1,4]
2	pulse		5	(5) + Q[2,4]
3	engineer(valvedes)	−0.3583	5	(5) + 6(3)
4	valvedes*pulse		5	(5) + Q[4]
5	Error	8.6708		(5)

means

valve ty	N	grad
A	12	8.667
B	12	14.167
C	12	14.500
D	12	11.750

There is evidence of a difference between **valve designs** (P = 0.015). **B** and **C** give the highest flow gradients and **A** is substantially less. To calculate the *least significant difference with a five percent significance*, LSD (5%), we need the standard deviation of a **valvedes** mean. Each **valvedes** mean is obtained by averaging over 12 errors and two engineers. In terms of the estimated variance components this is 8.6708/12 + (−0.3583)/2, which equals the unbiased estimate of **engineer(valvedes)** MS in the analysis of variance for grad table, 6.521, divided by 12. But, as sometimes occurs, the formal estimate of the variance of **engineer (valvedes)** is negative. This is physically impossible and the estimate of the **engineer** variance should be reported as zero.

Nevertheless, proceeding with the unbiased estimate of the variance of the valve design means, the LSD (5%) is:

$$t_{4,\,0.025} \times \sqrt{6.521\left(\frac{1}{12} + \frac{1}{12}\right)} = 2.90$$

Using this criterion we have statistical evidence that **B**, **C** and **D** are higher than **A**. Therefore we prefer valve design **A**. We cannot confidently conclude that **B** and **C** are higher than **D**, but these comparisons are of less practical interest. **Pulse rate** has a strong effect on flow gradient. There is no evidence of an interaction between **valve type** and **pulse rate**.

Scenario B. Suppose two engineering companies (1 and 2) were asked to manufacture one valve of each of the four types (**A,B,C,D**). Now we have a split-plot design. Engineering company (**manu-facturer**) is the block and **valve type** is the main-plot treatment. **Pulse rate** is the sub-plot treatment. An interaction between **manufacturer** and **pulse rate** seems unlikely and we assume there is none. The model is now:

grad(*i*, *j*, *k*) = valvedes(*i*) + manufacturer(*j*) + main_plot_error(*i*, *j*)
    + pulse(*k*) + valvedes*pulse(*i*, *k*) + sub_plot_error(*i*, *j*, *k*)

We use **General Linear Model** and have edited the **valvedes*manufacturer** interaction to read **main-plot**.

ANOVA: flow versus valvedesign, manufacturer, pulse

Factor	Type	Levels	Values					
valvedes	fixed	4	1	2	3	4		
manufacturer	random	2	1	2				
pulse	fixed	6	1	2	3	4	5	6

Analysis of Variance for flow

Source	DF	SS	MS	F	P
valvedes	3	261.896	87.299	10.45	0.043
manufacturer	1	1.021	1.021	0.12	0.750
main-plot	3	25.062	8.354		
pulse	5	1192.354	238.471	27.50	0.000
valvedes*pulse	15	55.729	3.715	0.43	0.951
sub-plot error	20	173.417	8.671		
Total	47	1709.479			

	Source	Variance component	Error term	Expected Mean Square for Each Term (using unrestricted model)
1	valvedes		3	(6) + 6(3) + Q[1,5]
2	manufacturer	−0.30556	3	(6) + 6(3) + 24(2)
3	main-plot error	−0.05278	6	(6) + 6(3)
4	pulse		6	(6) + Q[4,5]
5	valvedes*pulse		6	(6) + Q[5]
6	Error	8.67083		(6)

There is evidence of a difference between valve types (P = 0.043). **B** and **C** give the highest flow gradients and **A** is substantially less. Each **valvedes** mean is obtained by averaging over 12 sub-plot errors and two main-plot errors. In terms of the estimated variance components this is 8.6708/12 + (−0.05278)/2, which equals the **main-plot** error MS, 8.354, divided by 12. Again, we have a physically impossible estimate of a component of variance, this time the **main-plot** error. However, as before, we choose to use the value given as our **main-plot** MS on the ground that it is an unbiased estimator of the sub-plot error variance plus six times the **main-plot** error variance.

The LSD (5%) is now

$$t_{3,\,0.025} \times \sqrt{8.345\left(\frac{1}{12} + \frac{1}{12}\right)} = 3.76$$

The means are unchanged. Using this criterion we have statistical evidence that **A** is lower than **B** and **C**. But, we cannot confidently conclude that **D** is lower than **B** and **C** or that **A** is lower than **D**. We would prefer design A though we might wish to carry out further tests between design **A** and **D**, especially if **D** had any other advantages such as being easier to manufacture. **Pulse rate** has a strong effect on flow gradient. There is no evidence of an interaction between **valve type** and **pulse rate**.

Other scenarios. Suppose one engineering company makes all the valves. An engineer makes the two valves of each type consecutively. This would be convenient for the manufacturing, but it is not ideal for the experiment because the two valves are likely to be more similar than valves of the same type produced at different times. It would be better to manufacture the valves at different times, resetting all the machines for each one. The analysis would be the same as for scenario A, except that '**engineers** nested within **valve design**' would be described as '**valves** nested within **valve design**'.

△ Review of Case 14.4

It seems that the variability between different valves manufactured to the same design is negligible. Ishmael was relieved that he had read about split-plot designs. It would have been embarrassing to have insisted on manufacturing six valves of each type.

The flow gradient depends on pulse rate and is lowest within the range of typical human pulse rates.

Design A had the lowest flow gradient. We are quite confident it is better than B or C in this respect, but rather less confident that it is better than D.

15

Mixture designs

Many processes in the food and drinks industry, and in the pharmaceutical and chemical industry, involve blending ingredients. The proportions of ingredients in the mixture are crucial, and blenders rely on experiments to obtain the best results. The designs we have studied so far may need modification because of the constraint that the sum of the proportions must equal one.

If one component of the mixture predominates, we can use the designs we have already discussed, provided we do not attempt to include the proportion of the predominant component as a variable in the analysis. For example, if we add a water softener and a detergent to water, then we have a mixture of three components but the water is such a large component that we can ignore it and plan an experiment, such as a 2^2 factorial, in terms of the two additives. Usually, steel alloys contain so much iron that we can use a non-mixture design in terms of the alloying elements.

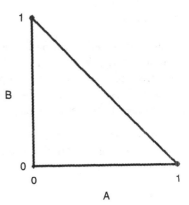

Figure 15.1

In other cases, there are much more efficient designs, *mixture designs*, that should be used. MINITAB Pro provides routines for generating and analysing these. If you are relying on Student, you will have to follow the general principles, that we explain below, and construct your own design.

Suppose we have a two-component mixture, and let A and B be the proportions of each. If we draw a graph with A and B as axes (Figure 15.1) then the constraint that $A + B = 1$ limits our choices for A and B to points on the diagonal line. We can simplify the diagram by looking only at the straight line (Figure 15.2). A point on the line has coordinates (A, B) such that $A + B = 1$.

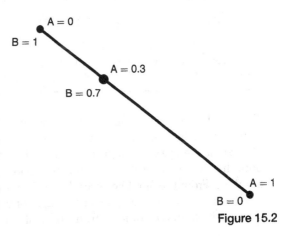

Figure 15.2

If we have a mixture of three components, with proportions A, B and C, the constraint $A + B + C = 1$ limits our choices for A, B and C to lie on an equilateral triangle in three dimensions (Figure 15.3). The triangle is a two-dimensional object in our three-dimensional design space. The coordinates of the vertices are shown. Values of A, B and C are restricted to points on or within the triangle. Any point on the triangle has coordinates (A, B, C) such that $A + B + C = 1$.

The triangle is called a *simplex* because it has the least number of vertices of any planar object in three dimensions constructed with

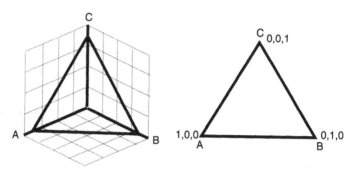

Figure 15.3

straight lines. Planes are flat, so they are two-dimensional objects in a three-dimensional space. If we have four components (A, B, C, D), the simplex is a three-dimensional object in a four-dimensional space. It is a tetrahedron in which any point has the coordinates (A, B, C, D) and $A + B + C + D = 1$. Figure 15.9 is a picture of this. Generally, we have an L-simplex with L components. A 5-simplex is called a pentachoron; it has five cells, each of which is tetrahedral.

There are three common forms of design: *simplex centroid*; *simplex lattice*; and *extreme vertices* (see Figure 15.12). You can use MINITAB to design (**Stat > DOE > Mixture**) and analyse all of them. For both simplex centroid and simplex lattice designs, you can add points to the interior of the design space. These points provide information on the interior of the response surface, thereby improving coverage of the design space.

Control variables in a mixtures experiment are called *component* variables rather than *process* variables. Designs with both component variables and process variables are possible. We present such a design in Case 15.4.

First we illustrate the principles with a simple case.

▽ Case 15.1 (AgroPharm)

Recently a blend of the red wine, Shiraz, with small amounts of the white wine, Viognier, has become popular in Australia and the USA. The winemaker at the AgroPharm vineyard, Dion Darkred, decides to experiment by blending Shiraz with Viognier in 0.01 increments from 0.00 to 0.10.

Table 15.1 shows the average scores for each blend from a panel of five tasters. The mixture design here is the same as in Figure 15.2, where factor A represents Viognier and factor B represents Shiraz. Dion plotted the average scores against the proportion of Viognier as in Figure 15.4. Now he cannot include both the proportion of Viognier and the proportion of Shiraz in a regression because of the constraint **S + V = 1**. From the plot Dion sees that he needs to allow for a quadratic response. A regression using a centred variable **(V - 0.05)**, and the square of it, confirms that the quadratic term is statistically significant. Note that he omits **S** from the analysis.

Viognier	Shiraz	taste
0	1	88.6
0.01	0.99	88.4
0.02	0.98	89.3
0.03	0.97	89
0.04	0.96	88.5
0.05	0.95	89.1
0.06	0.94	88.4
0.07	0.93	87.3
0.08	0.92	86.8
0.09	0.91	85.3
0.1	0.9	85.2

Table 15.1

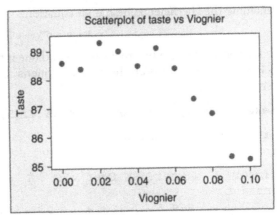

Figure 15.4

Regression Analysis: taste versus V-0.05, (V-0.05)^2

The regression equation is
taste = 88.6 − 36.7 (V-0.05) − 790 (V-0.05)^2

Predictor	Coef	SE Coef	T	P
Constant	88.5993	0.1832	483.73	0.000
V-0.05	−36.727	3.834	−9.58	0.000
(V-0.05)^2	−790.2	137.3	−5.76	0.000

S = 0.402127   R-Sq = 94.0%   R-Sq(adj) = 92.5%

Dion concludes that a small amount of Viognier may improve the panel score slightly but amounts above 5% will reduce it. Formally, if he differentiates the expression for taste with respect to V and sets the derivative to zero, he obtains a predicted maximum score of 89.03 when V equals 0.027.

He intends to present the results to his colleagues and wonders if the results might be more intuitive if he did have both the proportion of V and the proportion of S in his formula. How can he get round the constraint of S + V = 1?

In any equation of the form $y = a + bx + \ldots$, we could include a constant factor, $x_0 = 1$, attached to the intercept. Then the fitted equation is $y = a*x_0 + b*x + \ldots$. So Dion can replace x_0 by S + V, because S + V = 1. Also, since V = 1 - S, he can replace V^2 by V(1 - S) = V - VS. Substituting in the equation

$$\text{taste} = a + b*V + c*V^2,$$

Dion has

$$\text{taste} = a*(S + V) + b*V + c*V - c*V*S$$
$$= (a + b + c)*V + a*S - c*V*S$$

So, renaming the coefficients, he has an equation:

$$taste = d*V + e*S + f*V*S$$

which is an equation with two main effects and one interaction but with no constant (intercept). In **Stat > Regression > Regression**, he goes to **Options** and unchecks **Fit intercept**. The results of the analysis are as follows:

```
Regression Analysis: taste versus V, S, VS

The regression equation is
taste = − 659 V + 88.5 S + 790 VS

Predictor      Coef  SE Coef        T       P
Noconstant
V            −659.5    123.8    −5.33   0.001
S          88.4601   0.3064   288.74   0.000
VS           790.2    137.3     5.76   0.000


Analysis of Variance

Source          DF     SS      MS          F       P
Regression       3  84835   28278  174874.65   0.000
Residual Error   8      1       0
Total           11  84836
```

This appears to be almost a perfect fit because the ANOVA breakdown is of $\sum y^2$ rather than $\sum(y - \bar{y})^2$. It should be interpreted with caution, and the equation presents him with another complication. If Dion wishes to find the values of V and S that give a maximum from this equation, he needs to add the constraint (V + S - 1) multiplied by a Langrange multiplier and then set the partial derivatives with respect to V and S equal to zero. This is equivalent to using the constraint to replace one variable and then differentiating with respect to the single remaining variable, but requires more advanced mathematics.

He now thinks it might be simplest to present results using V and V^2, rather than with the centred variable V, or with both V and S.

```
Regression Analysis: taste versus V, V^2

The regression equation is
taste = 88.5 + 42.3 V − 790 V^2

Predictor      Coef  SE Coef        T       P
Constant   88.4601   0.3064   288.74   0.000
V            42.29    14.25     2.97   0.018
V^2         −790.2    137.3    −5.76   0.000

S = 0.402127   R-Sq = 94.0%   R-Sq(adj) = 92.5%

Analysis of Variance

Source          DF      SS      MS       F       P
Regression       2  20.195  10.098   62.44   0.000
Residual Error   8   1.294   0.162
Total           10  21.489
```

He notes that the coefficient of the squared term is the same as for the centred variables although the other coefficients are different. All three equations can be shown to be equivalent by making appropriate substitutions.

The negative coefficient of V^2 in the equation from this analysis

$$\text{taste} = 88.5 + 42.3V - 790V^2$$

indicates that a maximum value of taste can be estimated. As before, the highest predicted value of **taste** is **89.03** when **V** is **0.027**. The corresponding value of **S** is **0.973**. This is in accord with the data plot and will be easy to explain to the AgroPharm directors.

Dion is also aware that the data are consistent with a model that has a constant response up to a proportion of about 0.05 Viognier, followed by a decline. He will resolve this issue with a further tasting experiment to compare single-variety Shiraz with the blends of 0.973 Shiraz and 0.027 Viognier and of 0.95 Shiraz with 0.05 Viognier. They will certainly sell single-variety Shiraz; the marketing decision is whether also to sell some Shiraz and Viognier blend.

△ Review of Case 15.1

Although Dion will be interested in the results of the further tasting experiment, he thinks he already has enough data to make a marketing decision. There is no substantial change in the taste scores over a range of 0–5% Viognier. Single-variety Shiraz is the safer option, but he does not want to ignore fashion. If he is going to market a blend, he thinks it should have at least 5% of the second variety. So, he decides to blend one-tenth of the Shiraz with 5% Viognier.

▽ Case 15.2 (SeaDragon)

Sadah Saffron works in the division of SeaDragon that manufactures industrial machinery. She wishes to investigate the elongation properties of fibre made from up to three components: polyethylene (x1); polystyrene (x2); and polypropylene (x3). The fibre will be spun into yarn for the garment industry. The elastic elongation of the fibre is important because higher elongation reduces the tendency of the clothing to crease and tear. However, higher elongation makes the yarn harder to spin, and her aim is to investigate the likely range of fibre elongation that SeaDragon spinning machines may be expected to handle.

She chooses MINITAB's *simplex centroid* design by following **Stat > DOE > Mixture > Create Mixture Design**. She sets the number of components to 3 and clicks **Designs**, checks the box to include axial points and asks for two replicates (Figure 15.5). Remember that two replicates implies that she will have to make two test batches of each blend of fibre. It would not suffice to test the same batch twice, as this would not allow for variation between batches. Note that in this window, MINITAB labels **point types: (1) vertex; (2) double blend; (0) center point; (-1) axial point**. If you refer back to Figure 15.3, the axial points are on lines joining a vertex to the mid-point of the opposite side. These lines are referred to as the axes, and the centre point is where they meet. The numeric values chosen to distinguish the points are arbitrary and they are simply labels. Back in the **Create Mixture Design** window, Sadah clicks on the **Components** tab and names the components **polyeth, polysty** and **polyprop**.

MINITAB generated a design as in Table 15.2. Note that the run order has been randomised. We present the data here and on the website in the standard order because, each time you use this procedure, MINITAB produces a different run order. When you repeat the procedure, you should use **Data > Sort** and sort by standard order before copying and pasting the values of elongation. Note, in Figure 15.6, that all columns of the data are sorted by the column headed **StdOrder** in ascending order and are stored in the original columns. If you were to choose **New worksheet** instead of **Original column(s)**, MINITAB would treat the new worksheet simply as a table of values and not as a mixture design. Sadah plotted the points of the design using **Stat > DOE > Mixture > Simplex Design Plot** and

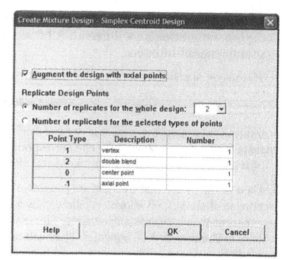

Figure 15.5

StdOrder	RunOrder	PtType	Blocks	polyeth	polysty	polyprop	elongation	FITS1
1	2	1	1	1	0	0	13.98	13.10
2	11	1	1	0	1	0	8.36	8.70
3	18	1	1	0	0	1	15.57	15.07
4	8	2	1	0.5	0.5	0	16.38	16.13
5	15	2	1	0.5	0	0.5	16.21	16.64
6	17	2	1	0	0.5	0.5	10.07	10.26
7	10	0	1	0.333	0.333	0.333	14.28	15.03
8	1	−1	1	0.666	0.167	0.167	15.06	15.97
9	5	−1	1	0.167	0.666	0.167	11.69	12.38
10	12	−1	1	0.167	0.167	0.666	15.15	14.67
11	19	1	1	1	0	0	12.62	13.10
12	16	1	1	0	1	0	8.85	8.70
13	4	1	1	0	0	1	13.97	15.07
14	13	2	1	0.5	0.5	0	15.53	16.13
15	3	2	1	0.5	0	0.5	16.34	16.64
16	9	2	1	0	0.5	0.5	9.11	10.26
17	6	0	1	0.333	0.333	0.333	14.62	15.03
18	7	−1	1	0.666	0.167	0.167	16.61	15.97
19	20	−1	1	0.167	0.666	0.167	14.60	12.38
20	14	−1	1	0.167	0.167	0.666	16.90	14.67

Table 15.2

choosing **Proportions** (Figure 15.7). Each of the points on the plot represents two replications. For example, run order point 19 is a replicate of run order point 2.

Do not copy and past the values of the component variables because rounding will prevent them from summing exactly to one. Use, instead, the values generated by MINITAB.

Also note that we have added a column headed **FITS1**. We shall explain this later.

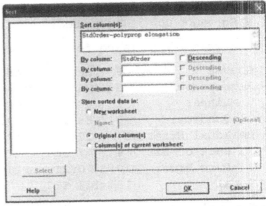

Figure 15.6

Sadah made 20 fibres with mixtures specified in the designed experiment and then tested the elongation of each. She added the elongation values to the table. Case 15.2 at www.greenfieldresearch.co. uk/doe/data.htm has two spreadsheets: the full Table 15.2 and the column of elongation values. Select only the elongation values. Be sure that you have already sorted the table in the MINITAB worksheet into standard order before pasting in these values. Sadah followed **Stat > DOE > Mixture > Analyze Mixture Design** Note that in Case 15.1 Dion used MINITAB's standard regression procedure but, in this case, Sadah uses the procedure specific to mixture designs.

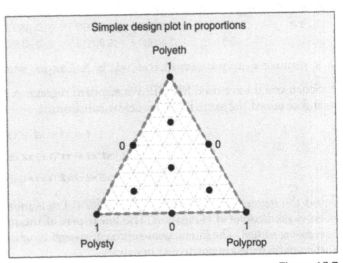

Figure 15.7

Sadah checked **Mixture components only** and **Proportions**. She clicked on **Terms** and asked for main effects **(A, B, C)** and first-order interactions **(AB, AC, BC)**. Actually, these terms are specified automatically because at the top right of the **Terms** window she asked for a **Quadratic** model. This does not mean that squared terms are included explicitly; the constraints of a mixture make squared terms implicit if first-order interactions are included. The analysis results were as follows:

Regression for Mixtures: elongation versus polyeth, polysty, polyprop

Estimated Regression Coefficients for elongation (component proportions)

Term	Coef	SE Coef	T	P	VIF
polyeth	13.098	0.7565	*	*	1.964
polysty	8.700	0.7565	*	*	1.964
polyprop	15.065	0.7565	*	*	1.964
polyeth*polysty	20.916	3.4865	6.00	0.000	1.982
polyeth*polyprop	10.240	3.4865	2.94	0.011	1.982
polysty*polyprop	−6.496	3.4865	−1.86	0.084	1.982

```
S = 1.10926      PRESS = 30.6733
R-Sq = 88.00%    R-Sq(pred) = 78.63%    R-Sq(adj) = 83.71%
```

Analysis of Variance for elongation (component proportions)

Source	DF	Seq SS	Adj SS	Adj MS	F	P
Regression	5	126.286	126.286	25.257	20.53	0.000
Linear	2	67.127	46.727	23.364	18.99	0.000
Quadratic	3	59.159	59.159	19.720	16.03	0.000
Residual Error	14	17.226	17.226	1.230		
Lack-of-Fit	4	7.015	7.015	1.754	1.72	0.222
Pure Error	10	10.212	10.212	1.021		
Total	19	143.512				

Unusual Observations for elongation

Obs	StdOrder	elongation	Fit	SE Fit	Residual	St Resid	
14	20	16.901	14.673	0.393	2.227	2.15	R
20	19	14.604	12.381	0.393	2.223	2.14	R

R denotes an observation with a large standardized residual.

Sadah could have used MINITAB's standard regression procedure, x1, x2, x1*x2, x1*x1 and x2*x2 and obtained the same fitted model by substituting

$$1 = x1 + x2 + x3$$

$$x1*x1 = x1*(1 - x2 - x3)$$

$$x2*x2 = x2*(1 - x1 - x3)$$

but the regression for mixtures of MINITAB Pro is more convenient. The fitted model provides good predictions of elongation. The linear part of the predicting equation is known as the *linear blending portion*. The interactions represent *synergistic* or *antagonistic blending* depending on whether the coefficients are positive or negative.

A term that appears in the output of this analysis is **VIF**. MINITAB's analysis detects multi-colinearity or correlation among predictors and summarises this with a *variance inflation factor* **(VIF) VIF = 1.0** indicates no relation among predictors; **VIF > 1.0** indicates that the predictors are correlated; if the **VIF** lies between **5.0** and **10.0** then the estimators of the regression coefficients are highly correlated so they will be poorly estimated unless we have a very large sample. The predictors are correlated in this analysis but a correlations matrix shows that there are no correlations so high as to indicate that a predictor should be removed from the analysis.

	polyeth	polysty	polyprop
polysty	−0.5		
polyprop	−0.5	−0.5	
elongation	0.433	−0.675	0.242

MINITAB omits *t*-ratios and P-values for main effects because a hypothesis test, that each main effect is zero, makes little sense when there is no intercept in the model. The standard errors of the coefficients are small compared with differences between them, indicating that the elongation properties of the components differ. The interaction terms are all statistically significant, at least at the 10% level.

Sadah's objective was to analyse the likely range of fibre elongation. She runs the analysis again and, in the **Analyze Mixture Design** window, she clicks the **Storage** tab and selects only **Fits** to be stored. These appear in an extra column on the worksheet. This is headed **FITS1** in Table 15.2. They range from 8.7 up to 16.6. She looks down the columns of the spreadsheet, including the fitted values, and notices that the highest (16.64) corresponds to 0.5 polyethylene and 0.5 polypropylene. She then makes some predictions around these proportions and obtains a prediction of 16.73 with a 95% prediction interval of (14.0, 19.5) when the blend is 0.4 polyethylene and 0.6 polypropylene. The quickest way to do this is to enter the proportions – examples being 0.4, 0.0, 0.6; 0.6, 0.0, 0.4; 0.45, 0.05, 0.50; 0.45, 0.10, 0.45 – into columns **C11** to **C13**, say. Then follow **Predictions. . .Components c11–c13**.

Alternatively, you can obtain a more reliable estimate of the maximum using Excel Solver with

$$D1 = 13.098^*A1 + 8.7^*B1 + 15.065^*C1 + 20.916^*A1^*B1$$

$$+ 10.24^*A1^*C1 - 6.496^*B1^*C1$$

$$F1 = A1 + B1 + C1$$

and constraints

$$F1 = 1$$

$$A1 >= 0$$

$$B1 >= 0$$

$$C1 >= 0$$

You still need to use **Predictions . . . 0.4 0.0 0.6** to obtain the prediction interval.

An informal estimate of the likely upper limit of the range of elongation is about 20. Yarns with low values of elongation are easier to spin and she is less concerned about the lower limit.

△ Review of Case 15.2

Sadah recommends that SeaDragon spinning machines for polyethylene, polystyrene, and polypropylene and blends of these should be capable of handling fibre with elongation up to at least 20%.

▽ Case 15.3 (AgroPharm)

Dion Darkred's Darkred Label blended red wine has always been popular. It is a blend of Cabernet Sauvignon, Shiraz, Merlot and Grenache but the proportions vary from year to year. This year, the members of his panel of tasters are enthusiastic to continue their work, and he decides to perform a mixture experiment. However, he thinks the success of Darkred Label is due partly to its being a blend of all four grape varieties and he wants to retain at least 0.1 of each in the mixture.

An extreme vertices design is appropriate when some components of the mixture cannot go as low as zero or as high as one. MINITAB can handle this with ease. Dion chooses the extreme vertices design by following **Stat > DOE > Mixture > Create Mixture Design** checks **Extreme Vertices** and sets **Number of Components** to **4**. He goes to **Designs** and selects **Degree of, design 1**, keeping the augmentation with centre and axial points and the 1 replicate. He clicks **OK** and then, in the **Components** window (Figure 15.8), he enters the variety names and specifies a range of 0.1 up to 0.7 for each variety. MINITAB generated a design as in Table 15.3. Note that the order has been randomised, but Dion has sorted the table into standard order.

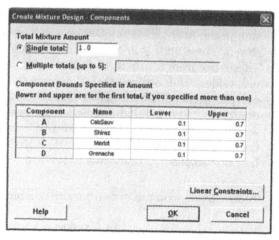

Figure 15.8

Tables 15.3 and 15.4 are on separate spreadsheets in Case 15.3 at www.greenfieldresearch. co.uk/doe/data.htm.

StdOrder	RunOrder	PtType	Blocks	CabSauv	Shiraz	Merlot	Grenache	taste
2	1	1	1	0.7	0.1	0.1	0.1	88
4	2	1	1	0.1	0.1	0.7	0.1	91
5	3	0	1	0.25	0.25	0.25	0.25	86.9
9	4	−1	1	0.175	0.175	0.475	0.175	89.3
6	5	−1	1	0.175	0.175	0.175	0.475	85.4
1	6	1	1	0.1	0.1	0.1	0.7	82
8	7	−1	1	0.175	0.475	0.175	0.175	89.8
7	8	−1	1	0.475	0.175	0.175	0.175	89.6
3	9	1	1	0.1	0.7	0.1	0.1	92

Table 15.3

A simplex representing the design space of four component variables is a tetrahedron as in Figure 15.9. It is the simplest three-dimensional object with straight edges. The design points are on the edges or on the faces of the tetrahedron or inside it; these are not shown in Figure 15.9.

It is hard to visualise points in three-dimensional space. MINITAB helps with **Graph > 3D Scatterplot**. Also follow **Tools > Toolbars > 3D Graph Tools**. You can use the graph tools to rotate the scatterplot until all three faces and all nine points of the design are visible as in Figure 15.10.

Figure 15.9

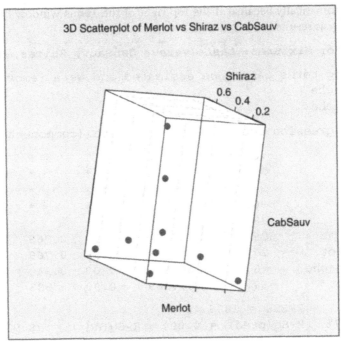

Figure 15.10

The extreme points are at the vertices of a tetrahedron. The other five points are inside the tetrahedron.

Stat > DOE > Mixture > Simplex Design Plot is restricted to plotting plane figures as in Figure 15.11. These do not show the interior points. Use **Select four components for a matrix plot** and **Proportions**.

Click on the **Settings** tab to choose set values for the fourth component in each of the three-component plots. In Figure 15.11, these are all set to 0.1.

Dion prepares the nine blends according to the designed experiment and submits them to the five tasters. He records their average scores in the table. He follows **Stat > DOE > Mixture > Analyze Mixture Design**. He defines **taste** as the response variable and checks **Mixture components only** and **Proportions**. He clicks on **Terms** and asks for main effects **(A, B, C, D)** and first-order interactions **(AB, AC, AD, BC, BD, CD)**. These terms

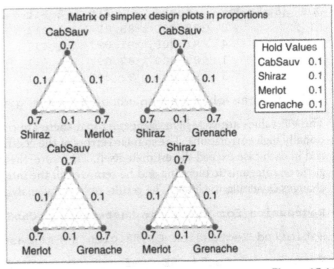

Figure 15.11

are specified automatically because at the top right of the **Terms** window he asks for a **Quadratic** model. The analysis results are as follows:

Regression for Mixtures: taste versus CabSauv, Shiraz, Merlot, Grenache

```
The following terms cannot be estimated and were removed:
Shiraz*Grenache
Merlot*Grenache
```

Estimated Regression Coefficients for taste (component proportions)

Term	Coef	SE Coef	T	P	VIF
CabSauv	83.10	7.700	*	*	26.66
Shiraz	97.76	7.700	*	*	26.66
Merlot	96.10	7.700	*	*	26.66
Grenache	78.43	7.700	*	*	26.66
CabSauv*Shiraz	26.67	70.303	0.38	0.769	73.09
CabSauv*Merlot	26.67	70.303	0.38	0.769	73.09
CabSauv*Grenache	−6.25	66.169	−0.09	0.940	64.75
Shiraz*Merlot	−59.58	66.169	−0.90	0.533	64.75

```
S = 1.41481     PRESS = 1806.48
R-Sq = 97.37%   R-Sq(pred) = 0.00%   R-Sq(adj) = 78.93%
```

Analysis of Variance for taste (component proportions)

Source	DF	Seq SS	Adj SS	Adj MS	F	P
Regression	7	74.0139	74.01386	10.57341	5.28	0.323
Linear	3	72.1695	7.84593	2.61531	1.31	0.554
Quadratic	4	1.8444	1.84436	0.46109	0.23	0.894
Residual Error	1	2.0017	2.00170	2.00170		
Total	8	76.0156				

Unusual Observations for taste

Obs	StdOrder	taste	Fit	SE Fit	Residual	St Resid	
1	2	88.000	88.097	1.411	−0.097	−1.00	X
2	4	91.000	91.097	1.411	−0.097	−1.00	X
6	1	82.000	82.097	1.411	−0.097	−1.00	X
9	3	92.000	92.097	1.411	−0.097	−1.00	X

X denotes an observation whose X value gives it large influence.

The **VIF** values are very high, indicating that there are colinearities in the data leading to exceptionally high correlations. The standard errors of the coefficients of interaction terms are very large and in each case exceed the estimate itself. Therefore, there is no substantial evidence of any synergistic or antagonistic blending and he removes all the interaction terms. In the **Terms** window, he changes **Quadratic** to **Linear**. The results of his new analysis are as follows:

Regression for Mixtures: taste versus CabSauv, Shiraz, Merlot, Grenache

Estimated Regression Coefficients for taste (component proportions)

Term	Coef	SE Coef	T	P	VIF
CabSauv	88.61	1.169	*	*	1.600

```
Shiraz      94.07     1.169   *   *   1.600
Merlot      92.41     1.169   *   *   1.600
Grenache    77.81     1.169   *   *   1.600
```

The regression is now highly significant (P = 0.001). The linear blending portion indicates that panel would rate a blend with 0.7 Shiraz the highest, with an estimated taste score of 88.61*0.1 + 94.07*0.7 + 92.41*0.1 + 77.81*0.1 = 91.7. But the Merlot is not statistically significantly worse and he could justify increasing its proportion. His final choice will depend on how much good quality wine he has of each variety. He can use the fitted equation to predict values.

If you do not have Pro you can plan your mixture design using **Calc > Make Patterned Data**. For a suitable design for Case 15.3, head four columns with the component variable names: **CabSauv, Shiraz, Merlot** and **Grenache**. Make patterned data for **CabSauv** using the range 0.1 to 0.7 in steps of 0.3, with **Number of times to list each value = 1** and **Number of times to times to list the sequence = 9**. Make patterned data for **Shiraz** using the range 0.1 to 0.7 in steps of 0.3, now with **Number of times to list each value = 3** and **Number of times to list the sequence = 3**. Make patterned data for **Merlot** using the range 0.1 to 0.7 in steps of 0.3, with **Number of times to list each value = 9** and **Number of times to list the sequence = 1**. This gives you **27** rows of data. But now you must add values into the **Grenache** column.

Use **Calc > Calculator** and compute **Grenache = 1.0 – (CabSauv + Shiraz + Merlot)**. Insert the brackets using the double brackets tab **( )** to the left of the **Not** tab. Do not use the symbols < and > as brackets. Now you have some negative values for **Grenache**. Use **Data > Sort**. Sort all columns by **Grenache**. This moves all the negative values together. Select these rows and use **Edit > Delete cells**. You now have ten rows (Table 15.4) that correspond closely to the design produced by Pro. Suppose Dion did this experiment with a different panel of tasters and obtained the average taste scores shown in Table 15.4.

You can analyse the results of this experiment, and of any mixture experiment, using regression. There are several approaches to this but the easiest is to omit the intercept term. If you do not omit the intercept term, MINITAB detects that Grenache is highly correlated with other variables and removes Grenache from the analysis. MINITAB then provides VIF values. When the intercept is omitted, no VIF values are given.

CabSauv	Shiraz	Merlot	Grenache	taste
0.4	0.4	0.1	0.1	90
0.4	0.1	0.4	0.1	89
0.1	0.4	0.4	0.1	91
0.7	0.1	0.1	0.1	88
0.1	0.7	0.1	0.1	92
0.1	0.1	0.7	0.1	91
0.4	0.1	0.1	0.4	85
0.1	0.4	0.1	0.4	87
0.1	0.1	0.4	0.4	86
0.1	0.1	0.1	0.7	82

Table 15.4

The results for this follow-up experiment, without the intercept, are:

Regression Analysis: taste versus CabSauv, Shiraz, Merlot, Grenache

```
The regression equation is
taste = 88.9 CabSauv + 93.9 Shiraz + 91.7 Merlot + 77.8 Grenache
```

Predictor	Coef	SE Coef	T	P
Noconstant				
CabSauv	88.933	1.227	72.48	0.000
Shiraz	93.933	1.227	76.56	0.000
Merlot	91.711	1.227	74.75	0.000
Grenache	77.822	1.227	63.43	0.000

S = 1.00554

We could include the four interaction terms that were selected by the Pro **DOE > Mixture** procedure.

```
The regression equation is
```

$$\text{taste} = 88.2\ \text{CabSauv} + 95.0\ \text{Shiraz} + 93.4\ \text{Merlot} + 77.5\ \text{Grenache}$$
$$-0.16\ \text{CabSauv}*\text{Shiraz} - 4.60\ \text{CabSauv}*\text{Merlot} + 0.79\ \text{CabSauv}*$$
$$\text{Grenache} -4.76\ \text{Shiraz}*\text{Merlot}$$

Predictor	Coef	SE Coef	T	P
Noconstant				
CabSauv	88.190	1.021	86.34	0.000
Shiraz	95.0000	0.8054	117.95	0.000
Merlot	93.4444	0.8054	116.02	0.000
Grenache	77.4762	0.6137	126.25	0.000
CabSauv* Shiraz	−0.159	3.927	−0.04	0.971
CabSauv*Merlot	−4.603	3.927	−1.17	0.362
CabSauv*Grenache	0.794	3.888	0.20	0.857
Shiraz*Merlot	−4.762	3.888	−1.22	0.345

S = 0.292770

There is no evidence of any synergistic or antagonistic blending from the data in Table 15.4. We can usefully fit four interactions out of six possibilities only because of the identification with the four squared terms.

△ Review of Case 15.3

Dion estimates that a blend of 70% Shiraz with 10% wine of each of the other grape varieties would have a taste score of 91.7. However, given his stocks of wine and his wish to declare a higher percentage of Cabernet Sauvignon than Grenache, he decides to blend 40% Shiraz, 30% Merlot, 20% Cabernet Sauvignon and 10% Grenache, with an estimated taste score of 90.9.

15.1 Mixtures generally

We have introduced several types of mixture designs through the cases in this chapter. Figure 15.12 illustrates some of the designs with three components. We recommend a careful study of MINITAB's help and manuals for fuller information.

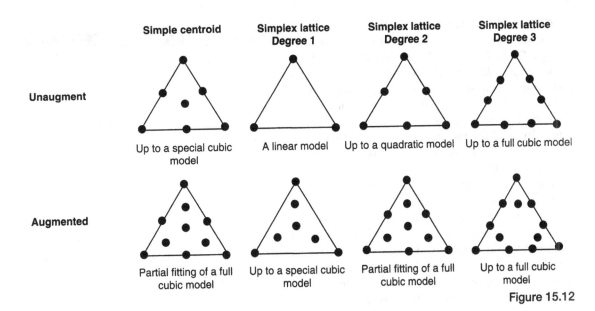

Simple centroid

Unaugment

Up to a special cubic model

Simplex lattice Degree 1

A linear model

Simplex lattice Degree 2

Up to a quadratic model

Simplex lattice Degree 3

Up to a full cubic model

Augmented

Partial fitting of a full cubic model

Up to a special cubic model

Partial fitting of a full cubic model

Up to a full cubic model

Figure 15.12

Each of the triangles in Figure 15.12 represents a two-dimensional plane in a three-dimensional space, as shown in Figure 15.13 and as we described at the beginning of this chapter.

An extreme vertices design uses tighter constraints within the natural limits of proportions (0, 1), as illustrated in Figure 15.14.

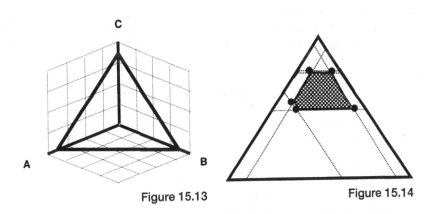

Figure 15.13

Figure 15.14

▽ Case 15.4 (AgroPharm)

Dion now considers a blend of Shiraz and Merlot, the grape varieties for which the vineyard is renowned. He and two colleagues will taste the trial blends and award each a score. There are also two process variables that he wishes to consider: the vintage of the wine and the temperature at which it is tasted.

The vintages are 2005 and 2004. Older Shiraz is sold as a vintage single variety rather than blended, and the Merlot is not usually kept longer than two years. The tasting temperatures are

17°C and **19°C**. He uses MINITAB's DOE Mixture routine for the design, following **Stat > DOE > Mixture > Create Mixture Design**, specifying **Number of components 2** and choosing **Simplex lattice**. Clicking on **Designs** (Figure 15.15), he specifies a degree of 2 to ensure that a quadratic model can be fitted. He also asks for the design to be augmented with a centre point and with axial points. Although Dion wants to replicate the design so that each taster will be presented with all combinations of component variables and process variables, in this window he leaves the number of replicates as 1. This is because he intends to add dummy variables for the tasters and, in this particular procedure, the replication order related to the tasters will be confusing. There is a better way to add replications that we shall explain later.

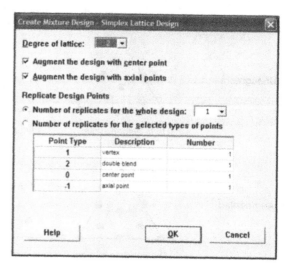

Figure 15.15

Returning to the **Create Mixture Design** window, Dion clicks on **Components** and names the variables **Shiraz** and **Merlot**. He clicks on **Process Vars** and specifies two process variables, naming these **Vintage** and **Temp**. Returning to the **Create Mixture Design** window, he chooses **Options** and removes the tick from **Randomize runs**. He will use another MINITAB procedure to randomise the runs after replicating the design for each taster. He clicks **OK** and the design of 20 runs appears in the worksheet: the basic lattice design has five points (Table 15.5) and is repeated for all four combinations of the full factorial design for vintage and temperature. The coding of vintage is –1 for younger (**2005**) and **+1** for older (**2004**). The coding for temperature is –1 for **17°C** and +1 for **19°C**.

Shiraz	Merlot
1	0
0	1
0.5	0.5
0.75	0.25
0.25	0.75

Table 15.5

To replicate the design, Dion follows **Stat > DOE > Modify Design** (Figure 15.16). He chooses **Replicate design**, clicks on **Specify** and then asks for two replicates to be added. This procedure ensures that there is a clear replicate for each of the three tasters. The worksheet has 60 rows. Fortunately, MINITAB numbers the replicates: **block 1, 2, 3**. These represent the tasters and can be referred to as tasters in the analysis.

Now he can randomise the runs. He returns to the **Modify Design** window (Figure 15.16), clicks on **Specify** and selects **Randomize entire design**. The full data, design and tasters scores are in Table 15.6. This is available in Case 15.4 at www.greenfieldresearch.co.uk/doe/data.htm.

Having set up the worksheet, he makes two bottles of each of the ten blends, which he labels with a code number. One bottle of each blend is stored in a temperature-controlled cabinet at **17°C**, and the other bottles are stored in a temperature-controlled cabinet at **19°C**. Three days later, they begin the tasting. Dion's daughter presents glasses to the three tasters in the MINITAB **Run Order**. The full data, design and tasters scores, are in Table 15.6. Note that values in the column headed **PtType** represent the type code of each point: type 0 is a centre point; type 1 is a vertex point; type −1 is an axial point. The column headed **Blocks** indicates the replication number. This corresponds with the taster number. Dion could change the column head to **Tasters** but MINITAB would ignore this: in the analysis, the results would still be labelled **Blocks**.

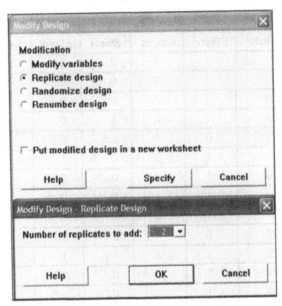

Figure 15.16

StdOrder	RunOrder	PtType	Blocks	Shiraz	Merlot	Vintage	Temp	Score
1	43	1	1	1	0	−1	−1	85
2	50	0	1	0.5	0.5	−1	−1	84
3	48	1	1	0	1	−1	−1	82
4	56	−1	1	0.75	0.25	−1	−1	90
5	59	−1	1	0.25	0.75	−1	−1	86
6	51	1	1	1	0	1	−1	86
7	60	0	1	0.5	0.5	1	−1	89
8	45	1	1	0	1	1	−1	82
9	46	−1	1	0.75	0.25	1	−1,	92
10	44	−1	1	0.25	0.75	1	−1	89
11	41	1	1	1	0	−1	1	85
12	52	0	1	0.5	0.5	−1	1	87
13	58	1	1	0	1	−1	1	83
14	54	−1	1	0.75	0.25	−1	1	89
15	49	−1	1	0.25	0.75	−1	1	86
16	47	1	1	1	0	1	1	92
17	57	0	1	0.5	0.5	1	1	94
18	42	1	1	0	1	1	1	82
19	55	−1	1	0.75	0.25	1	1	89
20	53	−1	1	0.25	0.75	1	1	92
21	31	1	2	1	0	−1	−1	82
22	36	0	2	0.5	0.5	−1	−1	87

Table 15.6

StdOrder	RunOrder	PtType	Blocks	Shiraz	Merlot	Vintage	Temp	Score
23	21	1	2	0	1	−1	−1	79
24	40	−1	2	0.75	0.25	−1	−1	88
25	25	−1	2	0.25	0.75	−1	−1	85
26	32	1	2	1	0	1	−1	86
27	29	0	2	0.5	0.5	1	−1	90
28	26	1	2	0	1	1	−1	80
29	38	−1	2	0.75	0.25	1	−1	87
30	34	−1	2	0.25	0.75	1	−1	86
31	27	1	2	1	0	−1	1	86
32	30	0	2	0.5	0.5	−1	1	88
33	22	1	2	0	1	−1	1	81
34	33	−1	2	0.75	0.25	−1	1	89
35	23	−1	2	0.25	0.75	−1	1	84
36	24	1	2	1	0	1	1	87
37	39	0	2	0.5	0.5	1	1	88
38	37	1	2	0	1	1	1	82
39	35	−1	2	0.75	0.25	1	1	94
40	28	−1	2	0.25	0.75	1	1	90
41	20	1	3	1	0	−1	−1	83
42	16	0	3	0.5	0.5	−1	−1	90
43	15	1	3	0	1	−1	−1	81
44	4	−1	3	0.75	0.25	−1	−1	88
45	7	−1	3	0.25	0.75	−1	−1	87
46	10	1	3	1	0	1	−1	91
47	9	0	3	0.5	0.5	1	−1	88
48	18	1	3	0	1	1	−1	85
49	2	−1	3	0.75	0.25	1	−1	93
50	5	−1	3	0.25	0.75	1	−1	89
51	14	1	3	1	0	−1	1	87
52	6	0	3	0.5	0.5	−1	1	89
53	17	1	3	0	1	−1	1	83
54	3	−1	3	0.75	0.25	−1	1	91
55	12	−1	3	0.25	0.75	−1	1	85
56	1	1	3	1	0	1	1	92
57	8	0	3	0.5	0.5	1	1	89
58	19	1	3	0	1	1	1	83
59	13	−1	3	0.75	0.25	1	1	86
60	11	−1	3	0.25	0.75	1	1	87

Table 15.6 (Continued)

A plot of **score** against proportion of **Merlot**, with all 12 combinations of vintage, temperature, and taster shown suggests that the there is a strong quadratic effect (Figure 15.17). There are no other striking features. The MINITAB analysis confirms this. There is also evidence that for **Shiraz**, at least, the older vintage improves the **score**. Another finding is weak evidence that **Shiraz** obtains a higher score at the higher temperature (P = 0.045).

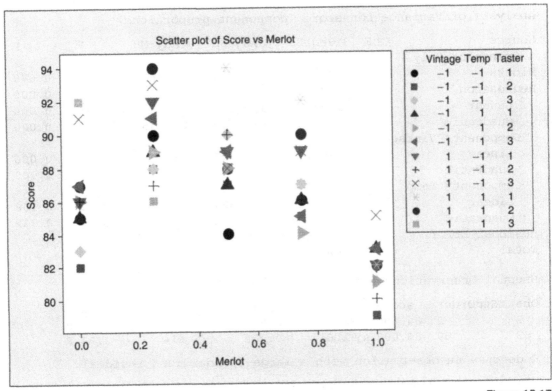

Figure 15.17

Regression for Mixtures: score versus Block, Shiraz, Merlot, ...

Estimated Regression Coefficients for score (component proportions)

Term	Coef	SE Coef	T	P	VIF
Block 1	0.367	0.3740	0.98	0.332	1.333
Block 2	−0.883	0.3740	−2.36	0.022	1.333
Shiraz	86.943	0.5566	*	*	1.661
Merlot	82.010	0.5566	*	*	1.661
Shiraz*Merlot	18.857	2.5289	7.46	0.000	2.429
Shiraz*Vintage	1.752	0.5566	3.15	0.003	1.661
Merlot*Vintage	0.819	0.5566	1.47	0.148	1.661
Shiraz*Merlot*Vintage	−0.952	2.5289	−0.38	0.708	2.429

Shiraz*Temp	1.143	0.5566	2.05	0.045	1.661
Merlot*Temp	0.476	0.5566	0.86	0.396	1.661
Shiraz*Merlot*Temp	−2.476	2.5289	−0.98	0.332	2.429

* NOTE * Coefficients are calculated for coded process variables.

S = 2.04868   PRESS = 295.407
R-Sq = 72.81%   R-Sq(pred) = 60.94%   R-Sq(adj) = 67.26%

Analysis of Variance for score (component proportions)

Source	DF	Seq SS	Adj SS	Adj MS	F	P
Component Only						
Blocks	2	23.633	23.633	11.817	2.82	0.070
Regression	8	527.043	527.043	65.880	15.70	0.000
Linear	1	182.533	182.533	182.533	43.49	0.000
Quadratic	1	233.357	233.357	233.357	55.60	0.000
Component* Vintage						
Linear	2	88.200	47.374	23.687	5.64	0.006
Quadratic	1	0.595	0.595	0.595	0.14	0.708
Component* Temp						
Linear	2	18.333	19.524	9.762	2.33	0.108
Quadratic	1	4.024	4.024	4.024	0.96	0.332
Residual Error	49	205.657	205.657	4.197		
Total	59	756.333				

Unusual Observations for score

Obs	StdOrder	score	Fit	SE Fit	Residual	St Resid	
2	2	84.000	88.319	0.806	−4.319	−2.29	R
59	59	86.000	91.614	0.728	−5.614	−2.93	R

R denotes an observation with a large standardized residual.

Dion notices that the analysis includes indicator variables **blocks 1** and **2**. These represent tasters 1 and 2. He and his colleagues are experts at wine tasting and they frequently make comparisons amongst themselves with the aim of having a common standard for taste scores. This is a large tasting experiment and Dion is pleased that he can check if there is evidence of any systematic difference.

There is some evidence (P = 0.022) that taster 2 may be giving lower scores than Dion, who is taster 3. The estimate of the average difference is −0.88. But Dion's main concern is to achieve the best possible blend. He repeats the analysis but rejects all terms except for the two component variables, the wines, and their interaction.

Regression for Mixtures: score versus Shiraz, Merlot

Estimated Regression Coefficients for score (component proportions)

Term	Coef	SE Coef	T	P	VIF
Shiraz	86.94	0.6640	*	*	1.661
Merlot	82.01	0.6640	*	*	1.661
Shiraz*Merlot	18.86	3.0168	6.25	0.000	2.429

The estimated coefficients are almost the same as shown with all terms included.

He uses these coefficients in Excel Solver to maximise:

$$\text{taste} = 86.94\,\text{Shiraz} + 82.01\,\text{Merlot} + 18.86\,\text{Shiraz*Merlot}$$

subject to

$$\text{Shiraz} + \text{Merlot} = 1$$

He estimates the optimum blend to be **0.63 Shiraz** and **0.37 Merlot** with a taste score of 89.5.

He uses these values and refers to the full regression analysis, for the coefficients of **Shiraz*Vintage**, **Merlot*Vintage**, and **Shiraz*Merlot*Vintage** to estimate the increase of taste score by using the older vintage:

$$1.752*0.63 + 0.819*0.37 - 0.952*0.63*0.37 = 1.18$$

This justifies a higher price for the older vintage blend.

△ Review of Case 15.4

Dion decides to blend 60% Shiraz with 40% Merlot. The predicted taste score is 89.5 for the more recent vintage of Shiraz and 90.7 for the older vintage.

16

Discrete response

16.1 Introduction

A discrete variable has distinct values, such as whole numbers or categories. In this chapter, we shall discuss the design and analysis of experiments in which the response variables are discrete. In particular, we introduce *logistic regression*. This is a versatile model for analysing the results of experiments in which the response is categorical. It is an extension of regression, and this is where you will find it in MINITAB (Figure 16.1). We shall also propose an informal method for estimating the sample size needed for a satisfactory experiment, or for assessing a proposed sample size. This method includes the simulation of an experiment and of the variation that we expect in an experiment. To do this, you will need to write a short program in MINITAB's macro language. An alternative would be to write a macro in Microsoft's Excel and to link the output of that to a MINITAB worksheet. In this chapter, we shall introduce MINITAB's macro language and demonstrate the simulation of an experiment in the context of logistic regression. The method, which will produce some plausible response data and estimates of standard coefficients in the model, is quite general and you can use it with other experimental models.

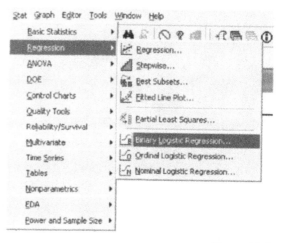

Figure 16.1

The simplest case of a discrete response is when there are just two categories that we usually call 'success' and 'failure'. Examples are potentially destructive tests that are used to assess the safety-critical features of products such as spectacle lenses, motorcycle crash helmets, vehicle petrol tanks and TV tubes. In a typical test, the experimenter will subject each item to a controlled impact and note whether or not it fails. Another class of manufacturing examples is whether or not a product satisfies all the clauses of some detailed specification. In such cases the 'yes' or 'no' response is a useful simplification. In agricultural trials, the response may be whether or not a seed germinates or a plant flowers. In sociological experiments, the response to some experimental situation, such as whether someone has abstained from smoking during a one-month period, may be 'yes' or 'no'. In medical experiments, the response may be 'recovered' or 'not recovered'.

Standard regression analysis is aimed at predicting the value, on a continuous scale, of a response. It is not suitable for predicting the value of a response that has only two possible and discrete values: 'yes' and 'no'. We need an analysis that will predict the probability of a 'yes' or 'no' response. We measure probability on a continuous scale but we constrain it to lie between zero and one. There are several approaches to this, but logistic regression has proved, generally, to be the most satisfactory. The theory behind this is briefly as follows. We have, as our response, a random variable Y that may be influenced by a set of predictor variables (X_i). But this variable (Y) can have values of only zero or one. Can we predict the probability (P_i) that a single value of Y will be one?

You might do this with a linear function of the X_i such as

$$P_i = \text{Prob } (Y_i = 1) = b_0 + b_1 X_1 + b_2 X_2 + b_3 X_3 + \cdots$$

You could try to fit this by ordinary linear regression but it would give unconstrained answers. That is, P could be any value from minus infinity to plus infinity, whereas probability is restricted to the range 0.0 to 1.0.

The first step of the logistic solution is to eliminate the upper bound of $P = 1$ by taking the odds: odds $= P/(1 - P)$. This ratio must be positive, since $0 < P < 1$, and there is no upper bound because, as P approaches 1.0, the ratio goes towards infinity.

The next step is to eliminate the lower boundary of $P = 0$ by taking the natural logarithm, $\log(\text{odds}) = \ln (P/(1 - P))$. This transformation is called the *logit* of the probability, and we can denote it as Z. The shape of the relationship is shown in Figure 16.2. You can see that, while the probability is constrained between zero and one, the range of the logit is from minus infinity to plus infinity, although 5 is close enough to infinity for most practical purposes.

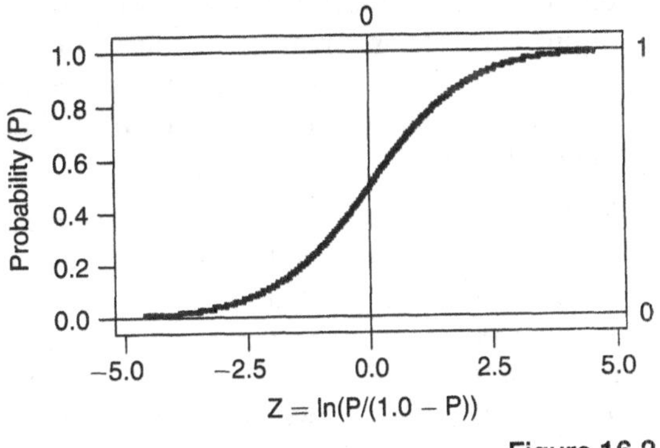

Figure 16.2

If we assume that the logit, Z, is linearly related to the predictor variables (X_i) as

$$Z = b_0 + b_1 X_1 + b_2 X_2 + b_3 X_3 + \cdots$$

MINITAB's logistic regression procedure will find the best fit, given a set of observed values of the predictor variables.

The reverse of the logit, known as *invlogit*, is the solution of $Z = \ln(P/(1 - P))$ for P:

$$P = \exp(Z)/(1 + \exp(Z))$$

After MINITAB has found the best fit, it will also provide the predicted values of probabilities for all cases in the data set, using the invlogit. Thus it can be used, predictively, as a diagnostic tool: given a set of values of the predictor variables for an item being tested, we can predict the probability that the item will succeed or fail; given a set of clinical test results for a patient, we can predict the probability that the patient has a particular condition.

▽ Case 16.1 (SeaDragon)

SeaDragon manufactures steel petrol tanks for a small company, Saturn Sports, that builds a popular sports car. Saturn is considering the use of a reinforced plastic petrol tank. This would save weight but the company is concerned that it would be more likely to shatter in a crash than a steel tank. AgroPharm could provide the plastic material if SeaDragon can produce suitable reinforcement material and moulds. Mould production is part of the work of the industrial machinery division, and Sadah Saffron has arranged for the manufacture of a prototype. She is looking forward to running an experiment with her friend Myfanwy Maroon at AgroPharm. They had been in the same class of mechanical engineering students at the University of Erewhon. From their general knowledge of strength of such materials, they anticipate that most tanks will survive an impact of 16 MJ/m², but expect a majority to fail under an impact of 21 MJ/m². They hope to be able to design a useful experiment using about 20 tanks, since a larger experiment would need approval from their divisional managers. But they decide to start with a non-destructive test: a deformation test in which deflection is measured at the centre of a tank.

Myfanwy is familiar with regression analysis when the response is a variable measured on a continuous scale. She can apply this to the deflections measured at the centre of the tank under impacts between 1.0 and 6.0 MJ/m², which should not be destructive. If she were sure that the relationship between impact and deflection over this range was approximately linear, the most efficient experimental design would be to test ten of the tanks at each end of the range. The limitation of this is that she would miss any evidence of curvature, such as was apparent in the plot of drying time of varnish against additive in Case 2.2. The same issue arose when we discussed two-level factorials and central composite designs. So, she should consider using the design with ten tanks at each end of the range only if she knows from experience that the relationship is approximately linear.

Myfanwy wonders how much better the design with ten tanks at 1.0 and ten tanks at 6.0 would be than a design with 20 tanks spaced at 0.25 intervals between 1.1 and 5.85. She can easily compare the alternative designs with MINITAB by a simple simulation. She sets up a column, **x1**, with ten 1s followed by ten 6s. She uses **Calc > Make Patterned Data > Simple Set of Numbers** with **Store patterned data in x1, From first value 1.0, To last value = 6.0, In steps of 5.0, Number of times to list each value 10; Number of times to list the sequence 1**.

Next to this she sets up a column, **x2**, with 20 numbers from 1.1 up to 5.85 in steps of 0.25. She uses **Calc > Make Patterned Data > Simple Set Of Numbers** to do this as in Figure 16.3. She then sets up a column of 20 random numbers from a normal distribution with zero mean and standard deviation 1.0, which she calls **nran**, using **Calc > Random data > Normal**, as in Figure 16.4.

She calculates **y1** as **x1** plus **nran**, and **y2** as **x2** plus **nran**. All these values are shown in Table 16.1.

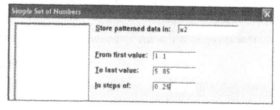

Figure 16.3

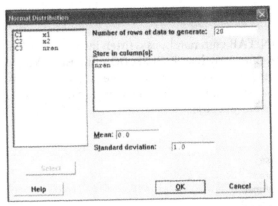

Figure 16.4

x1	x2	nran	y1	y2
1	1.10	−0.04	0.96	1.06
1	1.35	−0.95	0.05	0.40
1	1.60	0.81	1.81	2.41
1	1.85	0.11	1.11	1.96
1	2.10	1.48	2.48	3.58
1	2.35	1.18	2.18	3.53
1	2.60	−1.07	−0.07	1.53
1	2.85	−1.08	−0.08	1.77
1	3.10	−0.45	0.55	2.65
1	3.35	−0.13	0.87	3.22
6	3.60	1.04	7.04	4.64
6	3.85	−0.04	5.96	3.81
6	4.10	−0.32	5.68	3.78
6	4.35	−1.24	4.76	3.11
6	4.60	0.96	6.96	5.56
6	4.85	−0.53	5.47	4.32
6	5.10	−0.17	5.83	4.93
6	5.35	0.10	6.10	5.45
6	5.60	−1.15	4.85	4.45
6	5.85	2.80	8.80	8.65

Table 16.1

Using **Stat > Regression > Regression**, she regresses **y1** on **x1** and **y2** on **x2** and compares the estimates and standard errors of the estimators of the coefficients.

Regression Analysis: y1 versus x1

```
The regression equation is
y1 = −0.045 + 1.03 × x1

Predictor      Coef    SE Coef       T       P
Constant    −0.0455    0.4114   −0.11   0.913
x1          1.03153   0.09564   10.79   0.000
```

Regression Analysis: y2 versus x2

```
The regression equation is
y2 = −0.114 + 1.05 × x2

Predictor      Coef    SE Coef       T       P
Constant    −0.1137    0.6242   −0.18   0.857
x2          1.0514    0.1659    6.34   0.000
```

The more efficient design is the first because the standard error of the slope (0.096) is little more than half the standard error of the second (0.166). This is the result of just one simulation, but for a standard regression analysis the ratio of the standard errors of the slopes from the two designs (0.096/0.166 = 0.58) depends only on the choices for xs, and not on the values of the response variable. You can verify this using the numbers MINITAB gives you in **nran**. However, this is not so for discrete response designs and, if Myfanwy wants to investigate sample sizes for these designs using similar simulations, she needs to automate the procedure using a MINITAB macro.

16.2 Macros

A macro is a small program that uses mainly MINITAB commands, as written into or recorded in the session window. You can also add control statements such as **IF**, **ELSE**, **DO** and **WHILE**. You will write a macro into a text file, using a text processor such as Microsoft's Notepad. There are two types of macro: *global* and *local*. In this chapter, we shall illustrate just the global macro because it is simpler than the local macro and is suitable for our purpose.

The structure of a global macro is

> GMACRO
> *template*
> *body of macro*
> ENDMACRO

GMACRO and **ENDMACRO** mark the beginning and end of a global macro. **GMACRO** must be the first line because it labels the macro type as global, rather than local. **ENDMACRO** ends the macro command.

In a global macro, the word 'template' simply means the name of the macro and that can be anything you like. When you store the macro, you can use the same name for the file, but this is not essential. The file name extension is 'MAC', by default, but not necessarily so. When you invoke a macro you preface the file name with a percent sign (**%**). If the extension is not **MAC**, then you must include the extension. For example:

Template	File name	Invoked by
MyMacro	MYMACRO.MAC	%MYMACRO
MyMacro	MYMACRO.TXT	%MYMACRO.TXT

The body of a global macro contains MINITAB commands and control statements. It can also contain statements to invoke other global macros.

You can write a macro simply by recording a sequence of MINITAB commands.

1. Execute a series of commands using either menu commands or session commands.
2. Click on the **History** folder in the **Project Manager**. This folder displays the most recent commands (just commands, not output) executed in your session.
3. Highlight the commands you want to include in your macro, copy and paste into a word processor.
4. Change any commands if you wish and insert three lines to include GMACRO, the template and ENDMACRO.
5. Save the updated global macro file in text-only format, with a file name and the file extension MAC, to the Macros subdirectory of your main MINITAB directory. Note that, if you use Notepad, the **Save** window offers you a choice of two types of file: **Text Document (*.txt)** and **All Files**. If you choose the former and then add the extension. **MAC**, the file will be saved with two extensions, example.**MAC.txt**. MINITAB won't be able to find this when you invoke the name.

By following this procedure, you needn't worry about the fine details of program statements. MINITAB takes the worry away.

As a simple example of a macro, follow the procedure that Myfanwy used to generate the data of Table 16.1. Then open the **History** window (**Window > Project Manager**, then click on the **History** folder). Highlight and copy the text (see Figure 16.5). Now open **Notepad (Tools > Notepad)** and paste the text into it. Add **GMACRO**, a name (such as **example1**), and **ENDMACRO**. Ensure that, in the macro, you name the columns **c1, c2, …** then add the names at the end of the macro. The complete text will be

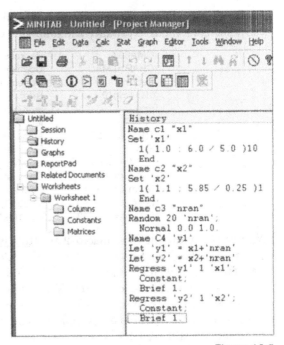

Figure 16.5

```
GMACRO
example1
Set c1
   1(1:6/5)10
   End.
Set c2
   1(1.1:5.85/.25)1
End.
Random 20 c3;
   Normal 0.0 1.0.
Let c4 = c1 + c3
Let c5 = c2 + c3
name c1 'x1'
name c2 'x2'
name c3 'nran'
name c4 'y1'
name c5 'y2'
ENDMACRO
```

Note that when a command, such as **Random 20 c3**, includes a sub-command, such as **Normal 0.0 1.0**, the command must be followed by a semi-colon and every sub-command except the last one must be followed by a semi-colon. The last one must be followed by a stop.

You can copy the file from **macro_1** on www.greenfieldresearch.co.uk/doe/data.htm. Paste it into Notebook and save the file as **example1.MAC** in the directory **Program Files > MINITAB 14 > Macros**.

Now run (invoke) the macro. Focus on the session window; follow **Editor > Enable commands**; type, after **MTB > %example1**. You will see columns in the worksheet fill with numbers.

▽ **Case 16.1 continued**

Myfanwy decides to verify that the **SE Coef** calculated from a single regression is plausible by writing a macro to simulate 100 runs of the experiment. She constructs most of it by working through a sequence of menu commands. Some of them are the same as in our short example.

Myfanwy copies the generated commands from the session window and pastes them into Notepad. Now she edits them to produce the following complete macro:

```
GMACRO                Regress c6; c1;       erase c6
CHAP16.MAC            Coefficients c6;      enddo
Set c1                Constant;             name c7 'enddint'
1(1: 6/5 )10          Brief 2.              name c8 'enddslo'
End.                  let c7(k1) = c6(1)    name c9 'disdint'
Set c2                let c8(k1) = c6(2)    name c10 'distslo'
1(1.1:5.85/.25 )1     erase c6              describe c7 c9 c8 c10;
End.                  Regress c5 1 c2;      mean;
do k1 = 1:100         Coefficients c6;      Stdev;
Random 20 c3;         Constant;             N.
Normal 0.0 1.0.       Brief 2.              ENDMACRO
Let c4 = c1 + c3      let c9(k1) = c6(1)
Let c5 = c2 + c3      let c10(k1) = c6(2)
```

You can copy the file from **macro_2** at http://www.greenfieldresearch.co.uk/doe/data.htm. Paste it into Notebook and save the file as **example2.MAC** in the directory **Program Files > MINITAB 14 > Macros**. As well as adding GMACRO and ENDMACRO, she makes other changes. She adds a loop, starting with **do k1 = 1:100** and ending with **enddo**. She introduces **Brief 2** statements to restrict the amount of output in the session window (see MINITAB **Help** for detailed description). In fact, she could use **Brief 0** to suppress all output since the data is stored and the coefficients and constants are still available. During each loop, the coefficients of each regression are put into **c6**. From there coefficients of the first regression are copied into **c7** and **c8**, using the loop index, **k1**; then the coefficients of the second regression are copied into **c9** and **c10**, also using the loop index, **k1**. Columns **c7** to **c10** are renamed. The name **enddint** combines **endd** for the end points design and **int** for the intercept. **slo** represents the slope. **disd** represents the distributed design. Finally, she orders descriptive statistics with **describe c7 c9 c8 c10**, specifying **mean**, **StDev** and **N**. Note that semicolons are used to separate sub-commands from commands and their list ends with a stop.

Myfanwy saves the macro in the MINITAB macros directory with the name **example2.mac**. Now she invokes the macro by typing **%example2** in the session window. The results from a run of this macro are as follows:

Descriptive Statistics: enddint, disdint, enddslo, distslo

Variable	N	Mean	StDev
enddint	100	0.0137	0.3576
disdint	100	0.0161	0.5496
enddslo	100	0.99517	0.08397
distslo	100	0.9945	0.1453

If you run the macro, you will get different random numbers and hence similar, but not identical, results. Notice that we have estimated the standard error of the estimators by calculating the standard deviations of the estimates from 100 simulations, rather than from a single sample as in a single regression analysis. The single regression gave estimated standard errors of the estimators of slope as 0.096 and 0.166 for the end-point and distributed design, respectively. The simulation gives 0.084 and 0.145, and the estimates from the single regression are well within sampling error of these values. The ratio 0.084/0.145 = 0.58 is the same as in the single regression. The estimator of the intercept is also more precise with the end-point design, because the intercept depends on both the mean deflection and the slope. If we used a centred variable for impact, the precision of estimators of the intercept would be identical.

The practical conclusion of the exercise is that the end-point design, with ten points at each end, is substantially more efficient than a distributed design. Since standard errors are inversely proportional to the square root of the sample size, she would need nearly 80 tanks for a distributed design to provide the same efficiency as the end-point design.

Myfanwy remains sceptical about end-point designs. She wonders if a compromise design, such as eight points at each end and four in the middle, would result in much loss of efficiency. It would be easy to adapt the macro to investigate this. But it is not necessary to do so. The four points in the middle provide no information about the slope. The standard error of the slope would increase by a factor of $\sqrt{(20/16)}$, about 13%, but she'd have some evidence to support or dismiss a straight line model.

That was the deformation experiment. The design issue for the destructive testing experiment that Myfanwy and Sadah have proposed is more subtle.

They expect all tanks to survive an impact of $14\,MJ/m^2$, and would be rather surprised if any could withstand an impact of $23\,MJ/m^2$. However, it is certainly not a good idea to test one half of the tanks at $14\,MJ/m^2$ and the other half at $23\,MJ/m^2$. If all the former survive and all the latter fail they would have no idea if the cut-off between surviving and failure is steep, for example all tanks surviving 18 and all failing at 19, or more gradual, with half the tanks surviving 17 and half failing at 20. The situation could be improved by testing some tanks in the middle of the range, because this would help to locate the impact at which half the tanks are expected to fail, but they would still have little idea about the steepness of the cut-off. In such cases, the binary logistic regression algorithm will:

1. tend to make the cut-off as steep as is possible, while remaining consistent with the data; display a warning about failure to converge; and,
2. quite reasonably, give very large standard errors for the coefficients in the relationship which relates the impact to the probability a tank survives.

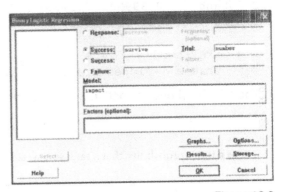

Figure 16.6

Their common sense suggests that they would do better to test tanks at, at least, six points within a range from 16 to 21, and Myfanwy suggests the following design requiring 24 tanks: two at each of 16 and 21, four at each of 17 and 20, and six at each of 18 and 19 MJ/m². She has made up some test results, so that they can try the logistic regression procedure.

Sadah types the test **impact** in column 1, the **number** of tanks tested at each impact in column 2 and the imagined number of tanks surviving in column 3, **survive**. She now selects **Stat > Regression > Binary Logistic Regression** (see Figure 16.6). MINITAB models the probability of success, for a response that must consist of just 0s and 1s, with 0 corresponding to failure and 1 corresponding to success. However, the results of an experiment can be summarised by the *number* of successes, or the *number* of failures, out of a particular number of trials at different factor settings, as in Table 16.2. Sadah's response variable is **survive**.

impact	number	survive
16	2	2
17	4	3
18	6	4
19	6	4
20	4	2
21	2	1

Table 16.2

She has the choice of description categories. She chooses the first, **Success**, for which she enters **survive** and, after **Trial**, she enters **number**. The model is **impact**. Thus, MINITAB will model the probability of a tank surviving an impact. MINITAB **Help** describes the alternatives, although you may think that the screen is self-explanatory. She clicks on **Storage** and chooses to store coefficients. After the analysis, these appear in the fourth column of the worksheet. She runs the logistic regression with commands enabled because they plan to incorporate the syntax into a macro.

Binary Logistic Regression: survive, number versus impact

```
Link Function: Logit

Response Information

Variable     Value    Count
survive      Success     16
             Failure      8
number       Total       24

Logistic Regression Table
                                               Odds    95%     CI
Predictor       Coef    SE Coef      Z      P Ratio  Lower  Upper
Constant     8.39958    6.38894   1.31  0.189
impact      -0.413714  0.339978  -1.22  0.224   0.66   0.34   1.29


Log-Likelihood = -14.472
Test that all slopes are zero:  G = 1.609,  DF = 1,  P-Value = 0.205
```

The MINITAB output continues, but this suffices for the moment. The fitted model is

$$\text{logit}(p) = 8.40 - 0.414\ \text{impact}$$

where p is the probability that a tank survives. We rearrange this to make p the subject of the formula:

$$p = \exp(8.40 - 0.414\ \text{impact})/(1 + \exp(8.40 - 0.414\ \text{impact}))$$

Sadah used the fitted model to calculate the logits and the predicted values of probability of survival, as in Table 16.3, where the equation for predicted probability is abbreviated to **EL/(1 + EL)**. Predicted probabilities of success are plotted against impact in Figure 16.7. They are also plotted against the logits in Figure 16.8, which can be compared with the graph in Figure 16.2.

impact	number	survive	logit	exp(logit)	EL/(1 + EL)
16	2	2	1.78	5.91	0.86
17	4	3	1.37	3.90	0.80
18	6	4	0.95	2.58	0.72
19	6	4	0.54	1.71	0.63
20	4	2	0.13	1.13	0.53
21	2	1	-0.29	0.75	0.43

Table 16.3

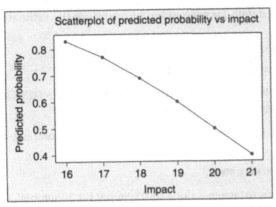

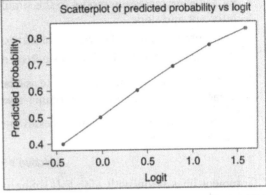

Figure 16.7   Figure 16.8

They will apply the same principle that they used when investigating the deflection and compare designs by adding random variation to some assumed model. However, in this case the random variation is binomial.

Sadah now writes the macro, copying most of the syntax from the session window. Comment lines begin with a hash (#). She needs to assign values to **k1** and **k2** before running the macro. She stores the simulated samples as well as the estimated coefficients.

Remember that an experimental design is the specification of the conditions at which experimental data will be observed.

In this case, Myfanwy and Sadah want an experiment that spans the range over which some tanks will fail and some will not fail. They also want the numbers of tests at all of the design points to be distributed so as to give precise estimates of the coefficients of the fitted function: the logistic regression function. They also need to specify how many tanks they might sacrifice so that they can estimate the cost of the experiment. They hope that 24 tanks will be enough. So their approach is as follows:

1. Specify the range of impacts over which the tests will be run. They choose: 16, 17, 18, 19, 20, and 21 MJ/m^2.

2. Specify three alternative designs: three sets of numbers of tests. They choose a uniform allocation, an allocation with weighting towards the middle of the range of impacts, and an allocation with weighting towards the ends of the range of impacts. Each of the alternative designs will have 24 tanks (Table 16.4).

Impact	16	17	18	19	20	21
Design 1	4	4	4	4	4	4
Design 2	2	4	6	6	4	2
Design 3	6	4	2	2	4	6

Table 16.4

3. Simulate a large number of possible outcomes for each experimental design. That is: each outcome will be a set of random numbers for the six points in the design; each random number can be 0, 1, . . . , N, where N is the number of tanks at that point. For example, if there are four tanks at 17 MJ/m^2, the random number can be 0, 1, 2, 3 or 4. The random number is drawn from a binomial distribution with N trials and probability of success p. She decides to simulate 1000 outcomes for each alternative design.

4. Fit a logistic regression model to the simulated data for each outcome and store the estimated intercept and slope.
5. Compute the standard deviation of all the stored intercept and slope values.
6. Compare the standard deviations between the alternative designs.

Paste it into Notebook and save the file as **example3.MAC** in the directory **Program Files > MINITAB 14 > Macros**. Invoke the macro by typing **%example3** in the session window.

The model is that the number of successes in N trials with probability of success p is binomial with probability of success p given by

$$\text{logit}(p) = \text{k1} + \text{k2} \times \text{impact}$$

Comment lines begin with a hash (#). She needs to assign values to **c1 'impact' c2 'number'**, **k1** and **k2** before running the macro. She stores the simulated samples as well as the estimated coefficients.

Two difficulties arise. First, the algorithm for logistic regression is not a direct calculation; it is iterative. That is, each time a set of coefficients is estimated, this set is used as the starting set for the next iteration. The default number of iterations in MINITAB is 20. This is reasonable, but, occasionally, a simulated set of numbers for success leads to non-convergence. This happens when there is a pattern of successes below one particular impact and no successes above it. Sadah allows for this in her macro by running a 21st iteration. She tests if the difference between the estimated intercepts on the 20th and the 21st iterations is small. If it is, then the algorithm has converged and she can accept the results; otherwise she rejects them. If non-convergence happens more than a few times, the sample size is either too small or the design points are too far apart near the middle of their range. She includes a counter to keep track of the number of non-convergences in the macro.

Second, the procedure generally prints out a lot of information in the session window that Sadah doesn't need. She avoids this by inserting **Brief 0** as a subcommand with each of the **Blogistic** commands. Also, since she wants to calculate the standard deviations of the 100 intercepts and slopes, she originally printed these in columns on the worksheet and then used MINITAB's **Describe** to calculate the standard deviations. This is slow so she speeded the process by using an updating procedure for calculations of mean and variance. This procedure uses the following recurrence relations:

$$m_i = [(i - 1)m_{i-1} + x_i]/i$$
$$Q_i = Q_{i-1} + (i - 1)(x_i - m_{i-1})^2/i$$

where m_i is the mean of first i observations and Q_i the sum of squares of deviations of the first i observations about their mean. Initial values are $m_1 = x_1$ and $Q_1 = 0$.

Sadah writes the macro and stores it with the name case16_1.mac. She enters, on the worksheet, data for the first design as in Table 16.5, leaving blanks in the survive column. Then, in the session window, she enters the commands

MTB > let k1 = 8

MTB > let k2 = −.4

MTB > %case16_1

impact	number	survive
16	4	
17	4	
18	4	
19	4	
20	4	
21	4	

Table 16.5

```
GMACRO
CASE16_1.MAC

# Assign values to k1
# and k2 before running
# k101 counts non-convergences

let k101 = 0
let c5 = exp(k1 + k2*c1)
let c6 = c5/(1 + c5)

let k3 = 0 # counter for while loop

let k11 = 0
# updating mean intercept
let k12 = 0 # used in updating
# standard deviation of intercept
let k14 = 0 # updating mean slope
let k15 = 0
# used in updating standard
# deviation of slope

#####
## use a WHILE loop to ensure that the
# number of runs rises to 1000 even
# if some runs are rejected
# because of non-convergence

while k3 lt 1000

## random number generation for survive
do k9 = 1:6
let k22 = c2(k9)
let k66 = c6(k9)
random 1 c100;
binomial k22 k66.
let c3(k9) = c100
erase c100
enddo

#####
## binary logistic regression with default
# of 20 iterations
## store coefficients in column 7
## Brief 0 suppresses output of results

BLogistic C3 C2 = C1;
ST;
Logit;
Coefficients c7;
Brief 0.

## run the 21st iteration
## store coefficients in column 8
BLogistic C3 C2 = C1;
ST;

Logit;
Coefficients c8;
Brief 0;
Iteration 21.

## k4 = difference between intercepts
# iterations 20 and 21

let k4 = abso(c7(1) − c8(1))

## if the difference is small, accept the
# result and add intercept and slope
# values to the running mean and
# variance calculations

if k4 < 0.05 # something small
let k3 = k3 + 1
else k101 = k101 + 1
endif

## calculate running mean and variance
## intercept
let k13 = c7(1) − k11
let k11 = ((k3 − 1)* k11 + c7(1))/k3
let k12 = k12 + k13*k13*(k3− 1)/k3

## slope
let k16 = c7(2) − k14
let k14 = ((k3 − 1)* k14 + c7(2))/k3
let k15 = k15 + k16*k16*(k3− 1)/k3

endif

endwhile

let c11(1) = k11 ## mean of intercept
let c12(1) = sqrt(k12/(k3− 1))
## SD of intercept
let c13(1) = k14 ## mean of slope
let c14(1) = sqrt(k15/(k3− 1))
## SD of slope

name c11 'intMean'
name c12 'intSD'
name c13 'slopeMean'
name c14 'slopeSD'

print 'number non-convergences in 1000'
print k101

ENDMACRO
```

Repeating the run with designs 2 and 3, she obtains a set of values as in Table 16.6.

Design 3 has smaller standard deviations than the other two designs but she is concerned that, even for this design, the standard deviations of the means and the slopes are almost as large as the means and the slopes. She wonders how the results would be affected if more tanks were tested. This should give smaller estimates of the standard deviations. She repeats the exercise with double the number of tanks tested at each impact and obtains a set of values as in Table 16.7. Again, design 3 has the smallest standard deviations.

	intMean	intSD	slopeMean	slopeSD
Design 1	9.160	6.994	−0.458	0.368
Design 2	9.868	8.324	−0.496	0.442
Design 3	9.544	6.486	−0.478	0.334

Table 16.6

	intMean	intSD	slopeMean	slopeSD
Design 1	8.659	4.142	−0.435	0.219
Design 2	8.650	4.872	−0.433	0.260
Design 3	8.450	3.577	−0.423	0.187

Table 16.7

The question now is whether Sadah and Myfanwy can persuade their managers to test the larger number of tanks. Their argument is that with the smaller number there is a risk that even the 90% confidence interval for a coefficient, which for physical reasons can only be negative, is so wide that it includes positive values. Then the whole trial and its cost could be wasted. The larger number gives the company a much greater chance of reaching a useful and statistically significant conclusion.

▽ Case 16.2 (UoE)

Ingrid Indigo would like to help people quit cigarette smoking. But what is the best way to do this? She decides to compare several strategies. She will recruit volunteers for the trial with the sole proviso that they smoke cigarettes and express a wish to stop smoking. She tells volunteers that she will randomly assign them to either of two treatment groups or to a control group. She gives each a brochure, commissioned by the Erewhon minister for community health, designed to encourage people to stop smoking, and she tells them that they will complete a questionnaire, with the help of an interviewer, after six weeks.

This is the extent of the help given to the control group (treat = 1). The first treatment (treat = 2) is an invitation to attend four classes that provide an invitation to various techniques, such as yoga, for breaking addictive habits. The second treatment (treat = 3) includes the offer of the same classes, but these are augmented by one month's supply of nicotine patches, with a decreasing dose of nicotine. For the first two weeks the patches contain 21 mg nicotine, and this is reduced to 14 mg and 7 mg for the third and fourth week. Participants will be followed up after six months. The sex, age and level of education will be recorded for all volunteers, and the number of classes attended will be recorded for those volunteers assigned to one of the active treatments.

While giving up smoking for a month is an excellent start, it is common for people to lapse into old habits. So, the participants will also be followed up after six months, but we do not present those data here.

It may be impractical to set precise sample sizes in studies, such as clinical trials, that rely on recruiting volunteers. Nevertheless, Ingrid must have some idea of the sample size she needs to make the study worthwhile. The cost of a treatment would be well justified, in cost–benefit terms, if the proportion of participants on the treatment who stop smoking is 0.5 while the proportion for the control group is only 0.2. She uses MINITAB's comparison of two proportions to investigate the width of the confidence intervals for different sample sizes if the results happened to be proportions of 0.5 and 0.2. She does this by using **Stat > Basic Statistics > 2 Proportions**. She clicks on **Summarized data** and enters **First: 10 trials, 5 events** and **Second: 10 trials, 2 events**. She repeats this several times with different numbers and obtains the following results:

Test and CI for Two Proportions (N = 10)

```
Sample   X   N   Sample p
1        2   10  0.200000
2        5   10  0.500000
```

Difference = p(1) − p(2)
Estimate for difference: −0.3
95% CI for difference: (−0.696862, 0.0968625)
Test for difference = 0 (vs not = 0): Z = −1.48 P-Value = 0.138

Test and CI for Two Proportions (N = 20)

```
Sample   X    N   Sample p
1        4    20  0.200000
2        10   20  0.500000
```

Difference = p(1) − p(2)
Estimate for difference: −0.3
95% CI for difference: (−0.580624, −0.0193759)
Test for difference = 0 (vs not = 0): Z = −2.10 P-Value = 0.036

Test and CI for Two Proportions (N = 30)

```
Sample   X    N   Sample p
1        6    30  0.200000
2        15   30  0.500000
```

Difference = p(1) − p(2)
Estimate for difference: −0.3
95% CI for difference: (−0.529129, −0.0708714)
Test for difference = 0 (vs not = 0): Z = −2.57 P-Value = 0.010

Test and CI for Two Proportions (N = 40)

```
Sample   X    N   Sample p
1        8    40  0.200000
2        20   40  0.500000
```

```
Difference = p(1) - p(2)
Estimate for difference: -0.3
95% CI for difference: (-0.498431, -0.101569)
Test for difference = 0 (vs not = 0):  Z = -2.96  P-Value = 0.003
```

Confidence intervals and P-values from these results are in Table 16.8. These are plotted in Figure 16.9: P-values in the upper part and the 95% confidence intervals in the lower part. If she wishes to halve the width of the confidence interval, she will have to increase the sample sizes by a factor of 4. This shows clearly in the figure.

N	LCI	UCI	P
10	−0.697	0.097	0.138
20	−0.581	−0.019	0.036
30	−0.529	−0.071	0.01
40	−0.498	−0.102	0.003

Table 16.8

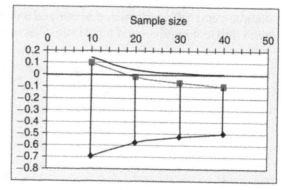

Figure 16.9

She would be happy if the results of her study turned out like the interval with samples of size 40. But such a difference in the population will not necessarily appear in the sample results. She decides to investigate the power, if she has two samples of size 40, more formally, using **Stat > Power and Sample Size > 2 Proportions**:

```
Power and Sample Size

Test for Two Proportions

Testing proportion 1 = proportion 2 (versus not =)
Calculating power for proportion 2 = 0.5
Alpha = 0.05

                Sample
Proportion 1    Size       Power
0.2              40     0.815532

The sample size is for each group.
```

Ingrid is somewhat reassured by this result. A power of 0.8 when testing at the 5% level is respectable. She thinks that it should be possible to recruit 120 volunteers, which would give 40 in each of the treatment groups.

She also needs to decide whether all volunteers need to start at the beginning of the study or whether they can join at any time during the period it is running. She is likely to attract more volunteers in the latter case, but there is a limit to the overall duration of the study. She has funding for one year, and decides to allow entry to the study over a ten-month period, hoping to recruit at least 120 people. Nevertheless, she expects the majority of participants will volunteer at the beginning of the study.

Although the aim of the treatment is that participants should give up smoking, and Ingrid based her sample size calculations on the basis of proportions of smokers who give up, she would like to account for how heavily they smoke at the time of entry to the study, and whether they achieve at least a substantial reduction in cigarette consumption, in the analysis. So, she decides to set up the following categories, corresponding to: not smoking; light smoking; moderate smoking; heavy smoking; and intensive smoking. The categories are chosen to correspond to numbers of packs of 20 cigarettes bought each week, as she doubts that participants can describe their smoking habits with any greater precision. The categories and codes are in Table 16.9.

Packs per week	Code	Category name
None	0	None
1	1	Light
2–4	2	Moderate
5–7	3	Heavy
8 or more	4	Intense

Table 16.9

The categories are described as *ordinal* because their order represents increasingly heavy smoking. The numeric coding follows this same order. The MINITAB ordinal logistic regression will enable her to model how the probabilities of moving between categories depend on the treatment and the other factors, such as age, that she has recorded. The initial allocation of participants to treatments was randomised subject to a restriction that there were equal numbers in each group. As more patients entered the study, they were randomised to groups subject to the condition that 12 consecutive entrants were split equally between the groups. You can find the complete table of results on the website (http://www.greenfieldresearch.co.uk/doe/data.htm). The first ten lines of the table are in Table 16.10.

ID	age	treat	enterstat	sex	education	attend	agecat	outstat
1	64	1	4	0	2		3	3
2	64	2	3	0	1	4	3	2
3	51	3	2	1	1	4	3	0
4	38	1	3	0	0		2	2
5	28	2	2	0	1	4	1	0
6	24	3	2	0	0	4	1	2
7	32	1	3	1	0		2	2
8	42	2	2	0	2	4	2	1
9	23	3	2	0	0	4	1	0
10	48	1	3	1	0		2	4

Table 16.10

Smoking status at entry to the trial is recorded in variable **enterstat** and status is at the end of the trial is in variable **outstat**. The treatment category is in variable **treat**. She also records **sex** (F = 0, M = 1), **age**, age category (**agecat** = 1 for ages under 30; **2** for ages 30–49; **3** for ages above 49), education level (**education** = 0 for up to school year 10; **1** for up to school year 12; **2** for tertiary), and number of attendances (**attend**) for volunteers assigned to the active treatments. The missing values in the attend column correspond to people assigned to the control group.

Ingrid starts the analysis with tables of the leaving state by entering state for the two treatments and control. She uses **Stat > Tables > Cross Tabulation and Chi-Square**; her categorical variable **For Rows** is enterstat, **For columns** is outstat, and **For layers** is treat:

Tabulated statistics: enterstat, outstat, treat
Results for treat = 1

Rows: enterstat   Columns: outstat

	0	1	2	3	4	All
1	2	6	0	0	0	8
2	2	1	4	0	0	7
3	0	4	7	5	2	18
4	1	0	1	4	1	7
All	5	11	12	9	3	40

Cell Contents: Count

Results for treat = 2

Rows: enterstat   Columns: outstat

	0	1	2	3	4	All
1	4	3	0	0	0	7
2	3	2	7	0	0	12
3	1	4	4	0	1	10
4	1	2	0	8	0	11
All	9	11	11	8	1	40

Cell Contents: Count

Results for treat = 3

Rows: enterstat Columns: outstat

	0	1	2	3	4	All
1	5	7	0	0	0	12
2	3	2	7	0	0	12
3	3	4	1	1	0	9
4	1	1	0	5	0	7
All	12	14	8	6	0	40

Cell Contents: Count

Her first impression is that participants on the active treatments have achieved rather better results than those on the control, since 21/80 stopped smoking at week 6 compared with 5/40. However, the proportion who gave up smoking is below the 0.5 hoped for.

She continues, using ordinal logistic regression analysis. But first she edits some of the data. The response is smoking status at the end of six weeks. She wishes to include: treatment; smoking status at entry to the trial; sex; education; and age category in the model. She would also like to account for the number of attendances at the classes, which turned out to be either three or four for all participants. She does this by copying the **attend** column to col-

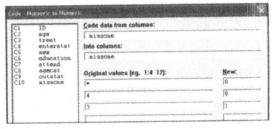

Figure 16.10

umn 10, which she renames as **missone**, and, using **Data > Code > Numeric to Numeric**, recoding **missone** as follows: from* to 0, from 4 to 0, and from 3 to 1 (Figure 16.10). The coefficient of this column is the change in response, for someone on an active treatment, due to attending three rather than all four classes.

She proceeds with **Stat > Regression > Ordinal Logistic Regression** as in Figure 16.11. Note that she specifies all the model variables as factors. Clicking on **Results** (Figure 16.12), she chooses the second option, which is recorded in commands as **Brief 1**. The results are as follows:

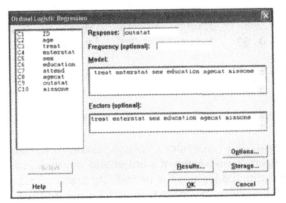

Figure 16.11

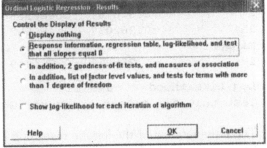

Figure 16.12

Ordinal Logistic Regression: outstat versus enterstat, treat, agecat

Link Function: Logit

Response Information

Variable	Value	Count
outstat	0	26
	1	36
	2	31
	3	23
	4	4
	Total	120

Logistic Regression Table

Predictor	Coef	SE Coef	Z	P	Odds Ratio	95% Lower	CI Upper
Const(1)	−0.216136	0.645568	−0.33	0.738			
Const(2)	1.57446	0.671381	2.35	0.019			
Const(3)	3.27793	0.710111	4.62	0.000			
Const(4)	5.91194	0.882316	6.70	0.000			
treat							
2	0.973744	0.466813	2.09	0.037	2.65	1.06	6.61
3	1.06877	0.445812	2.40	0.017	2.91	1.22	6.98
enterstat							
2	−1.52925	0.526861	−2.90	0.004	0.22	0.08	0.61
3	−2.07291	0.538527	−3.85	0.000	0.13	0.04	0.36
4	−4.06248	0.649337	−6.26	0.000	0.02	0.00	0.06
sex							
1	−0.0450646	0.359189	−0.13	0.900	0.96	0.47	1.93
education							
1	0.106939	0.418735	0.26	0.798	1.11	0.49	2.53
2	−0.0424102	0.448390	−0.09	0.925	0.96	0.40	2.31
agecat							
2	−0.818167	0.488556	−1.67	0.094	0.44	0.17	1.15
3	−0.262037	0.456952	−0.57	0.566	0.77	0.31	1.88
missone							
1	−0.785939	0.554349	−1.42	0.156	0.46	0.15	1.35

Log-Likelihood = −149.117
Test that all slopes are zero: G = 55.099, DF = 11, P-Value = 0.000

The interpretation of the logistic regression is easier if we consider a specific example: a female (sex category 1) in education category 2, age category 2, on treatment 3 but attended only 3 out of 4 classes, who starts in smoking category 3. The estimated logit of the probability that she ends in the first category (none, coded 0; see Table 16.9), is given by Const(1) plus the estimated coefficients for the predictor factors. For the predictors, unlike the way it treats the response category, MINITAB does use the same category numbers that are specified in the columns to identify coefficients, relative to the to the lowest numbered category. Thus, we have:

$$-0.216 + 1.069 - 2.073 - 0.045 - 0.818 - 0.786 = -2.869$$

The corresponding estimate of the probability that she ends in category 0 is 0.054.

The estimated logit of the probability that she ends in either the first or the second category is given by a similar expression with Const(1) replaced by Const(2), which is 1.5%. This gives −1.079 and the corresponding estimated probability is 0.2542. The estimated probability she ends in the second category, which we coded 1, can be found by subtraction: 0.254 − 0.054 = 0.200.

To find the estimate of the logit of the probability that she ends in the first or second or third category, replace Const(1) with Const(3), which is 3.278. This gives 0.625 and the corresponding estimated probability is 0.651. The estimated probability she ends in category 2 can be found by subtraction: 0.651 − 0.254 = 0.397.

In a similar way, the estimated probability that she ends in one of the first four categories, coded 0 to 3, is 0.9631. Hence, the probability that she ends in category 3 is estimated as 0.3112, and the probability that she ends up in the intense smoking category, worse than her entry category, is 0.037.

To summarise: the regression estimates cumulative probabilities. The numeric coding of the response categories is irrelevant, only the ordering of the numbers matters. The coefficients of the predictor variables estimate the change in the cumulative log-odds, attributable to the predictors. An example of an odds ratio is the ratio of the odds that someone on an active treatment, treatment 2 say, ends up not smoking to the odds they would end up not smoking if they were on the non-active treatment (generally referred to as the control). This is equal to exponential of the coefficient for treatment 2. Here it is estimated as $\exp(0.974) = 2.65$.

The coefficient of treatment 3 is 1.06877, so the estimated odds ratio for treatment 3 relative to the control is $\exp(1.06877)$ which is given in the sixth column of the logistic regression table as 2.91. The odds is the probability divided by the complement of the probability, and is close to the probability if the probability is small. So, another interpretation of the estimated effect of treatment 3 is that it increases a small probability of a participant giving up smoking by a multiple of 2.9.

Since her initial guess was a control proportion giving up of 0.2, Ingrid decides to calculate the estimated effect of treatment 3 on a participant with a probability of giving up, if on the control treatment, of 0.2. The odds corresponding to a probability of 0.2 is $0.2/(1 - 0.2) = 0.25$. She multiplies this by 2.91 to obtain an odds of 0.7275. Then the corresponding probability is obtained as $0.7275/(1 + 0.7275) = 0.42$.

A 95% confidence interval for the odds ratio corresponding to a factor is given in columns 7 and 8. For treatment 3 the 95% confidence interval is from 1.22 to 6.98. If all the interval exceeds 1.0, we are confident that the treatment has a beneficial effect. Since the multiplier is the exponential of the coefficient, an equivalent criterion is that the 95% confidence interval for the coefficient itself is all positive. This criterion is itself equivalent to the ratio of the coefficient to its standard deviation, Z, exceeding 2.0, and hence it has an associated P-value less than 0.05.

The two active treatments are estimated as having similar beneficial effects, and both are statistically significant. As expected, how far participants move towards giving up smoking (outstat = 0) depends on their state at entry. There is also some rather weak evidence (P = 0.094) that the middle **age** category **(2)** found it hardest to stop, or at least reduce, smoking.

There is no significant evidence that a participant's sex or level of education affects the probability of giving up smoking. There is a suggestion that missing one of the four classes reduces the treatment benefit, but this is not statistically significant at the 10% level (P = 0.156).

△ Review of Case 16.2

The cost of medical care for smoking-related illness is an increasing burden on Erehwon's health budget. Ingrid's study has provided significant evidence that the classes increase the probability that participants will quit smoking. This is encouraging, but the chief medical officer in the local area health authority needs to decide whether the increase is sufficient to justify the cost of providing the classes. Unfortunately, she doesn't have a sufficiently large budget to fund all schemes that can be justified by a cost–benefit analysis. She has to select the most promising.

Five out of 40 participants on the control treatment, a proportion of 0.125, had given up smoking at the end of the trial. Ingrid's best estimates are that the proportion would be increased to 0.249 for participants who attend classes and to 0.267 for participants who attend classes and use nicotine patches.

The medical officer is sufficiently impressed to continue funding the study of the effect of classes for another year, and to provide money to publicise the trial more widely. However, she decides that the possible additional beneficial effect provided by nicotine patches is too small, and equivocal, to pay for.

Ingrid will start a second phase of the study in which volunteers will be allocated to a treatment (classes) or a control group (brochure only). Nicotine patches are available by prescription in Erewhon, but most patients have to pay the full cost of prescription drugs. Ingrid will record whether or not volunteers have used nicotine patches and include this in the analysis as a concomitant variable. However, if nicotine patches are associated with a significantly increased probability of giving up smoking, it is possible that the increase arises because the more committed participants are the more likely to spend money on them. There is a possible confounding of nicotine patches and commitment to give up smoking. A similar issue arises with volunteers in the control group who attend yoga classes, or similar, that are not run by the health authority. However, this does not invalidate the comparison between a policy of the health authority providing a brochure and free introductory yoga classes, and the policy of providing a brochure only. The brochure may suggest people learn yoga from a book or attend any classes offered independently of the health authority.

It is notable that the statistical evidence of an association between smoking and many diseases is also subject to possible confounding. It is logically possible that gene X causes a desire to smoke cigarettes and a wide range of illnesses. However, there are many convincing scientific explanations for a causal link between cigarette smoking and disease.

Post-script

Have we justified our claim, in the Preface, that the methods we teach are just as applicable in social studies as they are in physics and chemistry; that they are as useful in linguistics, history and geography as much as they are in engineering and marketing? We believe we have.

You have worked, with us, through cases about school absence, metro noise levels, water fluoridation, diamond prospecting, wine tasting, compulsive gambling, prosthetic heart valves and many more. Perhaps you found no case that was quite like any research you want to do but surely, as you studied the methods we used, you will have thought: 'Yes, that is relevant to my work.'

You now understand the principles of the design and analysis of experiments and how MINITAB can help you. There are so many features in MINITAB that you are unlikely to remember all the procedures. Although many of those procedures are instinctive you may occasionally be stuck, as we have sometimes been, wondering how to do this or that. When this happens you may find the answer somewhere in the book. You will also find detailed guidance and more examples in MINITAB **Help** and in the manual. Don't ignore these. While you are working, refer repeatedly to **Help**: so easy, just one click away.

In every case, the researcher was responsible to a manager: in a company, a university or a government department. The researcher had to report the results so that the manager could understand them enough to make decisions. Research does not end with design and analysis. You, the researcher, must interpret and communicate the results. Unless you are able to describe and explain your results to people who do not share your analytic skills, your results will be worthless.

You have travelled a long road to acquire these new skills and you must be happy with your achievement. But that is not the end. The road has taken you to a peak from which you see a new landscape of opportunities: opportunities that were not previously visible. Introduce your new abilities to these new opportunities.

It is not enough to have a good mind; the main thing is to use it well.

Rene Descartes

Authors

Tony Greenfield was formerly head of process computing and statistics at the British Iron and Steel Research Association, Sheffield, and professor of medical computing and statistics at Queen's University, Belfast. He is a visiting professor to the Industrial Statistics Research Unit (ISRU), University of Newcastle-upon-Tyne, and is past President of the European Network for Business and Industrial Statistics (ENBIS). He is a fellow of the Royal Statistical Society and a Chartered Statistician. While at Queen's University, he established a course in research methods for the medical faculty. His publications include *Research Methods: Guidance for Post-graduates* (editor and co-author) and *The Pocket Statistician* (co-author), which people in business and industry, who are not highly qualified statisticians, use as a practical guide to the use of statistics for improving quality in their daily work. In 2004, he received William G Hunter Award presented by the Statistics Division of the American Society for Quality (ASQ). The citation reads: 'For excellence in statistics as a communicator, a consultant, an educator, an innovator, an integrator of statistics with other disciplines, and an implementer who obtains meaningful results.'

Contact: tony@greenfieldresearch.co.uk

Andrew Metcalfe is a senior lecturer in the discipline of statistics at the University of Adelaide. Before this, he was a senior lecturer in the Department of Engineering Mathematics at the University of Newcastle upon Tyne. He has enjoyed teaching statistics to diverse classes of students over thirty years. His industrial experience has included work in insurance and manufacturing, supervision of over fifty postgraduate students on placements, and a considerable number of consultancies. His research interests are engineering applications of statistics, particularly in the context of water resources. Amongst his publications are four other books: *Statistics in Engineering, Spectral Analysis in Engineering – Concepts and Cases* (with Grant Hearn), *Statistics in Civil Engineering and Statistics in Management Science.* He is a fellow of the Royal Statistical Society and a Chartered Statistician, and a member of the Statistical Society of Australia.

Contact: andrew.metcalfe@adelaide.edu.au

Glossary

In common with many other disciplines, there is sometimes more than one technical term for the same concept. Sometimes one particular term may seem more natural than another in a given context, but we have tried to be keep to a minimum of terms throughout this book. Here, however, we do include synonyms – for example, *concomitant variable* and *covariate*. We hope this glossary will assist you if you are reading other texts.

Accuracy – Lack of bias in an estimator. The closeness of an estimated value to the population value.

Alias – A characteristic of a process that cannot be distinguished from some different characteristic using the available data. In a factorial experiment when one comparison cannot be distinguished from another comparison, the comparisons are aliased.

ANOVA – A breakdown of the overall variability of a set of data into components, usually all but one of which can be attributed to the factors and a remaining component that represents random errors.

Balance – Each factor level appears in the same number of runs.

Bias – A systematic difference, which persists when averaged over imaginary replicates of the experiment, between the mean of the estimator and the parameter being estimated.

Bin – A grouping interval, or class interval, when summarising data.

Binomial distribution – Repeat some experiment with only two outcomes, success or failure, some fixed number of times. Suppose the probability of a success on any one occasion is also fixed. Then the binomial distribution gives the probabilities of total number of successes.

Case – A member of the sample. The information about the case may be values of several variables, in which case we have a multivariate datum. A run from an experiment is an example of such a case.

Centred – A variable is centred if some mid-value is subtracted. If the chosen mid-value is the mean it is also referred to as mean-corrected.

Code – A rule that assigns a discrete set of numbers to values, which can be categorical or numerical, taken by a variable.

Coefficient of variation – The ratio of the standard deviation to the mean for a non-negative variable.

Concomitant variable – A variable that may affect the response, and that cannot be set to a specific value although it can be measured. It is also commonly referred to as a covariate.

Confidence interval – A useful working interpretation is that a confidence interval is a range of values which is likely to include the parameter for which the interval has been constructed.

Confounding – When a characteristic of the process cannot be distinguished from some artefact of the design, such as a difference between replicates, using the available data. Sometimes used as a synonym for aliasing.

Contrast – See *effect*.

Control variable – A variable that affects the response and that can be set to specific values.

Correlation coefficient – A measure of linear association between two variables that is scaled to lie between -1 and 1.

Covariance – A measure of linear association between two variables with unit of measurement equal to the product of the units of measurement of those variables.

Crossed – The levels for one factor represent the same feature at different levels of each of the other factors.

Cumulative distribution function – A curve that gives the probability that a variable is less than some specific value.

Data – Information, especially numerical. Plural of datum but often treated as singular.

Degrees of freedom – The number of cases less the number of population parameters that have been estimated from the data.

Effect – In factorial experiments in which factors appear at two levels, the main effect of a factor is the difference between the response at its high level and the response at the low level. An interaction effect is the difference between the response when the product of factors is $+1$ and -1, when low and high levels are coded -1 and $+1$, respectively. Also called a contrast.

Errors – Differences between the deterministic part of the model and observations.

Estimate – A statistic calculated from a sample to provide the best single value, known as the point estimate, for some feature or parameter of the population. See below.

Estimator – The general form of an estimate. Usually a formula for estimating a constant (parameter or feature) of a population. We imagine it being used repeatedly in the same experiment and hence generating many different estimates. The distribution of these estimates is known as the sampling distribution of the estimator.

Expected value – A population average.

Experimental design – Specification of the values of the control variables at which runs will be made.

Exponential – The exponential of a number is the value of e, approximately 2.72, raised to that number. It is written as exp(.). It is the inverse function for the natural logarithm. If you take exponential of logarithm of some positive number you get that same number back.

F-value – An F-value is the ratio of two independent estimates of a population variance, if the null hypothesis of no effect is true. If there is an effect, the numerator estimate is expected to be larger. The P-value associated with the F-ratio is the probability of obtaining such a large, or larger, value if the null hypothesis of no effect is true.

Factor – A synonym for a control variable, usually used when the control variable is either physically restricted to a small set of levels or chosen to be limited to a small set of levels.

Fixed effect – The values taken by some factor are chosen by the experimenter.

Frequency – In statistics, this usually means the number of times that something has occurred. Relative frequency is the ratio of the frequency to the number of data, that is a proportion of the data, and estimates a probability. When drawing a histogram the relative frequency density, abbreviated to density in MINITAB, is the relative frequency divided by the bin width. Thus, the area of the rectangles equals the relative frequency.

Geometric mean – The geometric mean of n numbers is the nth root of their product. It is always less than their arithmetic mean. It is the exponential of the arithmetic mean of the logarithms of the data. An example of its use is averaging interest rates. The geometric mean of 1.10, 1.20 and 1.30 is 1.197. So investing at 10%, 20%, 30% per annum over each of three years is equivalent to investing at 19.7% per annum over three years.

Harmonic mean – The reciprocal of the arithmetic mean of the reciprocals. An example is calculating an equivalent single speed when travelling at different speeds over equal distances. Walking 1 km at 1 km/h and 1 km at 3 km/h takes the same time as walking 2 km at 1.5 km/h.

Hypothesis – In statistics a null hypothesis is a claim that some population parameter equals a single specified value. The evidence against the claim is based on the probability of obtaining results as extreme as, or more extreme, than those observed in the experiment if the claim is true. This probability is the P-value. The smaller the P-value the stronger the evidence against the null hypothesis

Independent – Two events are independent if the probability that one occurs does not depend on whether or not the other event has occurred.

Interaction – Variables interact if the effect on the response of one variable depends on the value of the others.

Indicator variables – Suppose a variable has c categories. There are c indicator variables, one for each category, coded 1 if the value of the variable is in that category and 0 otherwise.

Inverse cumulative distribution function – Provides a value of the variable such that the probability of being less than that value equals some specified probability.

Logarithm – The natural logarithm of a positive number is the power of e, approximately 2.72, that gives that number. The reason for mathematicians using e rather than 10 is that the derivative of $\exp(x)$ is $\exp(x)$. The natural logarithm is approximately 2.3 times the logarithm base 10.

Logit – Logarithm base e (natural logarithm) of the odds.

Mean – The mean of a set of data is their sum divided by their number. It is sometimes known as the arithmetic mean to distinguish it from other means such as: geometric mean, harmonic mean and weighted mean. A synonym for the mean is the average, but average tends to be loosely used for any typical measure. The mean is defined in the same way for a population but we cannot calculate it because we do not know all the population values.

Mean corrected – Data from which the mean has been subtracted to leave the deviations from the mean.

Median – The middle value when data are arranged into ascending order.

Mode – The most commonly occurring value.

Model – An algebraic description of the structure of data that represents the physical situation in some useful manner. It has a deterministic part and a random component.

Nested – The levels of one factor represent different features at different levels of other factors.

Normal distribution – A bell-shaped distribution. A standard normal distribution is a normal distribution with a mean of 0 and a standard deviation of 1.

Normal probability plot – A plot of data, sorted into ascending order, against their rank divided by the total number of data plus 1, which is an estimate of the probability of being less than that datum. The plot is on a graph with a probability scale adjusted so that the points appear to be scattered about a straight line if they are a random sample from a normal distribution.

Null hypothesis – In statistics a null hypothesis is a claim that some population parameter equals a single specified value. The evidence against the claim is based on the probability of obtaining results as extreme as, or more extreme than, those observed in the experiment if the claim is true. This probability is the P-value. The smaller the P-value the stronger the evidence against the null hypothesis. If the P-value, p, is reasonably small (less than 0.10 is

typical) we can say the result is statistically significant at the P $\times$ 100% level, though it is more common to give the smallest standard level (10%, 5%, 1%, 0.1%) that exceeds it.

Odds – The odds on an event occurring is the ratio of the probability that the event occurs to its complement.

One-sided alternative – The default alternative hypothesis to a null hypothesis that a parameter equals a specified value is that the parameter does not equal that specified value. P-values refer to this two-sided alternative hypothesis unless you specify otherwise. A one-sided alternative hypothesis is that the parameter is less than the specified value. The other one-sided alternative hypothesis is that it is greater than the specified value. P-values will be halved, relative to a two-sided alternative, if the experimental result is in the direction of the alternative hypothesis, but are not defined if the experimental result is in the other direction.

Order statistic – The ith smallest in a random sample is the ith order statistic, and i is the rank of the datum.

Orthogonal – Control variables are orthogonal if they are uncorrelated. The term is used because the values of the control variables are chosen by the experimenter. Correlation is usually reserved for random variables. A design is orthogonal if the control variables and their products, representing interactions, are uncorrelated.

P-value – The results of an experiment are summarised in one, or a few, statistics. These statistics are compared with their sampling distribution when some hypothesised null hypothesis is true. The P-value is the probability of obtaining a value of the statistic as far from, or further from, its average value in imaginary repeated sampling, as that obtained if the null hypothesis is true. The smaller the P-value the more is the evidence against the null hypothesis. If the P-value is p, we say the result is significant at the $100p\%$ level. If we decide to test the null hypothesis at some given level, c say, in advance we draw conclusions as follows. If $p < c$ we have evidence against the null hypothesis at the $100c\%$ level. If $p > c$ we have insufficient evidence to reject the null hypothesis at the $c*100\%$ level. The choice of c is subjective, but 0.05 is common. See *Hypothesis*.

Paired samples – Data in two samples are related, as pairs, across the two samples.

Parameter – Some constant numerical feature of a population. Examples are the population mean and the population variance.

Pilot sample – A small sample taken to estimate variability or to test the protocol for some planned experiment.

Point estimate – A single best estimate of a parameter.

Population – The collection of all possible experimental units from which those used in the experiment are assumed to have been drawn at random. We make inferences about this population from the results of the experiment.

Power – The probability of rejecting the null hypothesis if some alternative hypothesis holds. In general, it depends on the null hypothesis, but a useful choice is the technically significant difference.

Precision – A measure of how close replicate estimates are to each other. Formally defined as the inverse of the variance.

Predictor variable – Any variable used on the right-hand side of a regression model to help predict a response. Predictor variables are also known as an explanatory variables, the choice depending on the context. Yet another term for predictor variables is independent variables, which can be confusing. They are only independent inasmuch as the response is considered to depend on the values of the independent variables rather than vice versa.

Probability density function – A hypothetical histogram for the population. The areas under the curve between any two values equals the probability the variable will lie between those values.

Quadratic – Involving the second power, the square, of some variable.

Quartiles – The lower quartile is the value below which one quarter of the data lie. The upper quartile is the value below which three-quarters of the data lie, and hence above which one-quarter of the data lie.

Random – A simple random sample can be imagined as having been obtained by giving every member of the population a lottery ticket, and then drawing tickets such that each has the same chance of selection. A random ordering of runs in an experiment is one drawn at random from all possible sequences of the runs.

Random effect – The levels of a factor are assumed to have been drawn at random from some population.

Random numbers – Computer generated random numbers are usually pseudo-random because they are generated from some deterministic algorithm with a haphazard start. However, they appear random to the user. Random digits are a sequence of digits produced from an underlying process that generates one of the digits from 0 to 9 with equal probability of one-tenth at each draw. Uniform random numbers are drawn from the range 0 to 1 such that the probability of being between any two values, in this range, equals the absolute value of their difference. More generally, random numbers from any probability distribution are drawn so that the probability they are between any two values equals the area of the probability density function between those values.

Rank – If data are put into ascending order the rank of a datum is the number of its position starting with the smallest datum as 1.

Regression – In MINITAB's unmodified regression the response is modelled by a linear combination of predictor variables, or transformations of them, plus error. Commonly used transformations of predictor variables are squares and cross-products.

Replicate – A repeat of a designed experiment.

Residuals – Estimates of the errors. They are the differences between fitted values given by the estimate of the deterministic part of the model and the data.

Response variable – A response variable is a characteristic of a product or process that we wish to explain, control, or improve through adjustments of control variables.

Run – Operating the process once at a specified set of conditions to obtain an observation of the response. The design is made up of runs.

Sampling distribution – An estimate is thought of as a single value from the imaginary distribution of all possible estimates known as the sampling distribution.

Scaled – Data are scaled if they are multiplied, or divided, by some constant. More generally, a constant may be subtracted.

Significance level – The probability of a result as unlikely as, or more unlikely than, that obtained in the experiment if the null hypothesis is true. When set in advance it is the largest probability that will be claimed as evidence against the null hypothesis.

Six sigma – A business improvement strategy which aims to achieve near perfect quality through careful analysis of data followed by appropriate action. It is associated with lean manufacturing which aims to reduce waste and rework and to minimise inventory.

Skewed – A distribution is right-skewed or positively skewed if it has a longer tail to the right than to the left. It is left-skewed or negatively skewed if it has a longer tail to the left. A symmetric distribution is not skewed.

Standard deviation – The square root of the variance.

Standard error – The standard deviation of an estimator is also known as its standard error. Also, be warned – some software programs, such as Excel and R, refer to the estimated standard deviation of the errors in a regression model as 'standard error'.

Standard normal – A normal random variable with a mean of 0 and a standard deviation of 1. Commonly denoted by Z.

Statistic – Any numerical quantity calculated from a sample.

Technically significant difference – The smallest difference that is of practical importance.

Transform – A function of the data, commonly logarithm.

Trial – A single run out of an experimental design.

***t*-distribution** – The distribution of a standard normal variable divided by an estimate of its standard deviation with ν (nu) degrees of freedom.

***t*-test** – The ratio of the difference between an estimate and the null hypothesis value of the parameter, divided by the estimator's standard deviation which is estimated from the data. It is compared with a standard normal distribution.

Unbiased – An estimator for some population parameter is said to be unbiased if on average, in imagined repeats, it gives the value of that parameter.

Variance – The sample estimate of a population variance is the sum of the squared deviations of the data from their mean divided by their number less the degrees of freedom lost when estimating that mean. The variance of a population is the sum of the squared deviations of the data from their mean divided by their number.

Weighted mean – The weights are known constants. The weighted mean is the sum of products of data with weights divided by the sum of the weights. If the weights are equal this is the arithmetic mean.

Z-test – The ratio of the difference between an estimate and the null hypothesis value of the parameter, divided by the estimator's standard deviation which is assumed known. It is compared with a standard normal distribution.

References

Anderson VL and McLean RA (1974) *Design of Experiments: A Realistic Approach*. Dekker, New York.

Clarke GM and Kempson RE (1997) *Design and Analysis of Experiments*. Arnold, London

Close A, Hamilton G and Muriss S (1986) Finger systolic pressure: its use in screening for hypertension and monitoring. *British Medical Journal*, **293**, 775–778.

Frisby JP and Clatworthy JL (1975) Learning to see complex random-dot stereograms. *Perception*, **4**, 173–178.

Gorn GJ (1982) The effects of music in advertising on choice behaviour: a classical conditioning approach. *Journal of Marketing*, **46**, 94–101.

Graham G and Martin FR (1946) Heathrow. The construction of high-grade quality concrete paving for modern transport aircraft. *Journal of the Institution of Civil Engineers*, **26**(6), 117–190.

Greenfield Research (2003) Process Training CD.

Hanna J, Foster DJR, Salter A, Somogyi AA, White JM and Bochner, F (2005) Within- and between- subject variability in methadone pharmacokinetics and pharmacodynamics in methadone maintenance subjects. *British Journal of Clinical Pharmacology*, **60**(4), 404–413.

JFE 21st Century Foundation (2003) http://www.jfe-21st-cf.or.jp/chapter_6/6d_1.html (accessed 20 February 2006).

Kowalski SM, Cornell JA and Vining GG (2002) Split-plot designs and estimation methods for mixture experiments with process variables. *Technometrics*, **44**(1), 72–79.

Miller I, Freund JE (1977) *Probability and Statistics for Engineers*, 2nd edn. Prentice-Hall, Engelwood Cliffs, NJ.

Montgomery DC (2004) *Design and Analysis of Experiments*, 6th edn. Wiley, New York.

Morris VM, Hargreaves C, Overall K, Marriott PJ and Hughes JG (1997) Optimisation of the capillary electrophoresis separation of ranitidine and related compounds. *Journal of Chromatography* A, **766**, 245–254.

Norman P and Naveed S (1990) A comparison of expert system and human operator performance for cement kiln operation. *Journal of the Operational Research Society*, **41**(11), 1007–1019.

Sadras VO, Baldock JA, Cox JW and Bellotti WD (2004) Crop rotation effect on wheat grain yield as mediated by changes in the degree of water and nitrogen co-limitation *Australian Journal of Agricultural Research*, **55**(6), 599–607.

Singh SN, Mishra S, Bendapudi N and Linville D (1994) Enhancing memory of television commercials through message spacing. *Journal of Marketing Research*, **31**, 384–392.

Wu CFJ and Hamada M (2000) *Experiments*. Wiley, New York

Wudka J. (1998) http://phyun5.ucr.edu/~wudka/Physics7/Notes_www/node5.html (accessed 18 February 2006).

Index

Minitab procedures

Statistical methods